INTERNATIONAL

REVIEW OF CYTOLOGY

VOLUME 64

ADVISORY EDITORS

INTERNATIONAL

Review of Cytology

EDITED BY

G. H. BOURNE
St. George's University School of Medicine
St. George's, Grenada
West Indies

J. F. DANIELLI
Worcester Polytechnic Institute
Worcester, Massachusetts

ASSISTANT EDITOR
K. W. JEON
Department of Zoology
University of Tennessee
Knoxville, Tennessee

VOLUME 64

1980

ACADEMIC PRESS   *A Subsidiary of Harcourt Brace Jovanovich, Publishers*
New York   London   Toronto   Sydney   San Francisco

ACADEMIC PRESS, INC.
111 Fifth Avenue, New York, New York 10003

United Kingdom Edition published by
ACADEMIC PRESS, INC. (LONDON) LTD.
24/28 Oval Road, London NW1 7DX

LIBRARY OF CONGRESS CATALOG CARD NUMBER: 52-5203

ISBN 0-12-364464-X

PRINTED IN THE UNITED STATES OF AMERICA

80 81 82 83   9 8 7 6 5 4 3 2 1

Contents

List of Contributors vii

Variant Mitoses in Lower Eukaryotes: Indicators of the Evolution of Mitosis

I. Brent Heath

I.	Introduction	1
II.	Characteristics of Mitosis	5
III.	Evolution of Mitosis	55
IV.	Mitosis as an Indicator of Phylogeny	65
V.	Conclusions	69
	References	70

The Centriolar Complex

Scott P. Peterson and Michael W. Berns

I.	Introduction	81
II.	Organization of the Mitotic Centriolar Complex	82
III.	Centriolar Movements during Mitosis	84
IV.	Systems without Centrioles	85
V.	The Two Functional States of the Centriole	90
VI.	Indirect Immunofluorescence and Its Use in Mitotic Investigation	91
VII.	In Vitro Studies on Microtubular Nucleation	96
VIII.	The Role of Nucleic Acids in Centriolar Function	98
IX.	Multicentriolar Cells	100
X.	Conclusions	102
	References	103

The Structural Organization of Mammalian Retinal Disc Membrane

J. Olive

I.	Introduction	107
II.	General Aspects of the Organization of Photoreceptor Membranes	108
III.	Organization of the Rhodopsin Molecules into the Lipid Bilayer	124
IV.	Dynamic Aspects of the Disc Membrane	148
V.	Conclusion	156
	References	162

The Roles of Transport and Phosphorylation in Nutrient Uptake in Cultured Animal Cells

Robert M. Wohlhueter and Peter G. W. Plagemann

I.	Introduction	171
II.	Theoretical and Methodological Considerations	173

CONTENTS

III. Uptake of Nucleosides 192
IV. Uptake of Nucleobases 207
V. Uptake of Hexoses 213
VI. Uptake of Vitamins and Choline 230
VII. Concluding Discussion 232
References 234

The Contractile Apparatus of Smooth Muscle

J. Victor Small and Apolinary Sobieszek

I. Introduction 241
II. Architecture of the Contractile Apparatus 242
III. The Contractile Proteins 266
IV. Regulation of the Actin-Myosin Interaction via Ca 283
References 299

Cytophysiology of the Adrenal Zona Glomerulosa

Gastone G. Nussdorfer

I. Introduction 307
II. Fine Structure of the Normally Functioning Zona Glomerulosa 308
III. Fine Structure of the Hyperfunctioning and Hypofunctioning Zona Glomerulosa . 328
IV. Morphological-Functional Correlations in Zona Glomerulosa Cells 342
V. Zona Glomerulosa and the Maintenance of the Adrenal Cortex 353
VI. Concluding Remarks 359
References 360
Note Added in Proof 368

Subject Index 369
Contents of Previous Volumes 373

List of Contributors

Numbers in parentheses indicate the pages on which the authors' contributions begin.

MICHAEL W. BERNS (81), *Developmental and Cell Biology, University of California, Irvine, Irvine, California 92717*

I. BRENT HEATH (1), *Biology Department, York University, Toronto, Ontario, Canada*

GASTONE G. NUSSDORFER (307), *Department of Anatomy, University of Padua, 35100 Padua, Italy*

J. OLIVE (107), *Laboratoire de Microscopie Electronique, Institut de Recherche en Biologie Moléculaire du C.N.R.S., Université Paris VII, 2 Place Jussieu, 75221 Paris, France*

SCOTT P. PETERSON (81), *Developmental and Cell Biology, University of California, Irvine, Irvine, California 92717*

PETER G. W. PLAGEMANN (171), *Department of Microbiology, University of Minnesota, Minneapolis, Minnesota 55455*

J. VICTOR SMALL (241), *Institute of Molecular Biology of the Austrian Academy of Sciences, Billrothstrasse 11, 5020 Salzburg, Austria*

APOLINARY SOBIESZEK (241), *Institute of Molecular Biology of the Austrian Academy of Sciences, Billrothstrasse 11, 5020 Salzburg, Austria*

ROBERT M. WOHLHUETER (171), *Department of Microbiology, University of Minnesota, Minneapolis, Minnesota 55455*

INTERNATIONAL REVIEW OF CYTOLOGY, VOL. 64

Variant Mitoses in Lower Eukaryotes: Indicators of the Evolution of Mitosis?

I. Brent Heath

Biology Department, York University, Toronto, Ontario, Canada

I. Introduction 1
II. Characteristics of Mitosis 5
    A. General 5
    B. Membranes 20
    C. Matrix 25
    D. Nucleolus 26
    E. Chromatin 29
    F. Polar Structures 31
    G. Spindle 37
    H. Involvement of Cytoplasmic Microtubules 53
    I. Control Mechanisms 54
    J. Summary 55
III. Evolution of Mitosis 55
    A. The Prokaryote Ancestors 56
    B. Hypothetical Early Mitotic Systems 58
    C. The Phylogenetic Position of Extant Mitoses 63
IV. Mitosis as an Indicator of Phylogeny 65
    A. Advantages 66
    B. Disadvantages 66
    C. Tentative Relationships 67
V. Conclusions 69
    References 70

I. Introduction

In the present work mitosis will be defined as all those types of nuclear division that produce two, or rarely more, daughter nuclei, each containing a chromosome complement approximately similar (usually identical) to that of the original nucleus. The greatest range of variations by which mitosis is accomplished occurs in the protistan and fungal kingdoms, some members of which are probably most similar to the ancestors of higher plants and animals. Therefore, it is these two kingdoms that will be examined most extensively. The taxa considered are listed in Table I. This is not to imply that all mitoses in higher plants and animals are identical with one another; the converse is, in fact, true. However, the variations in the higher organisms are most probably secondarily derived

1

TABLE I

TAXA STUDIED

Phyla[a]	Lower taxa[b]	Genera studied	Trivial names used	
Myxomycota	Protostelidia	4	Protostelids	
	Dictyostelia	2		
	Myxogastria	7	Myxomycetes	
Acrasiomycota		0		
Plasmodiophoromycota		3	Plasmodiophoromycetes	
Labyrinthulomycota		3	Labyrinthulomycetes	
Oomycota	Saprolegniales	4	Oomycetes	
	Lagenidiales	3		
	Peronosporales	2		
Hyphochytridiomycota		1	Hyphochytrids	
Chytridiomycota	Chytridiales	2	Chytrids	
	Blastocladales	4		
	Harpochytriales	1		
Zygomycota	Mucorales	4	Zygomycetes	Fungi
	Entomophthorales	4		
Ascomycota	Hemiascomycetales	6	Hemiascomycetes	
	Plectomycetales	2		
	Discomycetales	3		
	Pyrenomycetales	6	Euascomycetes	
	Loculoascomycetales	1		
Basidiomycota	Heterobasidiomycetales	7	Basidiomycetes	
	Homobasidiomycetales	9		
Euglenophyta		5	Euglenids	
Dinoflagellata	Prorocentrales	2	Dinoflagellates	
	Gymnodiniales	2		
	Peridinales	9		
Cryptophyta		2	Cryptophytes	
Bacillariophyta	Pennales	6	Diatoms	
	Centrales	2		
Chrysophyta		1		
Haptophyta		1		Algae
Xanthophyta		1		
Chloromonadophyta		1		
Chlorophyta	Chlorococcales	7	Greens	
	Volvocales	3		
	Ulotrichales	6		
	Chaetophorales	3		
	Oedogoniales	2		
	Siphonacladales	4		
	Dasycladales	2		

(*continued*)

TABLE I (*continued*)

Phyla[a]	Lower taxa[b]	Genera studied	Trivial names used	
Zygnematophyta		4		⎫
Prasinophyta		4		⎪
Charaphyta		2		⎬ Algae
Phaeophyta		2	Browns	⎪
Rhodophyta		4	Reds	⎭
Sporozoa s.s.	Gregarinida	2	Gregarins	⎫
	Coccidida	3	Coccids	⎪
	Haemosporida	4		⎪
	Haplosporidia	2		⎪
Cnidosporidia		5		⎪
Sarcodina s.s. or		3	Amoebae	⎪
Rhizopoda				⎪
Actinopoda	Radiolarida	1	Radiolarans	⎪
	Acantharida	9	Acantharia	⎪
Foraminifera		3		⎬ Protozoa
Zoomastigina	Rhizomastigida	1	Flagellates	⎪
	Protomonadida	2		⎪
	Polymastigida	1		⎪
	Trichomonadida	1		⎪
	Hypermastigida	6		⎪
Ciliophora	Gymnostomatida	5	Ciliates	⎪
	Hymenostomatida	3		⎪
	Heterotrichida	2		⎪
	Oligotrichida	1		⎪
	Hypotrichida	2		⎪
	Peritrichida	3		⎪
	Suctorida	2		⎭

[a] Phyla largely after Whittaker and Margulis (1978).
[b] Lower taxa after Olive (1975); Chapman and Chapman (1973); Kudo (1966).

from the better known division patterns of "typical" plants and animals and thus have limited relevance to the present topic. Hence they will not be included. Because we have, at present, very little biochemical or genetical data on diverse mitotic systems, this article must necessarily concentrate on morphological characters. Since there is usually much more information with which to work at the ultrastructural level of analysis, it is this data that will receive primary coverage. However, much of the light microscope literature is relevant to a number of topics covered herein and should be consulted, perhaps via the reviews of Grell (1973), Olive (1953), Robinow and Bakerspigel (1965), and Godward (1966).

The essential design parameters of any mitotic system are (1) to produce at least two nuclei from one original, (2) to ensure that each daughter nucleus contains at least one copy of the total genetic information of the original nucleus, and (3) to accomplish (1) and (2) with a maximum efficiency that translates into the minimum expenditure of energy and cellular materials. As we shall see, these ends are achieved by a large variety of different means. The fundamental assumption underlying this, and similar works, is that some of these extant variations resemble the systems utilized by the extinct ancestors of present day organisms. Because no fossil record is ever likely to yield useful information on ancestral mitotic systems, and because it will be many years before sequence analysis of mitotic components will progress to the state where evolutionary patterns of the type constructed by Fitch (1976) and Schwartz and Dayhoff (1978) can be produced for mitotic systems, we can only attempt educated speculation on the past sequence of mitotic events. Such speculation is the essence of this article. However, it is hoped that this speculation will result in a better understanding of the mechanisms of mitosis and perhaps a clearer indication of the phylogeny of the organisms studied. Prior to embarking on this speculative trip, we must add one note of caution. There is no a priori reason why the mitotic system of an organism should have evolved in parallel with other components of that organism. Thus, it is not unreasonable to expect to find a "primitive" mitotic system in a cell that is otherwise somewhat "advanced," and vice versa. In other words, the selection pressures operating on the organism as a whole may well differ from those operating on the mitotic apparatus.

For the uninitiated reader, it should be pointed out that this is far from the only attempt at understanding how mitosis may have evolved. Earlier works that have made substantial contributions to this area include those of Pickett-Heaps (1969, 1972e, 1974b, 1975a,b), Margulis (1970), Ris (1975), Kubai (1975, 1978), Fuller (1976), Cavalier-Smith (1978), Oakley (1978), and Heath (1974a, 1975a). The serious student of this topic is referred to these reviews for alternate points of view and for further discussions of many of the topics dealt with here. For the reader who is unfamiliar with the state of the art in mitosis research, the following reviews provide useful background from various viewpoints: Mazia (1961), Forer (1966, 1978), Inoué and Sato (1967), Luykx (1970), Nicklas (1971), Bajer and Molé-Bajer (1972), Inoué and Ritter (1975), and Fuge (1977).

Before embarking upon the more speculative aspects of this work, it may be worthwhile considering the diversity of information that can be used to reconstruct the evolutionary process. Some topics have been considered frequently by previous authors and need only a brief review (e.g., the behavior of the nuclear envelope), but other characteristics of mitosis are less well considered and deserve mention, if only to point out our areas of ignorance. Most of the data used in this review is summarized in Table II. In most instances, reference in the text will only be made to the genus, the citations for each genus can then be obtained from Table II. Citation of individual papers will be used when only one of those listed in the table contains the relevant information.

II. Characteristics of Mitosis

A. GENERAL

1. *Time*

The time taken to undergo mitosis by different organisms varies by almost two orders of magnitude. For example, *Chilomonas* and *Fusarium* complete the process in about 5 minutes, whereas the higher plant *Tradescantia* (Barber, 1939) can take up to 340 minutes. A more extensive list of mitotic times is given in Mazia (1961), Hughes (1952), Milovidov (1949), and Heath (1978a). However, the value of such measurements in any phylogenetic argument is rather limited by the fact that, for example, the very distantly related insect *Drosophila* (Huettner, 1933) and the fungus *Fusarium* have similar times. Further detailed analysis of mitotic times is not profitable for a number of reasons. For example, it is difficult to accurately determine stop and start points in the process, especially in organisms with small nuclei such as many protists. Furthermore, as might be expected of a biochemical process, the time is temperature dependent. The optimal temperature varies from species to species as does the shape of the time versus temperature curve, which may have a sharp optimum extending over approximately 1°C for *Spirogyra* and over 20°C for *Sabellaria* eggs (Hughes, 1952). However, as noted by Mazia (1961) and confirmed by a perusal of the fungal data in Heath (1978a), there is a tendency for organisms with small spindles and chromosomes to have short mitotic times. This point emphasizes what is probably a generality, that mitotic times are secondarily derived from other features of the spindle and chromosomes, and are thus worthy of little consideration in an evolutionary context.

2. *Efficiency*

The efficiency of mitosis consists of two basic components; the frequency with which each daughter nucleus receives the necessary complete genome complement (genetic efficiency), and the amount of energy and materials expended in the synthesis and operation of the mitotic apparatus (energetic efficiency). Data on both aspects of the efficiency equation are too sparse to be conclusive. For example, whereas Day (1972) has noted that some fungi (where the spindle may be considered primitive, see later) have a high incidence of mitosis-derived aneuploidy and thus may be considered genetically inefficient, comparable data on other organisms growing in reasonably natural environments is lacking. (Clearly, data from cultured mammalian cells, for example, are inadmissible as a

TABLE II
CHARACTERISTICS OF MITOSIS[a]

Organism	References	Cell type[b]	Polar structures[c]	Polar structure migration[d]	Spindle development[e]	Nuclear envelope[f]	Perinuclear e.r.[g]	Telophase behavior[h]	Spindle vesicles[i]	Metaphase plate[j]	Chromosome pattern[k]	Kinetochores[l]	Microtubules/kinetochore[m]	Kinetochore location[n]	Central spindle[o]	Spindle elongation[p]	Framework m.t. pattern[q]	Anaphase chromosome migration[r]	Extranuclear spindle[s]	Nucleolus[t]
MYXOMYCOTA																				
Protostelidia																				
Cavostelium apophysatum	100	u	?	?	?	di P	–	–	?	+	M?	+	?	N	–	?	?	?	–	Ds
Ceratiomyxella tahetiensis	101	u	?	?	?	di P	–	–	?	?	?	+	?	N	?	?	?	?	–	?
Planoprotostelium auranitum	390 and pers. comm.	u	centrioles V	O	?	di P	–	–	+	+	s	+	1	N	–	+	?	+	–	Ds
Protostelium mycophaga	390 and pers. comm.	u	none	O	?	di P	–	–	+	+	s	+	1	N	–	+	?	?	–	Ds
Dictyostelia																				
Dictyostelium discoidium	257	u	plaque	D	2	pf	–	C	–	–	U	+	2-3	N	+	+	+	+	–	P
Polysphondylium violaceum	347,348	u	ring	D	2	pf	–	C	–	–	U	+	1	N	+	+	+	+	–	P
Myxogastrea																				
Physarum flavicomum plasmodium	7,8	c	none	O	?	pf	–	IE	–	+	s	+	1-2	N	–	+	?	+	–	Ds
Physarum polycephalum plasmodium	110,119,200,357, 358,392,405	c	none	O	?	pf	–	IE	–	+	s	+	1-2	N	–	+	+	+	–	Ds
Echinostelium minutum plasmodium	150	c	none	O	?	pf	–	?	–	+	s	?	?	?	+	+	?	+	–	Ds
Clastoderma debaryanum plasmodium	228	c	none	O	?	intact	–	?	–	+	s	?	?	?	+	?	?	?	–	?
Didymium iridis plasmodium	9	c	none	O	?	intact	–	?	–	+	s	?	?	?	–	?	?	?	–	Ds
Arcyria cinerea plasmodium	254	c	none	O	?	di T	–	–	–	+	s	?	?	?	–	?	?	?	–	Ds
Stemonitis virginiensis plasmodium	255	c	none ?	?	?	di T	–	?	–	+	s	?	?	?	?	?	?	?	–	Ds
Ceratiomyxa fruticulosa plasmodium	359	c	?	?	?	intact	–	–	+	+	s	+	?	?	–	?	?	+	–	?
Physarum flavicomum myxamoebae	8	u	centrioles V	B	?	di P	–	–	+	+	s	+	1?	N	+	+	?	+	–	Ds

6

Taxon	Ref.																		
PLASMODIOPHOROMYCOTA																			
Sorosphaera veronicae	31,32,78	c	centrioles--		B	2	pf	+	IR	+	+	M	+	many	N	–	+	–	P
Polymyxa betae	188	c	?		?	?	intact?	?	?	?	?	M	?	?	N	+	?	?	P
Plasmodiophora brassicae	102	c	centrioles--		?	?	pf	+	?	+	+	?	?	?	N	–	?	?	P
LABYRINTHULOMYCOTA																			
Labyrinthula sp.meiosis	294	u	centrioles V--		B	?	pf	–	?	+	?	?	?	?	N	–	?	?	Ds
Labyrinthula sp.vegetative	294,296,331	u	procentrioles ?		?	?	pf	–	?	+	?	?	1?	?	N	–	?	?	Ds
Sorodiplophrys stercorea	77	u	procentrioles V		B	?	intact	–	?	–	?	?	?	?	?	+	?	?	Ds
Thraustochytrium sp.	184	u	centrioles V		B	2	pf	–	IE	–	+	M?	+	1	N	–	+	+	Ds
OOMYCOTA																			
Saprolegniales																			
Saprolegnia spp. (3)[a]	18,135,137,138,161	c	centrioles--		D	1	intact	–	C	–	–	U	+	1	N	–	+	?	P
Thraustotheca clavata	130	c	centrioles--		D	1	intact	–	C	–	–	U	+	1	N	–	+	+	P
Achlya ambisexualis	82	c	centrioles--		?	?	intact	–	?	–	–	U	+	?	?	+	?	?	P
Aphanomyces spp. (2)	131,151	c	centrioles--		D?	1	intact	?	IE	–	–	U	+	?	N	–	+	+	P
Lagenidiales																			
Apodachlya sp.	132,135	c	centrioles--		B	?	intact	–	C?	?	–	?	+	?	N	–	+	?	P
Sapromyces elongatus	132,135	c	centrioles--		B	?	intact	–	C?	–	–	U	+	1	N	+	?	+	P?
Lagenisma coscinodisci	361	c	centrioles--		B	?	intact	–	?	–	–	U	+	?	N	–	–	?	P
Peronosporales																			
Phytophthora palmivora	81,142	c	centrioles--		D	?	intact	–	C	–	–	S	+	1?	N	–	+	+	P
Albugo candida	189	c	centrioles--		?	?	intact	–	?	–	–	U	+	?	N	?	+	+	P
HYPHOCHYTRIDIOMYCOTA																			
Rhizidiomyces apophysatus	98 and pers. comm.	c	centrioles V		B	?	pf	?	IE	+	+	?	+	1	N	?	?	?	Ds
CHYTRIDIOMYCOTA																			
Chytridiales																			
Phlyctochytrium irregulare	229	c	centrioles V		B	2	pf	+	IR	+	+	M	?	?	N	?	+	+	Di
Entophlyctis sp.	332	c	centrioles V		B	2	pf	+	IR	+	+	S	?	?	?	?	+	+	?
Blastocladiales																			
Blastocladiella emersonii	209	c	centrioles V		?	?	intact	–	IE	–	?	S?	?	?	?	?	+	?	P
Catenaria anguillulae	171	c	centrioles V		B	?	intact	–	IE	+	+	S	+	?	N	?	+	+	Di
Allomyces spp. (3)	286,287,343,424	c	centrioles V		?	?	intact	–	IE	+	+	S	?	?	N	?	+	–?	Di
Coelomomyces indicus	230	c	centrioles ?		?	?	intact	–	.	–	?	S	?	?	?	?	+	?	?
Harpochytriales																			
Harpochytrium hedinii	442	c	centrioles V		B	2	pf	–	C	–	+	M	+	1?	N	–	+	+	Ds

(continued)

7

TABLE II (continued)

Organism	References	Cell type[b]	Polar structures[c]	Polar structure migration[d]	Spindle development[e]	Nuclear envelope[f]	Perinuclear e.r.[g]	Telophase behavior[h]	Spindle vesicles[i]	Metaphase plate[j]	Chromosome pattern[k]	Kinetochores[l]	Microtubules/kinetochore[m]	Kinetochore location[n]	Central spindle[o]	Spindle elongation[p]	Framework m.t. pattern[q]	Anaphase chromosome migration[r]	Extranuclear spindle[s]	Nucleolus[t]
ZYGOMYCOTA																				
Mucorales																				
Mucor hiemalis	222	c	see Fig. 4	D	?	intact	–	C	–	?	U	+?	?	N	+	+	+?	+?	–	P
Phycomyces blakesleeanus	93	c	see Fig. 4	?	?	intact	–	C	–	?	U	+?	?	?	+	+	+?	+?	–	P
Pilobolus crystallinus	26	c	see Fig. 4	D	1	intact	–	C	?	?	U	+?	?	N	+	+	+	?	–	P
Absidia glauca	199	c	see Fig. 4	?	?	intact	–	?	?	?	?	?	?	?	+?	+	+	+	–	?
Entomophthorales																				
Conidiobolus villosus	342	c	see Fig. 4	D	?	intact	–	?	–	?	U	+	?	N	+	+	?	?	–	P
Basidiobolus ranarum	117,401,404	u	ring ?	B	?	di P	–	?	–	+	S	+	1	N	–	+	?	+	–	Ds
Ancylistes sp.	265	c	plaque	B	?	intact	–	C	?	?	U	+	?	N	–	+	?	?	–	P
Strongwellsea magna	168	c	ring	D	?	intact	–	C?	–	+	S	+	1	N	–	+	?	+	–	?
ASCOMYCOTA																				
Hemiascomycetales																				
Saccharomyces cerevisiae	39,40,118,246,260, 261,264,297,298,345, 407,456	u	plaque I	F	1	intact	–	C	–	–	U	+	1	N	+	+	?	+	–	P
Schizosaccharomyces spp.	13,219	u	plaque I	D?	?	intact	–	C	–	–	S	?	?	?	+	+	?	?	–	P
Wickerhamia fluorescens	346	u	plaque	D	?	intact	?	?	–	–	U	?	?	?	+	+	+?	?	–	?
Lipomyces lipofer	406 and pers. comm.	u	plaque	D	?	intact	–	IE	–	–	?	+	?	N	+	+	?	+	–	Di
Dipodascus sp.	407	u	plaque	?	?	intact	–	?	?	–	?	?	?	?	?	+	?	+	–	Di
Taphrina sp.	407	u	plaque	?	?	intact	–	?	?	–	?	+	?	N	?	?	?	?	–	P?

8

Plectomycetales																				
Aspergillus nidulans	197,281,344	c	plaque O	D	?	intact	–	C	–	–	S	+	1	N	+	+	?	+	–	?
Ceratocystis spp. (2)	5,399	c	plaque	D	?	intact	–	IE?	–	–	S	?	?	?	+	+	?	+	–	?
Discomycetales																				
Ascobolus spp. (2)	438,453,454,455	c	plaque	?	?	intact	–	C	–	–	S	+	5+	N	–	+	?	+	–	P
Pyronema domesticum	169,170	c	plaque O	?	?	intact	–	?	–	?	S	+	?	N	?	+	?	?	–	P?
Neotiella rutilans	440	c	plaque	?	?	intact	–	?	–	?	S	+	many	N	–	+	?	?	–	Di
Pyrenomycetales																				
Podospora spp. (2)	453,454,455	c	plaque	?	?	intact	–	C	–	–	S	+	?	N	–	+	?	+	–	P
Xylosphaera polymorpha	20	c	plaque	?	?	intact	–	?	–	–	S	+	2+	N	?	?	?	?	–	P
Neurospora crassa	427	c	sphere	?	?	intact	–	C	–	–	S	+	2+	N	+	+	?	?	–	P
Pustularia cupularis	363	c	plaque	?	?	di P?	–	?	–	?	S	+	?	N	?	?	?	?	–	?
Erysiphe graminis hordeii	226	c	plaque	?	?	intact	–	C	–	–	S	+	2+	N	–	+	?	?	–	P
Fusarium oxysporum	5,6,142	c	plaque	?	?	intact	–	C?	–	–	S	+	?	N	+	+	?	+	–	Ds
Loculoascomycetales																				
Cochliobolus sativus	163	c	plaque	?	?	intact	–	IE	–	–	S	+	?	N	–	+	?	+	–	Di
BASIDIOMYCOTA																				
Heterobasidiomycetales																				
Leucosporidium scottii	221	u	sphere	D	4	oc	–	C	–	?	U	+	?	N	+	+	?	?	–	Di
Rhodosporidium sp.	220	u	bar	D	4	oc	–	C	–	?	U	+	?	N	+	+	?	?	–	Di
Aessosporon salamonicolor	220	u	bar	D	4	oc	–	C	–	?	U	+	?	N	+	+	?	?	–	Di
Ustilago violacea	328,329	u	bar	D	4	oc? diA	–	C	–	–	S	+	1	N	+	+	?	+?	–	Di
Uromyces phaseoli var.vignae	139,140	c	plaque I	D	1?	intact	–	IE?	–	–	S	+	±3	N	+	+	+	+?	–	Di
Puccinia spp. (2)	123,124,450	c	plaque	?	?	intact	–	C	–	–	S	+	?	N	+	+	?	+?	–	Di
Gymnosporangium	256	c	plaque	?	?	intact	–	C	–	–	S	?	?	?	?	?	?	?	–	Di
Homobasidiomycetales																				
Coprinus spp. (5)	116,207,208,215, 216,337,351,411,414	c	sphere	D	4	oc di	–	?	–	–	S	+	2+	N	–	+	?	+	–	Ds
Poria latemarginata	379	c	sphere	?	?	pf	+	IE?	+	–	S	+	?	N	+	+	?	+	–	Di?
Armillaria mellea	266,267	c	sphere	?	?	di P?	–	–	–	–	S	+	?	N	–	?	?	+	–	Ds
Boletus rubinellus	227	c	sphere	?	?	di P	–	–	–	?	S	+	?	N	?	?	?	?	–	Ds
Trametes (Polystictus) versicolor	105,106,107	c	sphere	D	4–5	pf	–	?	–	?	S	+	?	N	?	+	?	?	–	Di
Schizophyllum commune	402	c	sphere	?	?	di A	–	?	–	?	S	?	?	?	+	?	?	?	–	?
Phanerochaete chrysosporium	378	c	?	?	?	pf	–	IE	–	–	S	+	?	N	+	?	?	?	–	?
Pholiota terrestris	439	c	sphere	D?	4–5	di M?	+	–	–	–	S	+	2–3	N	+	+	+?	+	–	Di?
Agaricus bisporus	413	c	sphere	?	?	intact	+	IE?	+	–	S	+	?	N	?	+	?	?	–	?

(*continued*)

TABLE II (continued)

Organism	References	Cell type[b]	Polar structures[c]	Polar structure migration[d]	Spindle development[e]	Nuclear envelope[f]	Perinuclear e.r.[g]	Telophase behavior[h]	Spindle vesicles[i]	Metaphase plate[j]	Chromosome pattern[k]	Kinetochores[l]	Microtubules/kinetochore[m]	Kinetochore location[n]	Central spindle[o]	Spindle elongation[p]	Framework m.t. pattern[q]	Anaphase chromosome migration[r]	Extranuclear spindle[s]	Nucleolus[t]
EUGLENOPHYTA																				
Astasia longa	27,52,400	u	none, see Fig. 4	?	?	intact	—	C	—	—	S	?	?	?	—	+	+?	+	—	P
Euglena gracilis	27,104,205,326	u	none, see Fig. 4	B	1	intact	—	IE	—	—	S	+	2+	N	—	+	+?	+	—	P
Scytomonas pusilla	251	u	?	?	?	intact	—	?	—	?	?	?	?	?	?	?	?	?	—	P
Isonema nigricans	367	u	centrioles V	?	?	intact	—	?	—	—	S	?	?	?	?	?	?	?	—	P
Phacus longicaudus	326	u	none, see Fig. 4	B	?	intact	—	IE	—	—	S	-?	O	O	—	+	+?	?	—	P
DINOFLAGELLATA																				
Prorocentrales																				
Prorocentrum spp. (2)	71,389	u	none	O	?	intact	—	C	—	—	S	?	?	?	O	+	?	?	+	?
Exuviella sp.	71	u	?	?	?	intact	—	?	?	—	S	?	?	?	O	+	?	?	+	?
Gymnodiniales																				
Amphidinium spp. (2)	72,278	u	none	O	?	intact	—	C	—	—	S	+	1	E	O	+	?	?	+	?
Crypthecodinium cohnii	195	u	none	O	6	intact	—	C	—	—	S	?	?	?	O	+	+?	?	+	P
Peridiniales																				
Woloszynskia micra	203	u	?	?	?	intact	?	?	—	—	S	+	?	E	O	+	?	?	+	P
Blastodinium sp.	387,388	u	sphere	D	6	intact	—	C	—	—	S	?	O	?	O	+?	+?	?	+	?
Amoebophrya acanthometrae	42	u	striate plaque	?	?	intact	—	?	—	—	M?	-?	?	O	O	+?	?	?	+	?
Amoebophrya rosei	42	u	striate plaque	?	?	intact	—	?	—	—	S	+	?	E	O	+?	?	?	+	?
Glenodinium foliaceum	70	u	?	?	?	intact	?	?	—	?	S	+	?	E	O	?	?	?	+	?
Oodinium fritillariae	43,44	u	sphere	D	?	intact	—	C	—	—	S	+	1	E	O	+	?	?	+	?
Haplozoon axiothellae	382	u	none	O	?	intact	—	C	—	—	S	+	?	E	O	+	?	+?	+	?
Syndinium sp.	153,339	u	centrioles V	D	6	intact	—	C	—	—	S	+	many	E	O	+	+	—	+	?

10

Solenodinium fallax	153	u	centrioles V	D	6	intact	—	—	—	—	S	+	many	E	O	+	+	?	—	?	?
Peridinium balticum	417	c	none	O	?	intact	—	?	?	?	S	?	?	?	O	+	+	+	?	+	?
Peridinium balticum eukaryo.nuc.	417	c	none	O	—	intact	—	C	—	—	S	truly	amitotic	?	O	+	+	—	—	—	P
CRYPTOPHYTA																					
Chroomonas salina	277,279	u	none	O	5	di P	—	—	+	+	M	+	?	N	—	+	+	—	+	—	Ds
Cryptomonas sp.	276,280	u	none	O	5	di P	—	—	+	+	M	+	1	N	—	+	+	—	+	—	Ds
CHRYSOPHYTA																					
Ochromonas danica	29,383	u	rhizoplast	B	5	di P	—	—	+	+	S	?	?	?	—	+	+	—	+	—	Ds
HAPTOPHYTA																					
Prymnesium parvum	235	u	centrioles V	?	?	di P	—	—	+	+	S?	?	?	?	—	+	+	—	+	—	Ds
XANTHOPHYTA																					
Vaucheria litorea	290	c	centrioles V	B	1	intact	—	IE	—	?	U	?	?	?	+	+	+	—	—?	—	Fr
CHLOROMONADOPHYTA																					
Vacuolaria virescens	145–148	u	none	O	?	pf di T	—	—	+	+	S	+	many	N	—	+	+	?	?	—	Fr
BACILLARIOPHYTA																					
Pennales																					
Diatoma vulgare	225,320	u	striate bar	D	5	di P	+	—	+	+	M	—	O	O	+	+	+	+	+	—	Ds
Surirella ovalis	418	u	striate bar	D	5	di P	+	—	+	+	M	C	?	N	+	+	+	+	+	—	Ds
Nitzschia sigmoideae	323	u	striate bar	?	?	di P	+	—	?	+	M	+	many	N	+	+	+	?	?	—	Ds
Synedra ulna	323	u	striate bar	D	5	di P	+	—	?	+	M	+	O	O	+	+	+	+	+	—	Ds
Pinnularia sp.	324,325	u	striate bar	D	5	di P	+	—	+	+	M	?	?	?	+	+	+	+	+	—	Ds
Fragilaria	419	u	striate bar	D	5	di P	+	—	+	+	M	?	?	?	+	+	+	+	+	—	Ds
Centrales																					
Lithodesmium undulatum	236–239	u	striate bar	D	5	di P	+	—	+	+	M	—	O	O	+	+	+	+	+	—	Ds
Melosira varians	416	u	striate bar	D	5	di P	+	—	+	+	M	—	O	O	+	+	+	+	+	—	Ds
CHLOROPHYTA																					
Chlorococcales																					
Chlorella pyrenoidosa	14,446	u	centrioles V	B	?	pf	+	?	?	?	?	?	?	?	?	?	?	?	?	—	?
Kirchneriella lunaris	302	c	centrioles ?	B	2	pf	+	IE	—	+	S	?	?	?	—	+	?	?	+	—	Ds
Hydrodictyon reticulatum	241	c	centrioles V	B	2	pf	+	IE	—	+	M	+	many	N	—	—	—	+	+	—	Ds

(continued)

TABLE II (continued)

Organism	References	Cell type[b]	Polar structures[c]	Polar structure migration[d]	Spindle development[e]	Nuclear envelope[f]	Perinuclear e.r.[g]	Telophase behavior[h]	Spindle vesicles[i]	Metaphase plate[j]	Chromosome pattern[k]	Kinetochores[l]	Microtubules/kinetochore[m]	Kinetochore location[n]	Central spindle[o]	Spindle elongation[p]	Framework m.t. pattern[q]	Anaphase chromosome migration[r]	Extranuclear spindle[s]	Nucleolus[t]
Tetraedron bitridens	306	c	centrioles V	B	2	pf	+	IE	-	+	S	?	?	?	-	+	?	?	-	Ds
Pediastrum boryanum	240	c	centrioles V	B	2	pf	+	IE	-	+	S	?	?	?	-	+	?	+	-	Ds
Sorastrum sp.	240	c	centrioles V	B	2	pf	+	IE	-	+	S	+	?	N	-	+	?	+	-	Ds
Scendesmus spp. (2)	273,322	c	centrioles V	B	2	pf	+	?	-	+	S	?	?	?	-	?	?	?	-	Ds
Volvocales																				
Chlamydomonas reinhardi	56,180	u	centrioles V	?	?	pf	+	?	-	+	S	?	?	?	-	+	?	+	-	Ds
Chlamydomonas moewusii	421	u	centrioles V	B	1	intact	-	IE	+	+	S	+	many	Z	-	+	?	+	-	Ds
Tetraspora sp.	309	u	centrioles V	?	2	pf	+	C?	-	+	M?	+	2+	Z	-	+	?	+	-	Ds
Volvox aureus	62	u	none	O	?	pf	-	?	-	?	?	?	?	?	?	?	?	?	-	Ds
Ulotrichales																				
Ulothrix fimbriata	87	u	centrioles V	?	?	di P	-	-	+	+	M	?	?	?	-	+	?	?	-	Ds
Ulva mutabilis	214	u	centrioles V	?	?	pf	-	?	-	+	M	?	?	?	-	+	?	+	-	Ds
Klebsormidium subtilissimum	305	u	centrioles	B	?	di P	-	-	+	+	M	?	?	?	-	+	?	-	-	Ds
Klebsormidium flaccidum	88	u	centrioles	?	?	di P	-	-	+	+	M	?	?	?	-	+	?	?	-	Ds
Microspora sp.	307	u	centrioles V	B	2	pf	-	IE	+	+	S	+	many	N	-	+	?	?	-	Di?
Stichococcus chloranthus	311	u	none?	?	?	di P	-	-	+	+	M	?	?	?	-	+	?	?	-	Ds
Cylindrocapsa involuta	319, N.B. also 152	u	centrioles V	B	?	pf	-	IE	-	+	M	?	?	?	-	+	-	-	-	Ds
Chaetophorales																				
Trentepohlia aurea	112	c	centriole	?	?	intact	-	?	-	?	?	?	?	?	-	+	?	?	-	Ds
Coleochaete scutata	242	u	centrioles V	B	2	di P	-	C	+	-	M	-?	O	O	-	+	?	?	-	Ds
Stigeoclonium helveticum	87	u	centrioles V	?	?	intact	-	?	+	-	M	?	?	?	-	+	?	?	-	Ds
Oedogoniales																				
Oedogonium spp. (2)	56a,56b,315,316	u	intranuc. sphere	B	3?	intact	-	IE	-	+	S	+	many	N	-	+	?	+	-	Fr
Bulbochaete hiloensis	310	u	none	O	?	intact	-	IE	-	+	S	+	many	N	-	+	?	+	-	P

12

Taxon	Pages																	
Siphonocladales																		
Cladophora spp. vegetative (2)	211,268	c	procentrioles	B	1	intact	IE	–	+	S	+	many	N	+	?	+	–	Fr
Cladophora spp. motile cells (2)	224,375	c	centrioles V	B	1	intact	IE	–	+	S	+	many	N	±	?	+	–	Fr
Pleurastrum sp.	263	c	none	?	5	?	?	–	+	S	?	?	?	+	?	+	–	Ds
Acrosiphonia spinescens	164	c	centrioles	B	2	pf	?	–	+	M	?	?	–	–	?	+?	–	Ds
Bryopsis hypnoides	38	c	centrioles	?	?	pf	?	?	?	S	?	?	?	?	?	?	–	P
Dasycladales																		
Batophora oerstedii	212	c	none?	?	?	intact	?	–	+	S	?	?	–	+	?	–	–	Ds?
Acetabularia mediterranea	449	c	?	?	?	intact	?	–	?	S	?	?	–	–	?	?	–	Ds?
ZYGNEMATOPHYTA																		
Spirogyra spp. (3)	92,181,268	u	none	O	2–5	di A	–	+	+	S	+	many	N	+	?	+	–	As
Mougeotia sp.	19	u	none	O	2–5	pf	IE	–	+	S	+	many	N	+	?	+	–	Ds
Closterium litorale	317	u	none	O	5	di P	–	+	+	S	+	1?	N	+	?	–	–	Ds
Cosmarium botrytis	304	u	none	O	5	di P	–	+	+	S	–?	O	O	–	?	+	–	Ds
PRASINOPHYTA																		
Pyramimonas parkeae	291	u	rhizoplast?	B	2	di ?	–	–	+	M	?	?	?	+	–?	?	–	Ds
Pedinomonas minor	321	u	centrioles V	?	1	intact	?	?	?	?	?	?	?	+	?	+?	–	?
Platymonas subcordiformis	398	u	rhizoplast?	?	?	?	?	–	+	M	?	?	?	+	?	+?	–	Ds
Heteromastix angulata	247	u	rhizoplast	B	2	pf	?	–	+	M	+	1	N	+	–?	+	–	Ds
CHAROPHYTA																		
Nitella missouriensis	425	u	centrioles V	B	?	di ?	?	+	+	S	+	many	N	+	?	?	–	Ds
Chara sp.	300	u	none	O	5	di P	–	+	+	S	+	?	N	+	?	+	–	As
PHAEOPHYTA																		
Zonaria farlowii	270	u	centrioles V	?	?	pf	?	+	+	?	?	?	?	+	?	?	–	Ds
Pylaiella littoralis	245	u	centrioles V	B	2	pf	IE?	+	+	M?	?	?	?	–	?	?	–	Ds
RHODOPHYTA																		
Griffithsia flosculosa	299	u	sphere	?	?	intact	?	–	+	M	?	?	?	+	?	?	–	Ds
Membranoptera platyphylla	223	u	ring	B?	?	pf	C	+	+	M	+	many	N	+	?	+	–	Ds
Porphyridium purpureum	36,362	u	granule	B	2	pf	IE	+	+	U	+	1	N	+	?	+	–	Ds
Polysiphonia spp. (2)	374	u	ring	B	2	pf	IE	?	+	S	+	7–8	?	+	?	+	–	Ds

(continued)

13

TABLE II (continued)

Organism	References	Cell type[b]	Polar structures[c]	Polar structure migration[d]	Spindle development[e]	Nuclear envelope[f]	Perinuclear e.r.[g]	Telophase behavior[h]	Spindle vesicles[i]	Metaphase plate[j]	Chromosome pattern[k]	Kinetochores[l]	Microtubules/kinetochore[m]	Kinetochore location[n]	Central spindle[o]	Spindle elongation[p]	Framework m.t. pattern[q]	Anaphase chromosome migration[r]	Extranuclear spindle[s]	Nucleolus[t]
SPOROZOA																				
Gregarinida																				
Stylocephalus longicollus	65	c	granule	D	6	Fig. 1	–	–	?	–	S	+	many	E	–	+	?	+	–	Ds?
Coccidia																				
Sarcocystis dispersa	377	c	cc + centrioles V	?	?	intact	–	?	–	–	S	?	?	?	?	+	?	?	–	P
Eimeria spp. (6)	57,121,159,185,186, 249,250,341,385	c	cc + centrioles V	D	?	intact	–	S	–	–	S	+	1	N	?	+	?	+	–	P
Eimeria aubernensis	122	c	cc + centrioles V	B ?	?	intact	–	S	–	?	?	+	?	?	?	?	?	?	–	?
Eimeria necatrix	73,74	c	cc + centrioles V	D	2	pf	–	S	–	–	S	+	1	N	?	+	?	+	–	P
Eimeria flaciformis	250	c	cc + centrioles V	?	2	pf?	?	?	?	?	S	?	?	?	?	?	?	?	–	?
Hepatozoon domerguei	433	u	ring	D	2	pf	?	?	–	?	S	?	?	?	?	?	?	?	–	?
Globidium gilruthi	330	c	centrioles V	B?	3?	intact	–	?	–	–	?	+	1	N	?	?	?	+	–	?
Haemosporida																				
Plasmodium spp. (4)	1,2,3,47,162,364,394, 410	c	plaque I	F	1	intact	–	C	–	–	U	+	1	N	?	+	–	+	–	P
Plasmodium elongatum	2	c	plaque	?	?	intact	–	?	–	?	S	+	1?	N	?	+	?	?	–	?
Plasmodium cynomolgi	409	c	plaque I	?	?	intact	–	?	–	?	S	+	?	N	?	+	?	+	–	?
Haemoproteus colombae	30	u	intranuc.ring+cents.	F	1	intact	–	C?	–	?	S	+	?	N	–	?	?	?	–	?
Parahaemoproteus velans	67	u	sphere	?	?	intact	?	?	–	?	?	+	?	N	?	?	?	?	–	?
Leucocytozoon spp. (2)	68,448	c	plaque	F	1	intact	–	?	–	?	U	+	1?	N	?	?	?	?	–	P

Taxon	Refs																
Haplosporida																	
Toxoplasma gondii	381	u	cc	?	?	intact	−	−		?	?	?	?	?	?	?	−
Minchinia spp. (2)	292,293,295	c	intranuc.plaque	?	?	intact	−	C	U	+	1.4+?	N	+	+?	+	?	P
CNIDOSPORIDIA																	
Stempellia mutabilis	66	c	plaque O	?	1	intact	−	−	S	+	1	N	−	?	+	?	?
Gurleya chironomi	213	u	plaque O	?	?	intact	−	?	S	+	many	N	+	?	+	?	?
Glugea weissenbergi	391	c	plaque O	?	?	intact	−	?	S	+	1?	N	+	+	+	?	?
Nosema vivieri	430	c	none ?	?	?	intact	−	?	U	+	1?	N	?	?	?	?	?
Thelohania bracteata	428	c	plaque I	?	?	intact	−	S	?	?	?	?	?	?	?	?	?
Metchnikovella hovassei	432	c	plaque O	?	?	intact	−	−	?	?	?	?	?	?	?	?	P
SARCODINA or RHIZOPODA																	
Pelomyxa spp. (2)	58,352	c	none	O	?	?	+	IE?	S	+	1?	N	−	?	?	?	Ds
Naegleria gruberi	99,366	u	none	O	?	intact	−	?	?	?	?	?	?	?	?	?	P
ACTINOPODA																	
Radiolarida																	
Collozoum pelagicum	154,338	u	plaque + cents. V	F	1	intact	−	C?	S	+	2+	N	+	?	+	?	?
Aulacantha scolymantha	204	u	?	?	?	di P	+	?	S	−	O	O	−	−	−	−	?
Acantharida																	
Acanthochiasma rubescens	84	u	plaque + cents. V	F	1	intact	−	S	S	+	1	N	+	±	+	?	Ds
Acantholithium stellatum	84	u	plaque + cents. V	F	1	intact	−	S	S	+	1	N	+	±	+	?	Ds
Astrolonche serrata	84	u	plaque + cents. V	F	1	intact	−	S	S	+	1	N	+	±	+	?	Ds
Gigartacon mülleri	84	u	plaque + cents. V	F	1	intact	−	S	S	+	1	N	+	±	+	?	Ds
Heteracon biformis	84	u	plaque + cents. V	F	1	intact	−	S	S	+	1	N	+	±	+	?	Ds
Conacon sp.	84	u	plaque + cents. V	F	1	intact	−	S	S	+	1	N	+	±	+	?	Ds
Amphilonche elongata	84	u	plaque + cents. V	F	1	intact	−	S	S	+	1	N	+	±	+	?	Ds
Lithoptera mülleri	84	u	plaque + cents. V	F	1	intact	−	S	S	+	1	N	+	±	+	?	Ds
FORAMINIFERA																	
Myxotheca arenilega	368,369,371	c	centrioles V	?	?	intact	−	?	S	?	?	?	−	?	+	?	Ds
Iridia lucida	51	?	sphere + centrioles	F?	1	intact	+	C	S	+	?	N	−	?	+	+	?
Allogromia laticollaris	370	u	large sphere	?	?	intact	−	?	S	?	?	?	−	?	?	?	Di

(continued)

15

TABLE II (continued)

Organism	References	Cell type[b]	Polar structures[c]	Polar structure migration[d]	Spindle development[e]	Nuclear envelope[f]	Perinuclear e.r.[g]	Telophase behavior[h]	Spindle vesicles[i]	Metaphase plate[j]	Chromosome pattern[k]	Kinetochores[l]	Microtubules/kinetochore[m]	Kinetochore location[n]	Central spindle[o]	Spindle elongation[p]	Framework m.t. pattern[n]	Anaphase chromosome migration[j]	Extranuclear spindle[g]	Nucleolus[i]
ZOOMASTIGINA																				
Rhizomastigida																				
Histomonas maleagridis	365	u	flagellar roots	?	?	intact	–	?	?	?	?	?	?	?	?	?	?	?	+	?
Protomonadida																				
Leishmania tropica	24	u	?	?	?	intact	–	?	–	?	?	?	?	?	?	?	?	?	–	?
Trypanosoma spp. (5)	64,149,173,429	u	none	O	?	intact	–	C	–	?	S	+	1	N	+	+	+?	?	–	a
Polymastigida																				
Dientamoeba fragilis	45	u	atractophores etc.	?	?	intact	?	?	?	?	?	?	?	?	?	+	?	?	+	?
Trichomonadida																				
Trichomonas vaginalis	37	u	atractophores etc.	D	6	intact	–	C	–	–	S	+	many	E	+	+	+	+	+	?
Hypermastigida																				
Barbulanympha ufala	156,158,174,175, 340	u	atractophores etc.	D	6	intact	–	C	–	–	S	+	many	E	+	+	+	?	+	?
Trichonympha agilis	155,156,192	u	atractophores etc.	D	6	intact	–	C	–	–	S	+	many	E	+	+	+	?	+	?
Staurojoenina caulleryi	156	u	atractophores etc.	D	6	intact	–	?	–	?	?	?	?	E	+	?	?	?	+	?
Spirotrichonympha psammotermitidis	156	u	atractophores etc.	D	6	intact	–	?	–	–	S	+	many	?	+	?	+	?	+	?
Holomastigotoides hemigynum	156	u	atractophores etc.	D	?	intact	–	?	–	?	?	?	?	E	?	?	+	?	+	?
Lophomonas striata	157	u	atractophores etc.	D	6	intact	–	?	–	–	S	+	many	?	+	?	?	?	+	?

CILIOPHORA

Gymnostomatida																	
Nassula spp. (2) μ	422	c	intranuc. vesicle	?	?	intact	—	—	S	+	1	N	—	+	+	+?	—
Nassula ornata μ	335	c	?	?	?	intact	C	—	S	+	?	?	—	+	?	?	—
Alloizona trizona μ	113	c	?	?	?	intact	IR	—	S	?	?	?	—	+	?	?	—
Dileptus anser μ	431	c	none	?	O	intact	IR	—	S	—?	?	O	—	+	?	?	—
Didinium nasutum μ	183	c	see Fig. 4	3	?	intact	IE	—	S	+	?	N	—	+	+	+	—
Hymenostomatida																		
Ichthyophtirius multifiliis μ	128	c	none	?	O	intact	?	?	S	?	?	?	?	+	?	?	?
Paramecium aurelia μ	182,396	c	none	?	O	intact	IE	+	S	+?	?	N	+	+	+	+?	—
Paramecium aurelia meiosis μ	395	c	none	?	O	intact	C	+	S	+?	?	N	—	+	—	—	—
Paramecium bursaria μ	210	c	none ?	?	O	intact	?	?	S	?	?	?	—	+	—	—	—
Tetrahymena pyriformis μ	60,177,201	c	?	?	?	intact	?	?	S	+	10	N	—	+	+	+	—
Heterotrichida																		
Blepharisma spp. (2) μ	178	c	none	?	?	intact	IR	—	S	—	O	O	—	+	+	+	—
Oligotrichida																		
Diplodinium sp. μ	354	c	?	?	?	intact	?	—	S	?	?	?	?	?	?	?	—
Hypotrichida																		
Gastrostyla steinii μ	434	c	none	?	?	intact	IR	—	S	?	?	?	—	+	+	?	—
Peritrichida																		
Epistylis anastatica μ	49	c	none ?	?	O	intact	?	?	S	?	?	?	?	?	?	?	—
Vorticella nebulifera μ	49	c	none ?	?	O	intact	?	?	S	?	?	?	?	?	?	?	—
Suctorida																		
Tokophrya infusionum μ	252	c	none	?	O	intact	C	—	S	?	?	?	+	+	?	?	P
Paracineta limbata μ	127	c	none	?	O	intact	?	?	S	?	?	?	—	+	?	?	Fr
Gymnostomatida																		
Nassula spp. (2) M	422	c	none	?	O	intact	C	—	S	—	?	?	+	+	+	?	?
Didinium nasutum M	183	c	none	?	O	intact	C	—	S	—	O	O	+	+	—?	—?	?
Loxodes magnus M	336	c	none	?	O	intact	IR	—	S	+	many	N	+	+	+	+	?
Trichostomatida																		
Isotricha sp. M	113	c	?	?	?	intact	?	?	S	?	?	?	?	?	+	?	?
Hymenostomatida																		
Paramecium aurelia M	182,397	c	none	?	O	intact	C	—	S	—?	O	O	—	+	—?	—?	P
Tetrahymena pyriformis M	60,83,445	c	?	?	?	?	?	—	S	?	?	?	—	?	?	—	?
Heterotrichida																		
Blepharisma spp. (3) M	178,179	c	?	?	?	intact	C	—	S	?	?	?	+	+	+	+	?
Protocrucia sp. M	355	c	?	?	?	?	?	—	S	?	?	?	—	+	?	?	?
Oligotrichida																		
Diplodinium sp. M	354	c	?	?	?	intact	?	?	?	?	?	?	?	?	?	?	?

(continued)

TABLE II (continued)

Organism	References	Cell type[b]	Polar structures[c]	Polar structure migration[d]	Spindle development[e]	Nuclear envelope[f]	Perinuclear e.r.[g]	Telophase behavior[h]	Spindle vesicles[i]	Metaphase plate[j]	Chromosome pattern[k]	Kinetochores[l]	Microtubules/kinetochore[m]	Kinetochore location[n]	Central spindle[o]	Spindle elongation[p]	Framework m.t. pattern[q]	Anaphase chromosome migration[r]	Extranuclear spindle[s]	Nucleolus[t]
Hypotrichida																				
Gastrostyla steinii M	435	c	none	O	7	intact	–	C?	–	–	S	–?	O	O	–	+	+	?	–	P
Stylonychia mytilus M	435	c	none	O	7	intact	–	C?	–	–	S	–?	O	O	–	+	+	?	–	P
Peritrichida																				
Campanella umbellaria M	49	c	none ?	O	?	intact	–	?	–	–	S	–?	O	O	–	+	+	?	–	?
Suctorida																				
Tokophrya infusionum M	252	c	none	O	?	intact	–	C	–	–	S	?	?	?	–	+	?	?	–	–
Acineta tuberosa M	17	c	none	O	?	intact	–	C	–	–	S	?	?	?	–	+	?	?	–	–

[a] The data presented in this table is derived from the numbered references, but in many cases, the interpretation is that of this reviewer. Such interpretation has been used where the original authors did not comment on a specific structure and where this reviewer's interpretation of the published pictures differs from that of the original authors. Taxa below the phyla are only designated separately when there is reasonable evidence for heterogeneity within phyla. Likewise, separate species are only listed when there is heterogeneity within genera. In genera in which a number of different species have been investigated, that number of species is indicated in brackets. In many cases, not all listed references contain all of the data for that organism or genus, thus each line represents the sum of the data from all references on that line. Among the dinoflagellates, all nuclei examined are the typical dinoflagellate type with one exception as noted. Among the ciliates, "μ" and "M" denote micro- and macronuclei, respectively. Throughout, (+), presence of the feature; (–) absence of that feature; (?) insufficient data for determination; O, denotes inapplicability of that character in view of preceding columns (e.g., one cannot describe polar structure migration time when such structures are absent); and a (?) after an entry denotes that that point is probably the case, but the evidence is not conclusive.

[b] This column differentiates between uninucleate cells (u) and coenocytes with more than one nucleus per cell (c).

c This column describes the morphology of the NAOs, most types of which are illustrated in Fig. 4. Basal bodies are not differentiated from centrioles in this table. In the case of centrioles, the suffix "V" denotes an orthogonal arrangement where the centrioles lie at approximately right angles to each other, whereas " – – " denotes the rarer 180° orientation; "c c" denotes a centrocone. In those cases where it is known, the position of a plaque with respect to the nuclear envelope is designated by "I" (in a close fitting pore in the envelope) and "O" (outside the intact envelope). In a number of cases (e.g., *Astasia, Euglena, Phacus*), basal bodies lie close to the spindle poles, but to one side. These cases are listed as having no centrioles at their poles.

d B, Denotes migration to the site of the spindle poles prior to spindle formation; D, denotes migration during spindle formation; F, denotes the formation of substantial adjacent fan-shaped arrays of microtubules that reorientate during spindle formation (see Fig. 6).

e Spindle (primarily nonkinetochore-associated components) development is described by the following number codes, which are further explained, in part, in Fig. 6: (1) forms inside the nucleus from differentiated regions of the nuclear envelope or from plaques set in the nuclear envelope; (2) pushes through polar fenestrae from extranuclear polar structures; (3) forms from structures that are permanently intranuclear and that are typically free of the nuclear envelope; (4) forms from NAOs, which enter the nucleus from the cytoplasm at prophase; (5) forms in the cytoplasm and sinks into the nucleus as the nuclear envelope disperses; (6) forms between extranuclear structures and remains external to the intact nuclear envelope; (7) forms inside the nucleus from numerous ill-defined sites.

f Intact, remains intact throughout mitosis; pf, develops polar fenestrae; di P, M, A, T, disperses at pro-, meta-, ana-, or telophase, respectively; oc, transiently opens and closes again to admit an extranuclear NAO. See Fig. 1.

g Refers to the presence of one or more perinuclear sheaths or endoplasmic reticulum outside of the nuclear envelope.

h As defined in Fig. 2, with (−) indicating that the nuclear envelope disperses prior to telophase and thus exhibits none of the listed behaviors.

i Refers to the presence (+) or absence (−) of vesicles within the spindle.

j Denotes the presence (+) or absence (−) of a well-defined equatorial array of chromosomes at metaphase.

k Refers to the arrangement of the metaphase chromosomes into distinguishable separate (S) chromosomes or a fused mass (M) of chromatin, or denotes the absence of sufficient chromatin condensation (U) to detect when the arrangement is.

l Denotes the presence of detectable structures at the point at which kinetochore microtubules intersect the chromatin. An absence (−) of structure does *not* imply an absence of kinetochore microtubules. c, Collar present on the fused mass of chromatin.

m Indicates, as accurately as possible, the number of microtubules per kinetochore. Numbers, such as 2+, denote a number close to, but a little higher than the listed number, as distinct from "many" which may be in the 20+ region.

n N, In the nucleoplasm; E, in the nuclear envelope.

o Denotes the aggregation of the framework, or nonkinetochore microtubules, into a discrete central bundle, at least by telophase.

p Indicates presence (+) or absence (−) of spindle elongation during anaphase–telophase.

q Describes the presence of close packing of the nonkinetochore microtubules in a configuration that would be predicted if intermicrotubule bridges in the range of 40 nm in length existed in the spindle. See Fig. 5.

r Indicates the presence (+) or absence (−) of chromosome-to-pole movements at anaphase as opposed to increase in inter-chromosome distance due to simple spindle elongation.

s Indicates those organisms that have a spindle lying external to a persistent or intact nuclear envelope.

t Describes the behavior of the nucleolus as outlined in Fig. 3 and Section II. D. P, Persistent; D, P, Persistent; Di, discardive; Fr, fragmentary; As, associative; Ds, dispersive; (−), absent from these nuclei; a, two persistent species and two dispersive species.

measure of mitotic accuracy.) However, at least in principle, useful data from organisms with diverse spindle types will become available. Such data could be a valuable indicator of the relative state of advancement of a particular mitotic variant. We might predict that relatively inefficient "relic" spindles could survive in organisms, such as coenocytes, where the penalties of a mitotic error may be light and thus would exert little selection pressure for increased genetic efficiency. It should, of course, be remembered that a number of organisms have evolved an alternate solution to the need for genetic efficiency by developing highly polyploid nuclei in which precise equipartitioning of the genome is not necessary and almost certainly does not occur. Examples of this are the ciliate macronuclei and probably the "eukaryotic" nuclei of certain dinoflagellates (Tippit and Pickett-Heaps, 1976). In the ciliates, such nuclei are secondarily derived from the micronuclei in response to specific cellular requirements and thus should not be considered representative of a primitive condition.

Measuring the energetic efficiency of various mitotic systems appears to be a daunting task that is unlikely to yield significant results in the near future. The reason for this is the apparently small amount of energy needed to accomplish mitosis. For example, Forer (1969) and Wolpert (1965) have calculated that only a few molecules of ATP could theoretically move the largest chromosomes, and Amoore (1963) has shown that, at least in pea roots, mitosis can continue normally in cells containing about 1.5% of their normal ATP complement. However, if the measuring difficulties can be overcome, the above points suggest that one may well find "relic" energy-inefficient mitotic systems since, with such a low percentage of total cellular energy involved, there could be relatively little selection pressure for greater efficiency.

B. Membranes

1. *Nuclear Envelope*

The behavior of the nuclear envelope during mitosis has received considerable attention from previous discussants of spindle evolution. There are three essential points that need to be emphasized.

1. As mentioned by most previous authors, great care is needed in interpreting the data on the persistence (closed mitosis) or dispersal (open mitosis) of the nuclear envelope because of the potential for artifact. In many organisms, the envelope undoubtedly disperses during prophase as it does in higher plants and animals; and, equally clearly, in many organisms it unambiguously remains completely intact throughout mitosis. However, in a number of other cases, the situation is ambiguous. For example, in the basidiomycetes, it is commonly said that the envelope disperses; yet Thielke (1974) has shown that, in the same genus

and species in which previous workers found dispersal, we can, in fact, demonstrate persistence. Similarly, in one species, Gull and Newsam (1976) illustrate considerable variability in degree of persistence of the envelope. It is clearly much more probably to find artifactual dispersal rather than artifactual persistence. Given this result, we must then also question the other reports of dispersal of the nuclear envelope in the other basidiomycetes, including the claimed transient openings in the envelope reported for the basidiomycetous yeasts by McCully and Robinow (1972a, b). (However, in this latter group, transient opening is quite likely since that is also what apparently occurs in *Coprinus* [Thielke, 1974].) There is also no a priori reason why the nuclear envelope may not vary in its sensitivity to fixation-induced disruption at different stages of the mitotic cycle. Other reports in which the apparent disruption of the nuclear envelope seems particularly questionable due to the overall poor quality of fixation include those on *Pyramimonas, Platymonas, Pleurastrum,* and *Pelomyxa.* However, as emphasized by Fuller (1976), the common observation of envelope disruption in, for example, the basidiomycetes following fixation schedules that preserve the envelope in other cells, may well indicate some fundamental difference in the envelope of these organisms. This difference, in itself, may have evolutionary significance.

2. The range of demonstrated behavior patterns of the nuclear envelope is much greater than often realized, as illustrated in Fig. 1. Examination of Table II shows that the three most common patterns are complete dispersal at prophase (Fig. 1a), complete persistence with no discontinuities other than the normal nuclear pores (Fig. 1c), and the formation of polar fenestrae in an otherwise intact envelope (Fig. 1f). However, there are intermediate types of behavior in which the membrane remains intact until it disperses at anaphase or telophase (Table 2). Similarly, in the polar fenestrae category, there is variability from the rather large fenestrae of, for example, the Chytridiales and Chlorococcales to the discontinuities that seem to completely seal to the edges of the polar structures (see Section II, F, 1 for a discussion of the terminology of these structures) of *Saccharomyces, Uromyces,* and *Plasmodium* (Fig. 1e). Because this latter group has such a close seal between the nucleus associated organelle (NAO) and the nuclear envelope during mitosis, there is effectively an obvious, continuous barrier between cytoplasm and nucleoplasm. Thus, these organisms are listed as having an intact nuclear envelope (Fig. 1e). In addition to these fairly common patterns of behavior, there are two more restricted patterns known. In *Stylocephalus,* the envelope persists only in the form of individual envelopes around each chromosome (Fig. 1b). The other variation appears to occur in *Coprinus* (Thielke, 1974), *Leucosporidium, Rhodosporidium, Aessosporon,* and possibly *Ustilago* (this interpretation differs from that of the authors and assumes fixation problems of the type found by many workers in *Coprinus,* as previously discussed), where the nuclear envelope transiently opens to permit entry of the

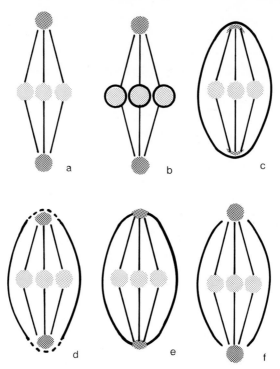

FIG. 1. Behavior of the nuclear envelope during mitosis. Light stippled circles represent the chromosomes, heavy stippled circles the polar structures, straight lines the spindle, and the heavy line the nuclear envelope. In (a) the envelope disperses, usually at prophase, to give an open spindle (see "di P, M, A, or T" in Table II); (b) represents a rare situation, only reported for *Stylocephalus,* in which the nuclear envelope disperses, but forms individual envelopes around each chromosome throughout mitosis; (c) and (e) are closed divisions with intranuclear spindle organizing or polar structures and membrane inserted NAOs respectively (see "intact" in Table II); (d) illustrates the "oc" behavior of Table II in which the membrane seems to transiently open to admit the NAO, then reseals around it during mitosis; (f) represents the polar fenestrate ("pf" in Table II) type of behavior where the cytoplasmic NAOs lie in large openings of the nuclear envelope through which the spindle was formed.

polar structures, then closes again for a further part of mitosis. This latter variation may prove to be more common, given the problems of fixation previously discussed.

One of the aspects of nuclear envelope behavior which has previously been somewhat neglected, is its behavior at telophase. In those organisms in which the membrane disperses sometime prior to telophase, it typically reforms around the separated chromosomes by apparent fusion of largely endoplasmic reticulum-derived cisternae. This system clearly requires the evolution of a totally different control system from the intuitively simpler enlargement and median constriction of the persistent envelope (Fig. 2a) in those organisms that have an intact envelope and no telophase interzone expulsion. However, in addition to these two types of behavior, those organisms with the intact or polar fenestrae type of envelope behavior may (1) constrict the nuclear envelope in two places, thus expelling a portion of the nucleus and/or spindle from the daughter nuclei (interzone expulsion, Fig. 2b), (2) form a completely new envelope around the chromosomes inside the persistent nuclear envelope, which subsequently disintegrates (internal reformation, Fig. 2c), or (3) produce a septum that grows across the nucleus as an extension of the old envelope (septation, Fig. 2d). The evolutionary aspects of these variations will be returned to in Sections II, F, 1, III, B, and III, C.

3. The behavior of the nuclear envelope is not entirely constant within an organism, as might be hoped for in a phylogenetically useful marker. As shown by Aldrich (1969), the envelope of *Physarum* nuclei remains largely intact in the coenocytic plasmodial mitoses, but disperses at prophase in the uninuclear myxamoebae. Steffens and Wille (1977) have reinforced the earlier arguments of Ross (1968), that this dimorphic behavior is not a consequence of different ploidy levels (*2n* versus *n*, respectively) by documenting comparable plasmodial mitoses in haploid plasmodia. These observations mean that there is presumably either some cytoplasmic control over the fate of the nuclear envelope or there are circumstances that favor the evolution of different behaviors at different life cycle stages, a point returned to in Section II, G, 5. This suggests that there is physiological significance to the different types of nuclear envelope behavior and

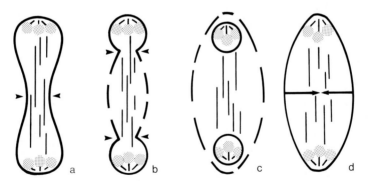

FIG. 2.   Behavior of the nuclear envelope at telophase. (a) After the chromosomes (stippled circles) have moved to the poles, the nuclear envelope may become medianly constricted (''C'' in Table II), or (b) it may be constricted in two places so that the interzone is cut off from the daughter nuclei (''IE'' in Table II). (c) A new envelope may form inside of, and independently from, the old envelope so that a large part of the nucleus is excluded from the daughter nuclei (''IR'' in Table II) or (d) a septum may grow centripetally across the equator of the nucleus to produce a similar result to that of median constriction (''S'' in Table II).

thus introduces the distinct possibility of convergent evolution in this behavior, thereby reducing its effectiveness as a phylogenetic marker.

Apart from the above points, the only remaining observation that needs to be made about the nuclear envelope is that, in general terms, there is a correlation between nuclear envelope behavior during mitosis and the major taxa, as seen in Table II. This correlation even extends to the dimorphic behavior referred to previously in *Physarum,* since it does not seem to occur outside of the myxomycetes. Because of the preceding general correlation, one may believe that nuclear envelope behavior is reasonably stable and thus useful, with care, as a phylogenetic marker. Its possible behavior during evolution will be returned to in Sections III, B and C.

2. *Perinuclear Reticulum*

In addition to a persistent nuclear envelope, some organisms develop a cisternum of largely nonfenestrated endoplasmic reticulum around their mitotic nuclei (Table II). This feature appears to have a very limited distribution, occurring generally in the chytrids, Plasmodiophoromycota, and Chlorococcales; in advanced, but not, supposedly, primitive (*Porphyridium*) Rhodophyta, and in one member each of the basidiomycetes (*Poria*) and foraminifera (*Iridia*). The consistent occurrence in such a few taxa would suggest that the character has excellent promise as a phylogenetic marker, especially as no function has yet been ascribable to it (e.g., we cannot even correlate it with the coenocytic habit, see Table II). However, to suggest common ancestry of the above groups, and to make the concomitant suggestion that the Chlorococcales are only distantly related to the other chlorophytes, seems so outlandish in our current state of knowledge that it will probably be more profitable to look for common physiological factors leading to convergent evolution of this structure.

3. *Vesicles*

As seen in Table II, numerous organisms with both open and closed spindles contain variously large vesicles (or in some cases, tubular cisternae; the two are not distinguished here) in their mitotic spindles. The role, or roles, of these vesicles is totally unknown, although in some cases they may be new nuclear envelope that is developing at an early stage of mitosis (e.g., *Sorosphaera*). However, a likely role is as sarcoplasmic reticulum equivalents, i.e., as calcium ion concentration control organelles used in modulating spindle development and function. Harris (1975), Wick and Hepler (1976), and Hepler (1977) have developed arguments for such a role for the vesicles found in the mitotic systems of many higher organisms. There is no experimental data to support such a hypothesis for any of the vesicles reported in the organisms listed in Table II. However, if such a role does prove to be correct, it is interesting to note that the possession of such a control system located in vesicles appears to be most common in higher plants and animals and is restricted to some members of the fungi and algae (Table II), thus perhaps indicating a more advanced type of mitosis in these organisms. It must be remembered that the absence of vesicles does *not* mean absence of a hypothetical calcium ion-mediated control system, since a persistent nuclear envelope could well have a sarcoplasmic reticulum type of function. Thus the organization of such a system in vesicles lying free in the spindle is perhaps the only indication of evolutionary advancement.

C. MATRIX

1. *Composition*

The purpose of this brief discussion of the nuclear matrix is to indicate its existence and its potential as a component that may prove to have value as an indicator of mitotic evolution. It is quite clear, especially in those organisms with an intact nuclear envelope, that the obvious features of the mitotic apparatus are surrounded and permeated by a mass of largely unidentified nucleoplasm. The nucleoplasm is evidently important to the mitotic process, if in no other way than as the bathing medium in which the entire event occurs. However, in addition to this "role," it is obvious that the complex shape changes undergone by, for example, telophase nuclei could as well be a result, at least in part, of activities of the nucleoplasm as of the spindle and nuclear envelope. Recent studies have shown that nuclei contain an identifiable morphological entity with various mechanical properties. Such an entity has been termed the nuclear matrix and has been described in some detail by Comings and Okada (1976), Berezney and Coffey (1977), Wunderlich and Herlan (1977), and Herlan *et al.* (1978) (see also references in the latter paper). The only comment, based on current data, that can usefully be made concerning the possible evolution of this matrix is that it seems to be similar in rats and mice (Comings and Okada, 1976; Berezney and Coffey, 1977), but the matrix of *Tetrahymena* macronuclei has a substantially different composition (Wunderlich and Herlan, 1977). Given the small number of organisms sampled and the highly specialized nature of the ciliate macronucleus, such a comparison is not very useful, but it does point to the existence of variability and thus the potential for determining a pattern of evolution in this feature. Clearly the analysis of this matrix, especially in organisms where it remains presumably intact around mitotic spindles (i.e., those cells with intact nuclear envelopes), offers great potential for significant results in the contexts both of evolution of mitosis and mechanisms of spindle action.

2. *Fate*

There is much more data on the fate of the nucleoplasm, and thus the nuclear matrix, during mitosis than there is on its structure and composition. In those organisms in which the nuclear envelope disperses during prophase, the fate of the nuclear matrix is essentially unknown, since it has not yet been described in intact cells. However, there are essentially three different types of fate discernible in other types of division if one uses the nucleoplasm as a synonym for nuclear matrix. As seen in Fig 2b and c, a portion of the matrix is included in the daughter nuclei, and a variably large portion is discarded into the cytoplasm where it seems to rapidly disappear, presumably to be recycled into cellular pools of amino acids, etc. Alternately, all of the matrix may be included in the daughter nuclei (Fig. 2a and d). The third variation is somewhat restricted, only being reported in some fungi [e.g., Poon and Day, 1976a; Girbardt, 1968; Thielke, 1973, 1974; also suggested, but not proven (see Heath and Heath, 1978), by Harder, 1976a,b, and Wright et al., 1978]. In this instance, a portion of the nucleoplasm is discarded from the mitotic nucleus early in prophase. The distribution of these variations is seen in Table II. The functional significance of these forms of behavior is obscure. Clearly, discarding the nucleoplasm is not necessitated by the inability of the spindle to pull it apart, since in the examples shown in Fig. 2b and c, the nucleus has already elongated and thus performed the separation of the nucleoplasm. In the absence of any clear understanding of the functional significance of a feature, it is very unwise to attempt to differentiate between convergent evolution and homology. At present, all that can be suggested is that retention of the total intact matrix as a functional unit with simple equipartitioning at telophase is a system that is easier to control, and requires less synthetic energy, than the various levels of loss during mitosis. Thus, this system may be the primitive one. Since higher organisms appear to have evolved the mechanisms needed for complete dispersal and reformation at each mitosis, there must be some force driving evolution in that direction, but its nature is obscure. It is also worth remembering that many organisms that retain the intact matrix during mitosis have a short mitotic cycle time. A correlation between short mitotic cycle time and insufficient time for loss and reformation of the matrix may exist and should be considered in future studies.

D. Nucleolus

As shown in Table II and as previously discussed by Pickett-Heaps (1970b), the nucleolus exhibits a considerable range of behavior patterns during mitosis. At one extreme, the nucleolus persists in essentially its interphase form throughout mitosis and becomes constricted into two approximately equal halves at telophase. I shall use the term "persistent" to refer to this behavior in preference to "autonomous," which was used by Pickett-Heaps (1970b) and which is less descriptively accurate. At the opposite extreme, the nucleoli disperse during prophase, and their subcomponents are no longer recognizable. In agreement with Pickett-Heaps (1970b), I shall use the term "dispersive" for this behavior. Between these extremes, there are three fairly discrete patterns of behavior: (1) "discardive,"[1] in which the intact nucleolus is discarded (often with some nucleoplasm) into the cytoplasm either at prophase (e.g., *Ustilago*) or at telophase (e.g., *Catenaria*); (2) "fragmentary," in which the nucleolus breaks up into small fragments that persist in a scattered array around the nucleus (e.g., *Cladophora*); and (3) "associative," in which the nucleolus fragments into diffuse material that coats the chromosomes (e.g., *Spirogyra*). The fragmentary mode of behavior may culminate either in reaggregation within the daughter nuclei at telophase (e.g., *Cladophora*) or in loss of at least some of the fragments to the cytoplasm together with the expelled interzone of the nucleus (e.g., *Oedogonium*). As with any feature based on observations of fixed cells, the possibility of artifact must be considered. Because different cells react in different ways (i.e., show a different appearance) to a given fixation schedule, it may emerge that associative behavior is merely an artifactual association of fragmentary nucleolar material. At present, there is no unambiguous way of resolving these types of problems; thus, the observed types are presented for consideration as reported.

As seen in Table II there is, in general, a good level of uniformity of nucleolar behavior within major taxonomic groups, thus suggesting that nucleolus behavior may be a stable character with good phylogenetic value. However, it is clear that the nucleolus is a labile organelle whose morphology is rapidly responsive to the metabolic activities of the cell (e.g., Nilsson, 1976). Thus, observed nucleolar behavior is more likely to be the secondary result of differing metabolic patterns within the cell. For example, it seems most probable that persistent nucleoli are a response to a cellular economy in which ribosome synthesis does not cease during mitosis, a trait that may be expected in cells with a short generation time. The converse situation would be expected in cells with dispersive nucleoli, and indeed the norm in higher organisms is dispersive nucleoli and absence of ribosome synthesis during mitosis (Prescott, 1964). Presumably much of the nucleolus morphology is due to ribosome precursors "in transit," thus if production of the ribosomal RNA precursors is halted without a concomitant stoppage in ribosome precursor transport to the cytoplasm (a likely situation), then much of the nucleolus will be lost, which is then likely to show up as dispersive behavior. Associative behavior may be the result of a functionally insignificant, simple physicochemical interaction between residual nucleolar material and condensed chromatin. However, a rationale for the somewhat uncommon discardive behavior is at present elusive, to say the least! Two possible explanations come to

---

[1]The word "discardive" is introduced as the adjective derived from discard because no other word has the same meaning.

mind. The nucleolus may have evolved a level of structural complexity that renders it incapable of being divided by constriction or of being dispersed via total export of its components. Thus, it may be necessary to subject the nucleolus to the degradative activity of cytoplasmic lytic enzymes in order to disperse it. This hypothesis predicts structural characteristics of the nucleolus that should be measurable. An alternate, and perhaps more attractive, hypothesis (suggested by Dr. M. C. Heath during review) postulates that synthesis of the nucleolus is halted during mitosis and the mechanisms by which the products of the nucleolus are exported to the cytoplasm are also shut down. Thus, the nucleolus persists. Subsequently discarding it to the cytoplasm would be a simple way of rapidly injecting a burst of nucleolar products to the cytoplasm that had been deprived of its supply during mitosis. When more is known about the mechanisms by which ribosomes are transferred from the nucleolus to the cytoplasm, this hypothesis may also be tested.

In terms of determining a primitive versus an advanced type of nucleolus behavior, we can perhaps recognize a trend for the morphologically and nutritionally simpler organisms to possess persistent nucleoli, whereas the higher organisms are characterized by dispersive nucleoli (Table II). We could argue that the simplest arrangement was a structurally simple nucleolus, which remained functional throughout mitosis; two behavior patterns could diverge from this arrangement. In one direction, either increased structural complexity produced a nucleolus that could not readily be divided nor dispersed, or inhibition of ribosome export from the nucleus occurred during mitosis so that the nucleolus persisted and discardive behavior evolved. In the other direction, increased efficiency and complexity evolved together with, or were succeeded by, increased levels of control sophistication, so that dispersal and reformation of the nucleolus could be reliably and rapidly achieved. Shut down of nucleolar activity during mitosis could have been the result of complex chromatin condensation necessitated by the increased size of the chromosomes that have evolved in higher organisms. The ability of higher organisms to reliably resynthesize nucleoli may be correlated with the evolution of nucleolar organizer regions (NOR) on certain chromosomes. Such structures are seldom reported in organisms with nondispersive nucleoli (Smetana and Busch, 1974). On this proposed scheme (shown diagrammatically in Fig. 3), the fragmentary and associative types of behavior are seen as intermediates along the line of evolution to a totally dispersed system. Perhaps these organisms have not yet evolved NOR and thus find it necessary to carry partially organized portions (potential organizers?) of the nucleolus from one nucleus to the next.

This hypothesis is highly tentative, but does make a number of explicit predictions that future research should be able to verify or negate. Thus, it may be a useful working hypothesis in our current state of knowledge.

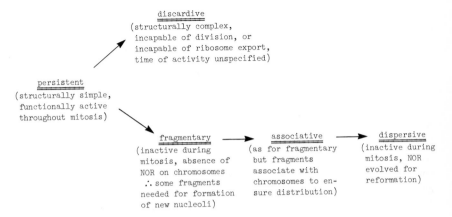

FIG. 3.   Hypothetical scheme for the evolution of nucleolar behavior.

E. CHROMATIN

1. *Condensation*

There are essentially two broad levels of chromatin organization, with inter-mediate levels still to be adequately described. At the molecular level, there is undoubted diversity in terms of both chromatin-associated proteins (e.g., his-tones) and nucleosomal organization. Such variations have been reviewed for the lower eukaryotes by Horgen and Silver (1978) and are beyond the scope of this work. However, at the other end of the scale, there is the interphase organization into heterochromatin and euchromatin (Frenster, 1974), and the reorganization of these states into the condensed states characteristic of the mitotic chromosomes of most organisms. Observations of this higher level of organization are fraught with danger in the form of preparation artifacts; both artifactual loss of organiza-tion and artifactual condensation (e.g., Dick and Johns, 1967; Svejda and Hor-nak, 1969; Poon and Day, 1976b) are likely. However, differences in behavior in response to essentially similar preparation procedures must presumably reflect underlying structural differences in the chromatin, so that even if the resultant pattern is an artifact, that artifact is likely to be an indication of real differences. With this tenuous justification in mind, and in the absence of better data, the observations available can be discussed at face value.

In the majority of organisms listed in Table II, the typical pattern of heterochromatin–euchromatin at interphase giving way to condensed chromo-somes at mitosis is observed. However, there is growing evidence that in at least some organisms, e.g., *Saprolegnia* (Heath, 1978b) and *Saccharomyces* (Peter-son and Ris, 1976), the chromatin does not condense at any stage in the nuclear cycle. Whereas the lack of condensation occurs in organisms with less DNA per

I. BRENT HEATH

chromosome than *Escherichia coli* (Table III), this size feature is not the only factor involved in this behavior, since *Coprinus,* for example, has a similarly small amount of DNA per chromosome, yet has clearly condensed mitotic chromatin. Assuming that the lack of chromosome condensation is verified, the implications for the molecular aspects of chromosome function are considerable, since in higher organisms, chromatin condensation seems to be a transcription regulating mechanism (e.g., Frenster, 1974). However, in the present context, it seems highly likely that the lack of chromatin condensation, especially at mitosis, is the result of two factors: (1) a short nuclear cycle time [e.g., less than 51 minutes for *Achlya* (Griffin *et al.,* 1974)], suggesting that it may be ineffi- cient, and perhaps impossible, to orchestrate chromatin condensation in a short nuclear cycle; and (2) efficient segregation of small chromosomes, without the need for condensation. Clearly, segregating DNA molecules (and any associated proteins) that differ in size by one or two orders of magnitude are likely to present inherently different design requirements. This point does not, of course, preclude the development of a chromosome condensation system in small chromosomes, but it may well mean that there is relatively little selection pressure for small chromosome condensation as compared with cells containing large chromo- somes. Since absence of chromosome condensation seems to be a feature of prokaryotes, it is likely that this is indeed a primitive feature. We would then

TABLE III
DNA Content of Selected Organisms

Organism	DNA per nucleus (picogram)	Chromosomes per nucleus	Average DNA per chromosome (femtogram)	Reference
Escherichia coli	0.0043	1	4.3	Prescott (1976)
Dictyostelium discoideum	0.050	7	7.1	Firtel and Bonner (1972)
Physarum polycephalum	0.50	40	12.5	Mohberg (1977)
"Oomycete"	0.046	21	2.2	[a]
Phycomyces blakesleeanus	0.032	14[b]	2.3	Dusenbery (1975)
Neurospora crassa	0.036	7[b]	5.2	Dusenbery (1975)
Saccharomyces cerevisiae	0.015	17[b]	0.9	Lauer *et al.* (1977)
Coprinus lagopus	0.042	12[b]	3.5	Dusenbery (1975)
Homo sapiens, sperm	2	23	87	Shapiro (1976)
Homo sapiens, somatic	4–19	46	87–413	Shapiro (1976)
"Average mammal"	—	—	100	Prescott (1976)

[a] These data are based on the DNA content of *Achlya bisexualis* given by Hudspeth *et al.* (1977) and prelimi- nary synaptonemal complex counts of *Saprolegnia ferax* obtained by Tanaka and Heath (unpublished). The two organisms are likely to be very similar to each other in these parameters. Complete data are not available on either one.

[b] These chromosome numbers were all derived from Altman and Dittmer (1972).

postulate that as the size of the genome and chromosomes enlarged, for whatever reasons, so the behavior of the chromatin also had to become more complex to permit efficient handling during mitosis.

2. *Arrangement*

We have just seen that chromosome condensation during mitosis is not a universal feature. In addition to this degree of variability, there is apparently further complexity among those organisms in which condensation does occur. As seen in Table II, there are many organisms in which chromatin condensation yields the well known arrangement of separate, independent chromosomes commonly observed in higher organisms. However, in a number of organisms, perhaps most clearly exemplified in the cryptophytes and diatoms, the metaphase chromatin forms an amorphous mass of material in which separate chromosomes cannot be differentiated. Again, we must be aware that artifactual dispersion of chromatin may be occurring, but equally, there must be differences in the degree of cohesiveness of the chromatin, otherwise there would not be differences in its appearance after similar fixation schedules. Thus, it is reasonable to discuss the phenomenon based on its symptoms, even if the cause is unknown. The biggest problem is that, while it is easy to differentiate between organisms exhibiting the extremes of behavior, numerous reports show more ambiguous chromatin arrangements in which the pattern of distribution of the chromatin mass is uncertain. Indeed, there is probably a gradation of behavior patterns among different genera. However, scrutiny of Table II does reveal a generally consistent distribution of mass versus separate chromosome organization among taxonomic groups, thus suggesting that there are consistent differences in organization. Translating these differences into any hypothetical evolutionary sequence is difficult. The only suggestion offered is that the mass arrangement is intermediate between the uncondensed situation and the highly condensed, individual chromosome pattern. Certainly the mass arrangement, in some cases at least (e.g., the cryptophytes), behaves as an integral unit during anaphase, thus suggesting that the evolution of a system for coordination of chromosome movements is unnecessary in this system. This then lends support to the proposal for the primitive nature of this arrangement, i.e., it evolved in response to the need for some mitotic chromatin condensation, but before a coordination system evolved.

F. POLAR STRUCTURES

1. *Morphology*

There is perhaps no other feature of mitosis that exhibits such a range of variation as the structures that lie at the poles of the mitotic spindle. These structures have been given a large number of different, but equivalent, names: nucleus associated organelle (NAO) of Girbardt and Hädrich (1975); spindle pole body (SPB) of Aist and Williams (1972); microtubule organizing center (MTOC) of Pickett-Heaps (1969); spindle or centrosomal plaque of Robinow and Marak (1966); and nucleus associated body (NAB) of Roos (1975a). In the present work, two generic names will be used; centrioles, for those structures that are clearly related to the familiar array of nine triplet microtubules and their associated linking structures, and nucleus associated organelles (NAOs) (Girbardt and Hädrich, 1975), for the range of other structures, including the intranuclear structures from which spindle microtubules of *Physarum, Oedogonium, Minchinia,* and *Sapromyces* seem to originate. Within these two generic terms, there are numerous specific morphological variations that are illustrated in Fig. 4 and will be discussed later. The distribution of these variations is listed in Table II.

In contrast to some of the other aspects of mitosis, there is no shortage of detailed data on the morphological range of centrioles and NAOs encountered in lower organisms. Thus, the problem is one of finding some pattern in this diversity. Perhaps the simplest feature of the diversity that we can deal with is the centriole and its variants. As emphasized previously by Pickett-Heaps (1969) and Friedländer and Wahrman (1970), the primary function of the centriole is almost certainly as a basal structure for flagellum development. The behavior of the centrioles during mitosis can then be explained on the basis of the need to ensure that each daughter nucleus and cell receives its essential complement of these possibly autonomous organelles. In other words, the centrioles are merely "going along for the ride," and we can thus add another function to the mitotic spindle, i.e., equipartitioning of genome *plus* centrioles. This concept is further supported by oservations on the coccids in which the centrocones probably have evolved for spindle organizing, with the centrioles also present outside of the intact nuclear envelope at the spindle poles.

Because a number of organisms do not retain a persistent set of centrioles during vegetative mitosis, but instead synthesize centrioles *de novo* when flagel-

---

FIG. 4. Examples of the diversity of types of polar structures found at the poles of mitotic spindles. The distribution of each type of structure can be seen in Table II. (a) Procentriole; (b) centrioles aligned end-to-end, 180° to each other; (c) orthogonally arranged centrioles; (d) orthogonally arranged basal bodies or kinetosomes; (e) plaque set into a close fitting pore in the nuclear envelope; (f) an intranuclear plaque, sphere, or bar; (g) a plaque lying on the exterior of the nuclear envelope; (h) a slight amount of electron-opaque material lying on each side of the nuclear envelope (e.g., some mucoralean zygomycetes); (i) a ring; (j) rhizoplast with associated kinetosomes (K) (from *Ochromonas*); (k) striate bar, characteristic only of prophase in diatoms (from *Surirella*); (l) centrocone with associated centrioles (C) (from *Eimeria*); (m) extranuclear sphere or atractophore (in the Zoomastigina the "etc." in Table II refers to assorted flagellar root structures that are associated with the atractophores). Scale bars are only intended to give an approximate indication of the size of the typical organelle, there is often some variability between species.

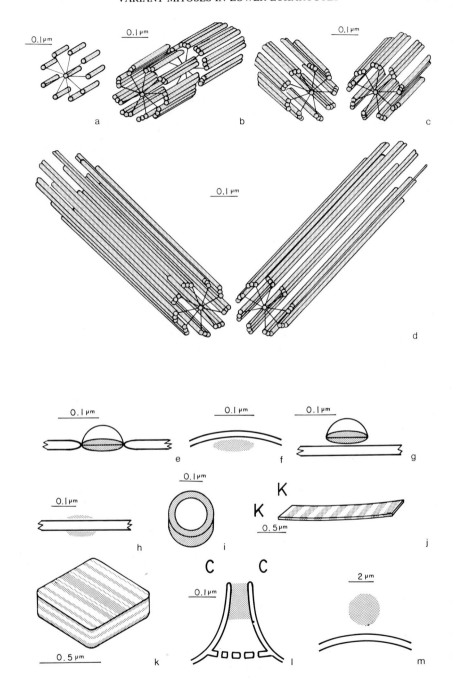

lum production is necessary (e.g., Myxomycetes, *Naegleria*), it is clear that retention during mitosis is not essential. It is tempting to suggest that the ability to obviate the energy and material required for retention of the centrioles during mitosis by evolving a *de novo* system is advanced with respect to the persistent system, but this seems unlikely since higher animals retain centrioles during their mitotic processes. It is thus more likely that the presence of the centrioles during mitosis correlates best with the frequency with which a cell synthesizes flagella. Thus it may be more "economical" to retain centrioles during all divisions if, in the normal environment, one nucleus out of twenty divisions normally forms a flagellate cell; and conversely, if only one nucleus out of twenty thousand divisions forms a flagellate cell, it may be more economical to form centrioles *de novo* as needed. The figures chosen are, of course, arbitrary; there is no evidence known to this author that suggests an accurate transitory figure at which one system would be favored over the other. However, the presence of centrioles during mitosis of nonflagellate cells (i.e., during divisions occurring in nonflagellate stages of the life cycle) is not an all-or-nothing phenomenon. A number of organisms (see Table II) synthesize, *de novo,* procentrioles that persist only during mitosis. These procentrioles are clearly related to centrioles, but are very short and may possess only singlet microtubules in place of the normal triplets (e.g., *Cladophora* [Lewis and Scott, 1979], *Labyrinthula* [Porter, 1972]). Intermediate lengths (interpretable as increasing levels of development) of centrioles are found up to the elongated basal body or kinetosome size of about 0.6 μm in length. The simplest interpretation of these variations in degree of development is that cells simply differ in the degree to which they proceed along the centriole maturation pathway at each mitosis. This variation then becomes essentially a simple extension of the all-or-nothing hypothesis discussed above. In other words, cells with intermediate frequencies of flagellate cell development per number of mitotic divisions may show intermediate degrees of centriole to basal body development. The prediction of this hypothesis is simple, but the test, i.e., number of total mitotic divisions per division leading to flagellate cell formation, is difficult, since it must be made strictly under the conditions in which the organisms evolved, i.e., in nature, where such data is hard to come by.

As noted above, the concept that the centrioles are distributed to daughter cells by the mitotic nucleus has been argued previously and now enjoys a fair degree of acceptance. What has not been discussed previously is the extension of this concept to other structures. This extension may help explain the diverse morphology of NAOs and is thus worth pursuing. In a number of organisms (e.g., *Thraustochytrium, Sorodiplophrys, Sapromyces*), the consistent positionings of the total cellular complement of Golgi bodies (dictyosomes) adjacent to the centrioles and poles of the mitotic nuclei indicates that they must replicate in synchrony with the centrioles and suggests that their even distribution to daughter nuclei and cells may well be effected by this linking with mitosis. This, then, would be another example of the mitotic system being used to distribute another organelle. While this process is somewhat restricted in its occurrence, it emphasizes the potential for the evolution of ancillary roles for the mitotic apparatus. What else may be utilizing the spindle in a comparable manner? The pole-associated microbodies of *Porphyridium* could be an example. Likewise, the virus particles associated with the pericentriolar material in certain mammalian cell lines (Wheatley, 1974; Gould and Borisy, 1977) could well be another example of the adoption of this very effective distribution, and thus, perpetuation, system; in this case, by an obligate parasite. Similarly the interphase microtubule center of *Surirella* seems to be distributed in this way.

What of the diversity of NAOs? It is quite evident that many NAOs function in some way to organize the spindle and also, in some cases, cytoplasmic microtubules. In these cases, there is a reasonable correlation between morphology and hypothesized function. However, in organisms such as *Basidiobolus* and *Polysiphonia,* where the NAO is small with respect to the large area of the spindle poles, it is hard to make a convincing case for a homologous function in spindle organization. In these cases, it may well emerge that these NAOs have some other cellular function *not* associated with mitosis and that their presence at the spindle poles is merely another example of the spindle acting to separate an independent cell organelle or infectious agent. It may well be argued that this hypothesis is a nonrigorous way of excluding observations that do not otherwise fit a particular hypothesis for correlating form, function, and evolution of the NAOs. However, as shown above, there are enough examples of structures distributed by the spindle to justify giving this position some consideration. Only the unambiguous elucidation of the function of these organelles will resolve the various possibilities.

In any organized but transient structure, such as a mitotic spindle, there must be some mechanism to effect the correct organization. Because the detectable components of most spindles develop from the vicinity of the NAOs or centrioles, it is usually assumed that spindle formation is one of their functions. We can make a good case for most of the listed NAO variants being microtubule organizing centers (MTOCs), as outlined by Pickett-Heaps (1969). The one ubiquitous characteristic of MTOCs seems to be that they are composed of variously arranged amorphous material that appears electron opaque after routine osmium-uranium-lead staining of thin-sectioned material. All of the polar structures listed in Table II contain various amounts of such material in various arrays. In those cells containing polar centrioles, the spindle, and usually the cytoplasmic, microtubules radiate from centriole-associated electron-opaque material. This point again emphasizes that it is not the centrioles themselves, but rather their associated material that is important in spindle organization. If spindle microtubule organization is the role of the NAOs, then we need to know if the observed morphological variants can be explained on either a functional or an evolutionary basis. In order to consider a functional explanation, we must more carefully consider the functional characteristics. There are two obvious ways in which an MTOC could function in controlling microtubule formation: (1) as a template, and (2) as a controller of local cellular environmental conditions such that self assembly could be induced. There is little reason to believe the latter role, especially since *in vitro* experiments on *Saccharomyces* NAOs (Borisy *et al.*, 1975; Byers *et al.*, 1978) show that they are capable of acting as MTOCs to give a predictable array of microtubules in a noncellular situation where it is hard to imagine their being able to exert any significant control on the local environment (i.e., the surrounding medium). However the vesicles that are frequently associated with the NAOs could have a role in controlling the local cellular environment (see Heath, 1978a, for a partial listing of organisms containing such vesicles, and Pickett-Heaps and Tippit, 1978, for further comment on this point). Thus a template role is favored; in which case, we might expect some structure–function correlations. The evidence is contradictory. Examination of the spindle poles of almost all of the organisms listed in Table II shows that all of the spindle microtubules, and probably all of the cytoplasmic microtubules, present in the polar regions do terminate in a portion of the NAO or pericentriolar material. Furthermore, in selected cases, e.g., *Saccharomyces* and *Ascobolus,* there is a good correlation between number of spindle microtubules and area of the discoidal NAOs. Similarly, diploid yeast cells have larger NAOs than haploid ones (Peterson and Ris, 1976). These two types of observation support the idea that at least part of the NAO morphology is attributable to a template function of the NAO. However, contrary evidence is found in a comparison between *Uromyces* and *Coprinus* for example, where one has very similar spindles yet considerably different NAO morphology. This suggests that there may be an evolutionary component in the observed diversity of NAO morphology. One could enumerate many more examples of both type of comparisons among the organisms listed in Table II, but such an exercise does not favor either proposal. Thus, on the basis of available information, one can only conclude that there are probably both functional and evolutionary components to the diverse range of NAO morphology.

   Can we discern any possible evolutionary pattern in NAO morphology? Up to a point, yes. As outlined in Section III, one can make arguments for the change in position of the NAO with respect to the nuclear envelope. Perhaps as a consequence of this, one would then predict that those NAOs that maintained a close contact with the nuclear envelope would have a flattened shape, whereas those that showed more independent behavior would be free to assume other shapes, e.g., spheres. Similarly, one could argue that the correlation between morphology and function is very obvious and thus indicative of a simple and therefore primitive organization in the case of the flattened disc-like NAOs, whereas the correlation is less obvious, and thus more complex and advanced in the spherical and nondiscoidal forms. In this context, one should emphasize that in those organisms that have intact nuclear envelopes at the spindle poles and extranuclear centrioles the NAO is probably discoidal in nature and is seen as the thin layer of electron-opaque material applied to the nuclear membranes in the polar region of the spindle. Thus, these organisms are probably at the "primitive" end of the scale. In those cells that have developed permanent complex flagellar root systems, the NAO mitotic activity seems to have finally been removed from a discrete organelle and applied to a part of the root system, e.g., a rhizoplast. Such a transfer may well have evolved as the most effective way to ensure equal distribution of flagellar root systems, as well as the divided nucleus, to daughter cells. However, the NAOs of the striate type found in diatoms, the centrocones of the coccids, and the massive sphere of *Allogromia* defy attempts at integration into the above scheme. There is no obvious functional requirements in these organisms that would help explain the NAO morphology. Therefore, at present, we may simply assume that these NAOs represent products of evolution driven by genetic drift rather than some strong functional selection pressure. If such an explanation is valid, it would indicate that these structures would have phylogenetic value, since convergent evolution would be unlikely. Unfortunately, their presently known distribution (Table II) only serves to confirm homogeneity within accepted coherent groups of organisms; it does not point to any interrelationships between groups.

2. Replication Cycle

In contrast to the range of morphological variations in mitotic polar structures, there is little diversity in the time of their replication. In most organisms, replication occurs shortly before spindle formation at late interphase or prophase. However, there are well documented exceptions to this pattern wherein replication occurs at late telophase and the interphase NAOs are double. Unfortunately, clear information on the time of NAO or centriole replication is rare; thus, the data is not included in Table II. A number of examples of both systems are given in Heath (1978a). There is generally consistency of replication time within major taxa, suggesting that it is a stable characteristic with some phylogenetic value, but unfortunately, since there are only two known alternatives, convergent evolution becomes inevitable. Nevertheless, it is a point of variability that deserves clarification in future investigations, so that anomalous situations can be investigated.

G. SPINDLE

While the mechanisms by which mitotic spindles generate the required force are unknown, it is clear that the spindle is the central component of mitosis, since it is the structure that has evolved to perform the primary task of mitosis, i.e., genome segregation. It is, therefore, surprising that the diverse variations in its detailed properties have been relatively neglected by those who have considered the evolution of mitosis. The following sections will attempt to redress this situation.

1. *Composition*

It can justifiably be argued that there is insufficient data on the composition of the spindles of any organisms to permit useful discussion of possible evolutionary relationships. However, there are a number of points that can be made. For example, the meaningful analysis of mitotic spindles has been hindered by the fact that they are either inactive after isolation or, if they do show continued movement, they are not truly isolated, but merely functioning in an unusually permeable cell (e.g., Cande *et al.*, 1974). In either case, the analysis is inaccurate, since it is based on less or more than the functional spindle. In principle, it should be easier to isolate functionally intact mitotic spindles from cells with intact nuclear envelopes at mitosis, especially in those cells in which there is evidence for an isolatable inducer of mitosis, e.g., *Physarum* (Oppenheim and Katzir, 1971). Unfortunately, such spindles would still be contaminated by nucleoplasm, but since this is arguably a component of the spindle in any case, it may be a less serious problem than those facing the isolators of open spindles. Clearly, analysis of primitive spindles, especially from a comparative point of view, is a worthwhile exercise.

The importance of compositional analysis of spindles as an aid to understanding their evolution is suggested by the increasing evidence that some components, e.g., tubulins, are generally conservative, but display some differences in their composition in different organisms (e.g., Snyder and McIntosh, 1976; Davidse and Flach, 1977). With the recent evidence that tubulin may have evolved in some prokaryotes (Margulis *et al.*, 1978), it becomes likely that one can construct phylogenies of these molecules upwards from the ancestral prokaryotes. However, when making such comparisons, it must be remembered that there is apparent heterogeneity of tubulins from different cellular locations within an organism (e.g., Bibring *et al.*, 1976; Stephens, 1978). Therefore, in order to obtain valid data on the evolution of mitosis, only mitotic spindle tubulins should be compared.

In addition to potential variations in tubulins, one might also expect to find variations in the microtubule associated proteins (MAPs of Snyder and McIntosh, 1976). As noted by Snyder and McIntosh (1976), ''It is likely that the MAPs are of considerable importance for the control of assembly and function of microtubules, ...'' Given the diversity of both organization and response to antimitotic agents of microtubules from different organisms (discussed later), it is highly likely that there will be diversity of MAPs also. Such diversity may help clarify the path of spindle evolution.

Apart from microtubule related components, there is growing evidence that mitotic spindles of at least some organisms contain muscle proteins such as actin, myosin, troponin, and α-actinin (critically reviewed by Forer, 1978, and Goldstein et al., 1977). Because there are a number of genes coding for actin (Tobin and Laird, 1977), it is likely that, for actin at least, there are spindle-specific muscle proteins, variations of which may indicate relationships among different spindle types. In addition to possible compositional variations of muscle proteins in spindles, it must be remembered that there may well be quantitative differences in such components in different spindles. Indeed, they may well be absent from some spindles. Such hypotheses will be hard to unambiguously prove but are worth considering as further indicators of evolutionary patterns in spindles.

The essential point about the composition of the spindle is that it is potentially as easy, or perhaps even more easy, to analyze functionally intact spindles from lower organisms as compared with higher ones. There is already a sufficient diversity of known components whose variations must yield useful data on spindle evolution. Thus, this area of investigation is a rich and largely untapped territory that is ripe for exploitation.

2. Structure

Before discussing the structure of the spindle, it must be made clear that the level of analysis to be used is highly susceptible to the perennial problem of fixation artifact. The primary component of the spindle that has been visualized is the microtubule, thus it is observations of these components that will be discussed as reported. However, Luftig et al. (1977) have presented evidence for artifactual length changes in microtubules during fixation. While Hardham and Gunning (1978) have made contrary observations in another cell type, it is clear that the potential for artifact exists. Likewise, there may also be positional changes as well as length changes. Because no technique other than serial reconstructions from electron micrographs of thin sections currently offers the required resolution together with the capability of three-dimensional reconstruction, we are forced to accept the available data at face value and assume that consistent differences following comparable treatments represent real in vivo differences.

a. Spindle Framework. Most mitotic spindles contain two basic components: a framework that runs in some way from pole to pole; and chromosome associated components that may interact with this framework in order to move the chromosomes during the active parts of mitosis. The roles of the framework seem to be (1) to establish a mechanical foundation against which chromosome movement can be generated, and (2) to widely separate the genomes and daughter nuclei at anaphase and telophase. The considerable morphological diversity found in the framework, together with examples of the organisms in which each type is found, is summarized in Fig. 5. Among these arrangements there appear to be three fundamental dichotomies. The framework may be either intranuclear (Fig. 5d–i) or extranuclear (Fig. 5a–c). It may be composed predominantly of either truly pole-to-pole (continuous microtubules, Fig. 5a–e) or of interdigitating microtubules that emanate from opposite poles and interdigitate with one another (Fig. 5f–i). Finally, it may form either a coherent central bundle around the periphery of which the chromatin is arranged, or a more dispersed array within which the chromosomes are scattered. The distribution of the central bundle versus the dispersed system is noted in Table II. Within the interdigitating type of organization, the microtubules from opposite poles may either lie close to one another (within ~ 50 nm) (Fig. 5f, h, i) or be more widely separated (Fig. 5g). Because these arrangements seem to be reasonably consistent within major taxa (as far as has been determined; unfortunately, relatively few organisms have been

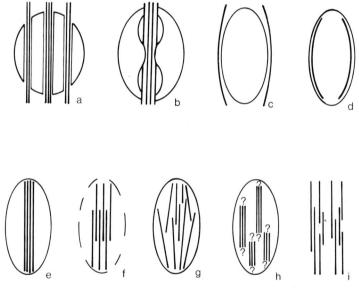

FIG. 5. Examples of the diversity or organization of the framework, or nonkinetochore, microtubules. (a) Bundles lying in extranuclear tunnels (e.g., *Cryptothecodinium*). (b) A single large bundle lying in the cytoplasm in a groove of the nucleus (e.g., *Syndinium, Amoebophrya, Barbulanympha*). (c) A layer of perinuclear microtubules adjacent to the nuclear envelope (e.g., *Blepharisma, Loxodes* macronuclei). (d) A layer of intranuclear microtubules adjacent to the nuclear envelope (e.g., *Tetrahymena*). In a to d, the microtubules are illustrated as being continuous from pole to pole. Such has not been proven, although it may be true in some or all cases. (e) A bundle of pole-to-pole microtubules (e.g., *Uromyces, Phycomyces, Saccharomyces*). (f) A bundle of closely spaced interdigitating microtubules (e.g., diatoms). (g) An interdigitating but widely spaced array of microtubles (e.g., *Saprolegnia, Thraustotheca*). (h) An array of bundles of closely spaced microtubules of undetermined length (e.g., ciliate macronuclei). (i) An array of variously overlapping and closely spaced microtubules of various lengths (e.g., *Cryptomonas*, higher plants and animals).

examined in sufficient detail), and because, at least in some cases, they can be partially verified by polarized light microscope observations of living cells (e.g., *Barbulanympha* and *Nitzschia*), it seems likely that they do represent real organizational differences rather than artifacts of fixation.

As will be argued later (Section III, B and C), the closed intranuclear spindles probably more closely resemble the primitive condition than the open and extranuclear ones. However, the relationships among the other variations is more complex. It has been argued before (Heath, 1975a) that there is no need to postulate different force generating mechanisms to explain spindle elongation in any of the variants shown in Fig. 5. For example, in those spindles where true continuous microtubules exist in close apposition to one another, intermicrotubule sliding mediated by dynein-type cross-bridges could generate spindle elongation without any detectable evidence for sliding if microtubule elongation were synchronous with the sliding process. Similarly, widely spaced interdigitating tubules could utilize a sliding mechanism if the microtubule based cross-bridges were able to interact with the nucleoplasm rather than another microtubule. It should also be remembered that the distinction between interdigitating versus continuous is not always simple. For example, in *Fragillaria,* prior to spindle elongation we clearly see a spindle composed entirely of pole-to-pole microtubules, but these slide apart into two interdigitating groups during spindle elongation. Thus, this organism has an interdigitating spindle, yet a contrary conclusion would be reached by analysis of prophase only. However, the simplest hypothesis, which should thus be favoured in the absence of contrary evidence, is that in those spindles composed primarily of continuous microtubules throughout spindle elongation, microtubule elongation via polymerization generates the force needed for spindle elongation as argued by Inoué and co-workers (Inoué and Sato, 1967; Inoué and Ritter, 1975). Since a system for controlling microtubule polymerization must have preceded any further development of a microtubule utilizing system, it is logical that spindles whose morphology is consistent with this type of force generation are the most primitive type. With increasing load, one might expect that larger spindles would evolve rigid cross-links to bind the elongating microtubules together for greater stability. All of the spindles of the types illustrated in Fig. 5a–e do have either cross-bridges between microtubules or at least inter-microtubule spacings small enough to permit cross-bridges to form. Another cellular example of the evolution of the use of cross-linked microtubules to produce rigid structures is in the axopodia of heliozoans (e.g., Edds, 1975; Roth *et al.,* 1970; Tilney and Porter, 1965) where elongation during development is undoubtedly at least partially due to microtubule polymerization. Within the organisms bearing continuous microtubules (Fig. 5a–e), we can detect a sequence of increasing potential for intertubule cross-bridging and thus presumably increasing strength on a per tubule basis. Thus, in types (c) and (d), the single layer permits the linking of any one microtubule to only two others, but in arrangements (a), (b), and (e) linkages to more than two are possible, the exact number depending on the type of packing (as well as the number of linkage sites around the circumference of a microtubule). Furthermore it is evident that, for a given number of microtubules, the large single bundle has the potential for more cross-bridges than numerous small bundles because the proportion of peripheral microtubules (with obviously reduced cross-bridging capacity) is reduced. Thus one can reasonably postulate (but not prove) that evolution went in the direction of increasing strength in the framework microtubule array in the order (c) and (d) to (a) to (b) and (e). We can support this hypothesis by the argument that undoubtedly the hypermastigote flagellates, such as *Barbulanympha,* have the most massive examples of type (b) and are undoubtedly some of the most highly evolved protists.

Because higher plants and animals seem to utilize a system most like Fig. 5, types (f), (g), and (i), it is likely that these arrangements are more advanced than the continuous types (a)–(e). This argument is supported by the necessity of evolving some form of mechanochemical cross-bridge to interact either with the adjacent microtubules or some component of the nucleoplasm. Why this system should have evolved over the alternative polymerization system is unclear, but obviously a means of generating an inter-microtubule sliding force was essential to the development of eukaryotic flagella, and possibly also for propelling organelles along cytoplasmic microtubules (Schmitt, 1968; Ochs, 1972). Possibly sliding "motors" evolved for these purposes and were subsequently utilized by the spindle framework system. Alternatively, increased efficiency may have been the driving force of evolution. It apparently requires hydrolysis of one molecule of GTP to add one tubulin subunit onto a microtubule (Olmsted and Borisy, 1975; Weisenberg and Deery, 1976). From the diagram in Snyder and McIntosh (1976), we can see that addition of thirteen 4 nm long subunits elongates a microtubule by the length of 4 subunits, thus giving an elongation of $4 \times 4/13 = 1.2$ nm per subunit or per GTP molecule. Conversely, hydrolysis of one molecule of ATP by a dynein cross-bridge in a flagellum axoneme can probably generate about 40 nm of movement (Brockaw, 1975). Clearly, sliding is a more energy efficient way of generating length change than polymerization, thus it seems likely that energy efficiency might have been a strong selection pressure leading to the development of sliding mechanisms for spindle elongation. We could therefore envisage the evolution of microtubule sliding as the result of both cytoplasmic and nuclear pressures. In fact, perhaps the principle evolved twice in the two locations, and thus we may find differences between the two systems.

Determination of relative advancement of the various sliding types of system (Fig. 5f–i) is difficult. Type (i) may be most like higher plants and animals, and thus, most advanced, with type (f) being morphologically more regular and simple, and thus, perhaps more primitive. Unambiguous demonstrations of type (g) are sufficiently rare and susceptible to fixation artefact to leave its reality an open question. Type (h) has not been studied enough to determine how the microtubules are arranged in the entire nucleus, thus it may in fact be an intranuclear version of type (a), i.e., bundles of continuous microtubules, or it may be basically type (f), arranged into numerous small units. Further discussion in the absence of obtainable data is not worthwhile.

The final aspect of the spindle framework that requires discussion is the tendency for the framework to form either a coherent bundle of microtubules with the chromatin arranged around its surface or to be interspersed with the chromatin. The distribution of these patterns is shown in Table II, where it can be seen that there is generally consistency of behavior within major taxa. Which pattern is primitive remains obscure, although, intuitively, it would seem simpler, and thus, perhaps primitive, to separate the two components of the spindle (i.e., central bundle versus peripheral chromatin). As with any character showing an either-or feature, convergent evolution is probable. Thus, the phylogenetic value of this pattern is very limited.

b. *Interactions between Chromatin and Spindle.* The dominant means by which chromatin interacts with the spindle is via kinetochores and kinetochore microtubules that typically intermingle with, or run adjacent to, the framework microcubules (Table II). However, there is considerable diversity within this system. Kinetochores may lie either in the nuclear envelope or free in the nucleoplasm (Table II). The number of microtubules per kinetochore varies from one to many. The morphology of the kinetochore ranges from very small funnel or disc-like structures that are almost indistinguishable from the nucleoplasm (typically in those cells listed with only one microtubule per kinetochore in Table II) to large, multilayered complexes such as those of *Oedogonium*.

Previous work on the evolution of mitosis has emphasized the similarity between the attachment of the bacterial chromosome to the cell membrane and the nuclear envelope-located kinetochores, thus suggesting that the latter represent the primitive type of organization in eukaryotes (Kubai, 1975, 1978; Ris, 1975). A contrary view, which will be returned to later, is that extranuclear spindles are not primitive, and in order for the kinetochores to interact with these spindles, nuclear envelope-located kinetochores became secondarily evolved. Thus, these types of kinetochore can be considered advanced with respect to intranuclear kinetochores. Some support for this view can be obtained from the fact that those cells with extranuclear spindles are typically ones that have complex morphologies and are assumed to be relatively advanced organisms.

With respect to intranuclear kinetochores, there are a number of observations that can be made. In general terms, there is a correlation between number of microtubules per kinetochore and apparent size of chromosome moved. Thus, the single microtubule bearing kinetochores are associated with poorly contrasted or small chromosomes, while the larger chromosomes that are very prominent tend to have more microtubules per kinetochore. This trend implies a load:

number-of-microtubules-per-kinetochore correlation. However, Moens (1978a, b) has recently analyzed this type of correlation in insect meioses and found that it did not hold. Instead, he suggests that increased microtubule-per-kinetochore numbers are derived from chromosomal fusions. Such a hypothesis is more attractive for recent fusions than older ones, where one might expect that redundant microtubules would be lost during evolution. However, whichever hypothesis we accept, it seems clear that low numbers of microtubules per kinetochore represent the primitive condition.

Because there is considerable variability in kinetochore morphology, this feature also deserves consideration as a phylogenetic indicator. As a general rule, the more complex morphology is positively correlated with increased numbers of microtubules per kinetochore. However, this correlation is not perfect, and in general, the use of kinetochore morphology as a marker for mitotic evolution is further discredited by reports of changes in morphology during different phases of mitosis (e.g., Alov and Lyubskii, 1977; Scott and Bullock, 1976). Furthermore, there are examples of supposedly closely related organisms in which there are considerable differences in kinetochore morphology (Table II). However, further examination of some of these apparent discrepancies, in fact, restores some confidence in the consistency of kinetochore morphology within major taxa. Thus, the kinetochores of *Polysphondylium* and *Dictyostelium* differ considerably from one another. While they are placed in the same family (Dictyostelidae) by Olive (1975), Dutta and Mandel (1972) have suggested a rather distant relationship on the basis of DNA base ratio analysis. Similarly, other examples of apparent heterogeneity within taxa in Table II, such as *Mougeota* versus *Closterium* and *Membranoptera,* and *Polysiphonia* versus *Porphyridium,* occur between different orders within the phyla. There does, however, seem to be some consistency of kinetochore structure within lower taxa and phyla. Therefore, this structure may indeed have value as an indicator of mitotic evolution and phylogeny. As with the number of microtubules per kinetochore, so presumably with kinetochore ultrastructure; greater complexity is an advanced feature. This point is supported by the generally complex kinetochore morphology of most higher organisms (see reviews of Alov and Lyubskii, 1977; Bajer and Molè-Bajer, 1972; Luykx, 1970), but again, a warning that there are certainly functional factors affecting morphology comes from the simple kinetochores of avian microchromosomes (Brinkley *et al.,* 1974) and the dimorphic kinetochores of certain insects where there are substantial differences in kinetochore morphology at meiosis and at mitotis (Comings and Okada, 1972; Ruthman and Permantier, 1973). At present, the only conclusion about kinetochore morphology is that it has some potential for use as an indicator of mitotic evolution, but very great care is needed in evaluating the data in this way.

While most organisms examined either possess some form of kinetochore and kinetochore microtubules, or have not been examined in sufficient detail to permit a definite statement for or against their presence, there are a number of organisms in which the existence of alternate chromatin–spindle interactions have been demonstrated with sufficient clarity to merit serious consideration. However, the earlier warnings of Kubai (1975) and Heath (1978a) concerning the artifactual loss of kinetochore microtubules must still be heeded. There appears to be two alternate systems in which kinetochore microtubules may not be involved in chromosome movement. In the mucoralean Zygomycetes, such as *Mucor, Pilobolus,* and *Phycomyces,* the available evidence would best fit a hypothesis postulating simple permanent attachment of the chromatin to the spindle poles, which are then separated by spindle elongation. If such proves to be true, this system comes close to what is hypothesized to be a very primitive type of mitosis (see Section III, B). The other system of interest is that found in most diatoms, with the exception of *Nitzchia,* in *Aulacantha,* and possibly in *Prymnesium,* where the behavior of the chromatin is consistent with a lateral sliding type of movement along the framework microtubules as discussed previously (Heath, 1974a). This suggestion is comparable to that of Fuge (reviewed in 1977) for the behavior of chromosomes and akinetochoric fragments in crane fly meiosis, where distinctive lateral interactions between chromatin and framework microtubules occur in addition to regular kinetochore-based microtubules. Such a system may account for the general movements of assorted particles and akinetochoric fragments that are frequently reported in many higher spindles (reviewed by, e.g., Lukyx, 1970; Bajer and Molè-Bajer, 1972; Nicklas, 1971). This type of movement is most probably a relatively advanced characteristic since, as will be argued in Section III, B, sliding activities associated with microtubules are probably a general phenomenon that evolved later than microtubule elongation-based activities.

The final type of chromatin–microtubule interaction that requires mention is that found in the ciliate macronucleus. Here there is at present little evidence of any specific interaction between the chromatin and microtubules other than the kinetochores of *Loxodes.* The observed morphology of most dividing macronuclei is consistent with the previously mentioned lateral type of interaction between chromatin and microtubules, but space constraints are such that it could easily be the product of random, functionally insignificant, juxtapositioning. At present there is no reason to believe that the chromatin is segregated by any more complex mechanism than simple passive division in the elongating nucleus. Such a system is probably sufficient in a highly polyploid nucleus where there is little need for accuracy in chromatin segregation. A similar situation probably occurs in the eukaryotic nucleus of *Peridinum* where even framework microtubules are lacking.

c. *Sensitivity to Antimitotic Agents.*   There is an extensive literature that shows that many, but not all, protists and fungi are either completely resistant to, or require unusually high levels of, various antimitotic and antimicrotubule agents such as colchicine, low temperature, and high pressure (e.g., Eigsti and Dustin, 1955; Wunderlich and Speth, 1970; Haber *et al.*, 1972; Kuzmich and Zimmerman, 1972; Jaekel-Williams, 1978; Tucker *et al.*, 1975; Turner, 1970; Cappuccinelli and Ashworth, 1976; Heath, 1975b,c, 1978a). In some cases, there is clearly a strong selection pressure for such resistance, as in the case of poikilothermic organisms residing in temperatures that depolymerize homiotherm microtubules. However other types of resistance, for example to chemicals, are hard to correlate with any known selection pressures. Therefore, a survey of such resistance has potential as an indication of evolutionary trends. Unfortunately, such data is currently totally inadequate, in part because interpretation of the results obtained are ambiguous. For example, resistance to colchicine could be due to enzymatic degradation of the drug, lack of permeability by all or part of the cell to the drug, absorption by nontarget cellular binding sites, masking of the target sites, or variation in the target (tubulin) itself. Only the latter cause is useful in the present context, and data on it is almost nonexistent. Thus, one can only conclude that antimitotic agent sensitivity is a potentially useful tool to aid in understanding the evolution of mitosis, but its time has not yet come.

3. *Formation*

There are two, and perhaps three, distinctive modes of spindle formation presently known. The major dichotomy is either (1) the polar structures migrate apart prior to spindle formation or (2) the spindle forms between the polar structures during their separation (Fig. 6). As far as is known these patterns of formation are listed in Table II. General consistency of behavior is seen at the subphylum level. Most inconsistencies occur in groups commonly thought to be heterogeneous on other characteristics (e.g., the Enteromophthorales). The only dubious group in which spindle formation heterogeneity does not correlate with organismal heterogeneity is the Coccidida, but here the problem is simply due to definition. The reported amount of centriole separation prior to spindle formation in *Eimeria aubernensis* and *Globidium* is very small, so that probably most of the separation occurs during spindle elongation as occurs in the other members of the group. Homogeneity of behavior among closely related organisms indicates stability and phylogenetic value for the character, but this value is limited by the fact that when there are very few options by which a process may occur, convergent evolution becomes very likely. However, it is probably safe to conclude that any organism that achieves spindle formation by a different means than the norm for the group should be carefully reexamined to determine the validity of its inclusion in that group. In terms of spindle evolution, this author favors migration during spindle formation as the most primitive type because then one process can contribute to the other, e.g., the force for centriole migration can largely be

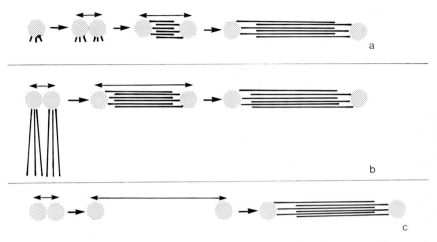

FIG. 6. Modes of spindle formation (a) The polar structures may develop very small arrays of perhaps only kinetochore microtubules prior to their separation. However the bulk of the spindle is formed as the polar structures migrate apart. (b) The spindle develops to approximately its mature number of microtubules, which grow to a considerable length as two adjacent fans, prior to their reorientation as the polar structures migrate. (c) The polar structures migrate to the distance apart that they will occupy at metaphase, *then* the spindle develops after this migration. (Designations D, F, and B, respectively, in Table II.)

derived from spindle elongation, whereas a system for generating migration as well as spindle formation must have evolved in the alternate case.

The third possible type of spindle formation is best seen in *Saccharomyces,* where two half spindles develop side by side, then reorient together when their poles separate. A similar situation seems to occur in the Dictyostelia, *Plasmodium berghei, Haemoproteus,* the actinopods, and possibly the cryptophytes. The difficulty of classification comes from the observation that, in those cells where pole migration and spindle formation are coincident events, there is a time when very small portions of half spindles lie side by side [e.g., *Saprolegnia* (Heath, 1978a,b)]. Clearly, it is then simply a matter of degree, rather than absolute difference, where one draws the line between the two systems. The lack of clarity between these two systems is further emphasized by observations in this author's laboratory which show that *Saprolegnia* can be induced to form extensive side by side half spindles by treatment with an antimicrotubule drug. Thus, at present, the *Saprolegnia* and *Saccharomyces* modes of spindle formation are best considered as minor variants of the same system.

The final point of variability observed in spindle formation is the behavior of the nuclear envelope. Three clear options have been described: (1) formation of the spindle entirely within the nuclear envelope, (2) formation largely outside of the nuclear envelope with entry following massive envelope breakdown, and (3) rupture of only the polar regions of the nuclear envelope as the spindle pushes through from the cytoplasm. The distribution of these alternatives is outlined in Table II, where the three types are indicated in the nuclear envelope column by intact, dispersed, and polar fenestrate entries, respectively. Within the intact type, there is, of course, some evidence for diversity. In some cases [e.g., *Minchinia, Oedogonium, Sapromyces, Saprolegnia, Physarum* (Tanaka, 1973; Sakai and Shigenaga, 1972)], formation seems to be nucleated by structures that are permanently intranuclear, whereas in other cases [e.g., *Coprinus* (Thielke, 1974), *Leucosporidium, Rhodosporidium,* and *Aessosporon*], the intranuclear spindle develops from cytoplasmic structures that enter the nucleus prior to spindle formation, with a subsequent resealing of the nuclear envelope after entry. These variations are discussed in more detail in Sections III, B and C because they involve spindle formation, NAO behavior, and nuclear envelope behavior.

4. *Dynamic Characteristics*

Apart from the overall time of the mitotic process, which is often largely made up of the prophase events and which has been discussed in Section II, A, there are three other dynamic aspects of mitosis that relate directly to spindle structure and function. These are (1) the rate of anaphase chromosome movement, (2) the rate of anaphase–telophase spindle elongation, and (3) the relative timing of these two events. Comparisons of the first two processes are hindered by temperature effects because, like any chemical process, the rate of movement is temperature dependent. This problem can only be overcome by comparison of rate versus temperature curves, which are unfortunately not available for a useful sample of organisms. Furthermore, there is some evidence for different rates at different stages of development in some organisms; thus, this too should be taken into account. However, an approximation to some degree of standardization can be obtained by comparing rates at physiological temperatures as given in Table IV. The general trend is toward faster rates for both (1) and (2) (as defined earlier) in the lower organisms. The extreme values vary by more than an order of magnitude. Because there is considerable variability in the data, with overlap between the ranges of various taxa, detailed comparisons are of little value. The only tentative conclusions that can be reached at present are that, in general terms, the spindles of primitive organisms generate faster rates of movement of both processes (1) and (2), and that the trend is thus toward slowing and perhaps better control of the process. The variability and overlap in the rates for different taxa make it unlikely that there are fundamentally different mechanisms operating in different organisms. However, given the crude nature of the measurement (i.e., time of the end produce of what must be a complex series of events), even this suggestion is very tentative.

TABLE IV
REPRESENTATIVE RATES OF CHROMOSOME MOVEMENT AND SPINDLE ELONGATION

Organism	Rate of chromosome to pole movement (μm/minute)	Rate of spindle elongation (μm/minute)	References
Myxomycota			
Diverse genera	0.26–1.85	?	Koevenig (1964)
Perichaena	3.3	?	Ross (1967)
Didymium	2.7–4.4	5–6	Kerr (1967); Ross (1967)
Chytidiomycota			
Phlyctochytrium	2.5	?	McNitt (1973)
Ascomycota			
Podospora			Zickler (1971)
Ascobolus	5	5	Aist (1969); Aist and Williams
Fusarium	9	5	(1972)
Sarcodina			
Pamphagus	10.4	?	Bêlar (1921)
Actinopoda			
Actinophrys	2.6	?	Bêlar (1922)
Zoomastigina			
Barbulanympha	0.4	1.5	Inoué and Ritter (1978)
MEAN	3.9	4.3	From above
RANGE	0.26–10.4	1.5–6	From above
Angiosperms			
Haemanthus	0.55	—[a]	Bajer in Mazia (1961)
Tradescantia	0.7	?	Barber in Mazia (1961)
Metazoans			
Homo (HeLa cells)	0.2	1.6	Oppenheim *et al.* (1973)
Dipodymus (Pt K$_1$)	2.7	2.0	Brinkley and Cartwright (1971)
Nephrotoma (Spermatocytes)	0.3	—	Forer (1966)
Tamalia (Spermatocytes)	0.34	2.1	Ris in Mazia (1961)
Psophus (Spermatocytes)	0.68	0.94	Jacquez *et al.* in Mazia (1961)
Popilius (Spermatocytes)	0.55	—	Roberts in Mazia (1961)
Triturus (Fibroblasts)	1.0	—	Mazia (1961)
Gallus (Osteoblasts)	2.2	1.4	Hughes and Swann in Mazia (1961)
MEAN	0.9	1.6	From above
RANGE	0.2–2.7	0.94–2.1	From above

[a] —, Denotes insignificant spindle elongation.

Irrespective of variations in the rates of the processes mentioned above, it is clear, in at least a few organisms, that the relative timing of the two events varies in different species. As shown in Table II, most spindles exhibit concomitant spindle elongation and chromosome-to-pole movement (+ in the column labeled "anaphase chromosomal migration"). However some organisms lack chromosome-to-pole movement because either the chromosomes are always closely adjacent to the poles (e.g., *Syndinium*), or because spindle elongation alone seems to generate adequate separation (e.g., *Coleochaete?*). The fourth situation seems to occur in *Closterium,* for example, where chromosome-to-pole movement occurs after spindle elongation. Because the anaphase stages of mitosis are typically rapid, the necessary data is often lacking, so that a detailed survey of these variations is not yet possible. In terms of spindle evolution, we might propose a primitive state for those cells lacking chromosome-to-pole movement entirely because the chromosomes never leave the poles (see Section III, B). However, it is hard to see much rationale for saying that there is much significance to the time of the process in the other organisms. It would seem very likely that coordination of the two processes could be very lax with little loss of efficiency.

5. *Dimorphic Nuclei*

One of the major problems in any discussion of the evolution of mitosis is the existence of organisms in which there is dimorphism of the mitotic system, either between nuclei within a cell, or between different life cycle stages. The first type of situation occurs in the ciliates, but this phenomenon presents few problems to the topic. Because the two types of nuclei have very different functions, the design requirements for their mitotic systems are also different. Therefore, it is highly probably that different mechanisms will have evolved. Thus, for the reproductive micronucleus, accuracy of division is essential for continuation of the entire genome, whereas for the highly polyploid vegetative macronucleus, accuracy is presumably unimportant, and the minimal apparatus capable of effecting approximately equal segregation is likely to be selected for. Thus, the presence of the rotary "mixing" movements at prophase (e.g., *Tokophrya*), and the subsequent formation of either an intranuclear (e.g., *Loxodes, Gastrostyla, Nassula,* etc.) or extranuclear (e.g., *Blepharisma, Didinium*) array of framework microtubules to effect nuclear elongation, are likely to be specially developed or retained features that have evolved for a very specific type of division. It is unlikely that this system has any relationship to the "mainstream" of mitotic evolution. As will be argued later (Section III, B), the intranuclear framework may well be primitive with respect to the extranuclear one, but both are likely to be derived from cells with more "conventional" spindles and cytoplasmic microtubule arrays.

A potentially more difficult problem is encountered among the myxomycetes, in which substantial mitotic dimorphism occurs at different life cycle stages where the functional requirements are presumably similar. For example, we cannot correlate the differences with ploidy level (Ross, 1968; Steffens and Willee, 1977) or type of cell division. If we uncritically utilize certain mitotic features (e.g., nuclear envelope behavior and centrioles) as indicators of mitotic evolution and aids to understanding phylogeny, we are forced to conclude that each species of myxomycete is polyphyletic, or that these characteristics of mitosis are controlled by the cellular environment and have no phylogenetic value. The former is clearly unacceptable, and the latter, undesirable for a writer of an article with the title of this one! However, if we can identify convincing functional differences between the two types of spindle, then we can more easily accept the dimorphism as a response to special selection pressures and, thus, secondarily derived from an organism with a stable monomorphic type of spindle. We can then also regain confidence in the value of spindle structure as a phylogenetic indicator. What, then, are the differences and possible explanations?

From Table II, a comparison of the data for *Physarum* myxamoebae and plasmodia shows the following differences.

1. Centrioles are present at the spindle poles in the myxamoebae, but not in the plasmodia. It has already been argued that centrioles only "go along for the ride" at mitosis (Section II, F), hence this is hardly a major difference. It may merely reflect the fact that the myxamoebae have progressed along the pathway of *de novo* centriole formation that is necessary for the impending production of flagella. A more serious difference would be the observation of different modes of spindle formation. For example, the plasmodia have a very characteristic intranuclear globule that plays a role in spindle production (Tanaka, 1973; Sakai and Shigenaga, 1972). If the two spindles are fundamentally similar, analysis of the myxamoebae should reveal a similar structure. Analysis of the appropriate stage of spindle formation has not yet been accomplished.

2. The nuclear envelope disperses during prophase in the myxamoebae but persists, with polar fenestrae, in the plasmodia. Because of variation in the behavior of the nuclear envelope in plasmodia of different genera (compare *Physarum* with *Arcyria*), this character is not very reliable, a point made earlier (Section II, B). It seems to be a matter of degree, whether the nuclear envelope disperses at prophase, anaphase, or only partially breaks at the poles. This interpretation gains extra weight when one considers that, in *Physarum* plasmodia, the spindle forms from an intranuclear granule (see earlier), eliminating the need for polar fenestrae that exists in organisms in which the spindle pushes through from cytoplasmic organizers (e.g., *Phlyctochytrium*). Thus, the polar fenestrae of *Physarum* may be an indication of an inherently labile nuclear envelope in this organism (cf. the discussion of *Coprinus* in Section II, B). This labile membrane may be "tipped over the brink" to dispersion by slightly different cellular factors in the myxamoebae.

The second factor that may explain the difference in membrane behavior is the speed of mitosis; it is faster in the myxamoebae (Kerr, 1967; Ross, 1968). A more rapid mitosis might be found in the absence of the nuclear envelope if cytoplasmic materials such as tubulin and ATP are required by the spindle. The selection pressure for a faster mitotic system in a myxamoeba may be as follows. If transcription is shut down during mitosis (as occurs in higher eukaryotes, and probably occurs in myxomycetes, since they have condensed chromosomes at mitosis), then a free-living single cell with a relatively small volume of cytoplasm might pay a bigger penalty for lack of transcriptional activity (and thus cellular activity?) during mitosis as compared with the plasmodia which presumably has a larger pool of "buffering" cytoplasm. Thus, in free-living cells, there may be a selection pressure for rapid mitoses permitted by a loss of the nuclear envelope at mitosis (cf. Oakley, 1978). Retention of the envelope in the plasmodia may be either a simple "relic" phenomenon, or it may be retained in order to prevent chromosome mixing on adjacent spindles in the coenocytic cells as suggested by Ross (1967). It may be that the price of a slower mitosis is worth paying for this latter reason.

The remaining two differences between plasmodial and myxamoebal mitoses are probably derived ones. Thus, interzone expulsion in the plasmodia is clearly impossible, by definition, when the nuclear envelope does not persist. The presence of vesicles in the myxamoebal spindle could be due to either the accidental entry of these vesicles from the cytoplasm in the absence of the nuclear envelope or, alternatively, to the need for an intraspindle calcium pump to replace the hypothetical one present in the intact nuclear envelope of the plasmodia (see Section II, B, 3).

Thus, the dimorphic myxomycete mitoses can be explained in a way that makes the differences due to function-specific selection pressure comparable to the phenomenon found in the ciliates. However, they do illustrate the need for care in discussing spindle evolution and its use as a phylogenetic indicator, especially with respect to centriole and nuclear envelope behavior.

6. *Polyspindles and Uninuclear Meioses*

In the vast majority of organisms, there is a tight correlation between genome separation and karyokinesis (division of the nucleoplasm and nuclear envelope), so that there is only one spindle per nucleus at mitosis and meiosis. However, there are two exceptions to this pattern: (1) cells in which numerous mitotic spindles coexist within a single nucleus (cells that I shall refer to as possessing polyspindles), and (2) cells in which the two sequential divisions typical of meiosis occur within a single nucleus without an intervening karyokinesis [a situation described by Moens and Rapport (1971) as uninuclear meiosis].

Currently, polyspindles are only known from the sporozoans such as *Sarcocystis, Plasmodium* (Canning and Sinden, 1973; Howells and Davis, 1971), and *Haemoproteus*. With such a limited distribution, it is unlikely that this organization represents a relic step from the main line of mitotic evolution. Rather, it probably evolved in response to some unknown specific feature of the biology of these parasitic organisms. However, these organisms do provide circumstantial evidence for a continuous connection between the spindle polar structures (probably mediated via kinetochore microtubules during mitosis) and the chromatin. Because of the many genomes in a single nucleus, it seems unlikely that a sequence of attachment at mitosis and detachment during interphase could function efficiently without resulting in considerable misconnections and serious results when karyokinesis does occur. It is interesting to note that the chromosomes typically do not appear to condense (or do so only to a limited extent) in these organisms during mitosis. We will return to this point in Section III, B and C.

Uninuclear meiosis occurs more widely among extant organisms than do polyspindles. It has been reported in some, but not all, hemiascomycetes (Moens and Rapport, 1971; Rooney and Moens, 1973), *Saprolegnia* (Howard and Moore, 1970), and some red algae (Scott and Thomas, 1975). This broad distribution, in relatively simple and primitive groups, makes it likely that such a process is a primitive type of meiosis that has been lost in most organisms. Because uninuclear meioses presumably require a type of continued interaction between chromatin and spindle components similar to that discussed above for polyspindles, it may be that loss of this type of meiosis coincided with the evolution of the means for controlling a more complex cycle of chromatin–spindle association–disassociation. Certainly among the oomycetes and hemiascomycetes, it is quite possible that there is a persistent connection between the chromatin and the spindle pole structures throughout the mitotic cycle (see Section III, C). Comparable data on the red algae is lacking at present. It is interesting to note that, on morphological grounds, numerous authors (most recently Kohlmeyer, 1975, and Demoulin, 1974) have postulated a close phylogenetic link between the asomycetes and red algae. The common possession of uninuclear meiosis may strengthen this suggestion and indicate common ancestry with the oomycetes also. We might then argue that uninuclear meiosis is a one-time evolutionary experiment in these organisms that has persisted, but not passed to other groups. However, we are reminded again of the dangers of ascribing any phylogenetic significance to a feature that can really only be accomplished by one or two ways, i.e., uninuclear or two separate divisions. Because these organisms only utilize uninuclear divisions at meiosis and do not possess polyspindles at mitosis, there may well be some, as yet unexplained, functional feature of these cells that has led to the evolution of uninuclear meiosis. In this event, uninuclear meiosis would have relatively little value as an indicator of mitotic evolution. Clearly a more extended account of its occurrence would be very useful.

H. Involvement of Cytoplasmic Microtubules

In addition to the various components of the spindle itself, many cells have a variable number of cytoplasmic microtubules associated with the spindle poles during part, or all, of mitosis. The extremes in the quantity of these associated cytoplasmic microtubules are represented by the massive arrays of astral microtubules typical of metazoan cells and the total absence of equivalent microtubules in the anastral angiosperm spindles. Most protists and fungi lie between these extremes and have a relatively small number of cytoplasmic microtubules present at least at some phases of mitosis. Two examples of the types of arrangements encountered are *Thraustotheca,* where a few (5–10) microtubules radiate from each pair of centrioles, predominantly into the cytoplasm away from the nuclei, throughout mitosis; and *Uromyces,* in which cytoplasmic microtubules are very rare during prophase and metaphase, but become abundant at anaphase and telophase. As discussed elsewhere (Heath, 1974a,b, 1978a; Heath and Heath, 1976), these cytoplasmic or astral microtubules must play some role in mitosis if their arrangement and numbers are modulated in this way. Presumably this role is to interact with the cytoplasm, in some way aiding in nuclear elongation and postmitotic nuclear movements, a role that clearly has impact on spindle design. Unfortunately, the variation in number of microtubules during mitosis makes comparison of this number between organisms possible only with detailed studies using serial sections of all phases. Such data is not available for many species. Comparisons are further complicated by seeming variability in behavior among closely related genera, e.g., *Uromyces* and *Puccinia.* Thus, while many of the organisms listed in Table II do possess some astral tubules at some stages of mitosis, there is not enough detailed data to make comparisons between different spindles a worthwhile project. However, these microtubules are clearly important to mitosis, and should be described in as much detail as possible in future studies so that a substantial comparative data base can be built up.

I. CONTROL MECHANISMS

The only realizable objective of this section is to draw attention to this facet of mitosis in order to encourage future investigators to consider their results in a comparative way. At present, we have very little data on what phenomena control the initiation of mitosis. However, there is evidence for the existence of cytoplasmic factors that can be isolated and used to induce premature mitosis (reviewed in general by Mazia, 1961, and more recently for some fungi and protists by Heath, 1978a). It seems quite likely that there will be intertaxa variability in the nature of these factors, just as there is variability between chemical messengers used during cellular slime mold aggregation (e.g., Olive, 1975). Thus, further comparative work on inducers of mitosis may give clues to understanding the evolution of mitosis, but useful results lie a long way in the future.

J. Summary of Characteristics

From the preceding discussion, three things are clear: (1) there are a large number of identified features of mitosis, the variations of which may help understand the way in which mitosis has evolved; (2) for many of these features, we have as yet too little data upon which to base worthwhile predictions and hypotheses; (3) by judicious use of some of these features one can make very tentative predictions of the design of a primitive mitotic apparatus, and from that, deduce directions of subsequent evolution.

A primitive mitotic system is likely to have had the following features:

1. Short duration (e.g., 5 minutes)
2. Energetic inefficiency
3. Genetic inefficiency (e.g., produces a high frequency of aneuploidy)
4. Accomplished within an intact nuclear envelope
5. Calcium ion control effected by a nuclear envelope-located pumping system, as opposed to vesicles lying free in the spindle
6. Simple median constriction of the nucleoplasm, nuclear envelope, and persistent nucleolus
7. Absence of chromatin condensation
8. Continuous connection between chromatin and spindle pole structures throughout the nuclear cycle
9. Flat disc-like polar structures closely associated with the nuclear envelope
10. Pole-to-pole framework microtubules
11. Few kinetochore microtubules per kinetochore (if kinetochores are present)
12. Simple kinetochore morphology
13. Rapid chromosome-to-pole and spindle elongation rates.

What I should now like to do is develop some of these points further into hypothetical pathways by which mitosis may have evolved, then consider how extant organisms may fit into that scheme.

III. Evolution of Mitosis

Since mitosis is essentially an eukaryotic phenomenon, and since eukaryotes undoubtedly arose from prokaryotic ancestors, it is important to consider the porbable changes that occurred in the genome distribution system during the transition from prokaryotes to eukaryotes. However, this discussion requires some prior consideration of the organization and segregation mechanism of the genome of prokaryotes.

A. The Prokaryote Ancestors

It is usually argued that the prokaryote ancestors of eukaryotes were bacteria, and that because there is good evidence for a connection between the bacterial genome and the cell membrane (e.g., Leibowitz and Schaechter, 1975; Nicolaides and Holland, 1978, and references therein), then the primitive eukaryote probably also had such an association (Kubai, 1975, 1978; Ris, 1975; Pickett-Heaps, 1974b, 1975a). Using this argument, it is often claimed that *Cryptothecodinium* has a mitotic apparatus most like that of the earliest primitive mitoses. However, recent data from the prokaryote kingdom suggests an alternative hypothesis. On the basis of a comparison of genome sizes of extant mycoplasmas and bacteria, Wallace and Morowitz (1973) argue that mycoplasmas are the most primitive prokaryotes, and that higher organisms evolved along two different lines from mycoplasma-like ancestors. In one line, genome doublings with end-to-end linkages gave rise to the typical large bacterial chromosomes. On the other line, genome doubling without end-to-end linkage gave rise to numerous separate, small chromosomes characteristic of lower eukaryotes. They argue that these numerous small chromosomes generated a strong evolutionary pressure for a new form of chromosome segregation, i.e., mitosis. We could also argue that evolution of a membrane bound compartment to keep these chromosomes together in a specialized environment, maximally conducive to their replication and transcription, would also be highly favored. Thus, we might more profitably look for the ancestral mitosis precursors among the genome distribution systems of mycoplasmas rather than the more highly specialized and evolved walled bacteria.

In most mycoplasmas, the genome is attached to the cell membrance, probably by its replication point (reviewed by Maniloff and Morowitz, 1972). However, in *Mycoplasma gallisepticum,* the genome seems to be attached to a 55 nm diameter disc that is attached to, but distinctly remote from, the cell membrane (Fig. 7) (Maniloff and Quinlan, 1973, 1974). These discs occur singly in nondividing cells, replicate prior to cell division, and lie at opposite poles of dividing cells (Fig. 8). Thus the morphology of a dividing cell is remarkably similar to that of a plaque-bearing mitotic nucleus without any microtubules. A clue to the causal agent of disc separation and/or cell division is found in the result that cytochala-

---

Figs. 7 and 8.   *Mycoplasma gallisepticum.* Figure 7 shows a cell that is probably undergoing division. It possesses two "blebs", each of which contains a small disc (arrows) which apparently represents the site of DNA attachment. Barring the absence of microtubules, this cell bears a striking resemblance to a mitotic nucleus with intranuclear NAOs! Courtesy of Dr. J. Maniloff and the American Society for Microbiology, from Maniloff and Morowitz, 1972. Figure 8 shows a variety of cells, one of which contains what may represent a replicated disc in one of the "blebs" (large arrow) and another which appears to have a face view of a disc (small arrow). Courtesy of Dr. J. Maniloff and the New York Academy of Sciences, from Maniloff and Quinlan, 1973.

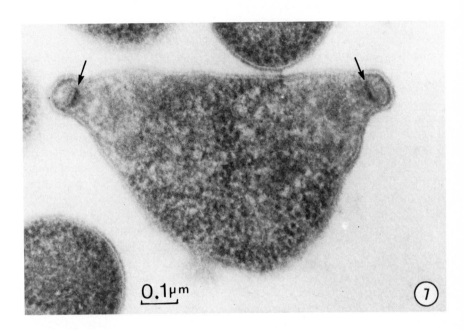

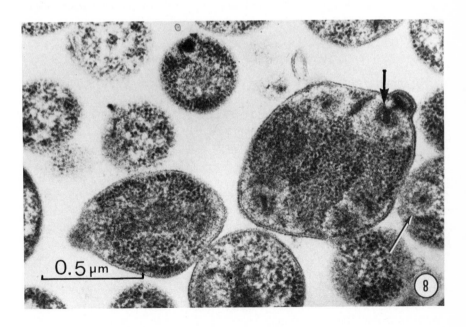

sin B blocks cell division, apparently by its action on an actin-like protein rather than any activity related to glucose or thymidine uptake (Ghosh *et al.*, 1978). Preliminary data cited by Ghosh *et al.* (1978) suggests that *M. gallisepticum* does indeed contain an actin-like protein as do other mycoplasmas (Neimark, 1977; Searcy *et al.*, 1978). These observations indicate that actin may well have evolved prior to tubulin. With the recently reemphasized possibility that actin can participate in cytoskeletal processes, as well as contractile ones (Pollard, 1976), it is very tempting to believe that *M. gallisepticum* has evolved a type of "mitotic spindle" in which the DNA is permanently attached to the discs that are pushed apart by the formation of a cytoskeletal array of actin filaments prior to median cell constriction. Unfortunately, such a hypothesis is strictly speculative, since the localization and organization of the actin is still unknown. Nevertheless, the important point is that *M. gallisepticum* has evolved an actin-based division system whose morphology is comparable to a mitotic system without microtubules. It is not important, in the present context, whether the mycoplasmas are truly primitive as suggested by Wallace and Morowitz (1972), or whether they are derived from other bacteria. The essential point is that *M. gallisepticum* does illustrate a level of organization achieved by a prokaryote. Such a level could well have been achieved by the ancestral prokaryotes. Thus, perhaps the behavior of the DNA of walled bacteria is less relevant to the evolution of mitosis than previously believed. For this reason an extended discussion of its possible segregation mechanisms (reviewed by Heath in 1974a and still unknown) is not given here.

B. HYPOTHETICAL EARLY MITOTIC SYSTEMS

For simplicity of cross referencing with later sections, the following discussion will be organized into a number of distinct steps that will be arranged in the most plausible order. However, it should be remembered that some of these steps may well have preceded, or occurred simultaneously with, steps listed later in the sequence, thus the numbering of the steps is not intended to be absolute and fixed. A diagrammatic interpretation of these steps is shown in Fig. 9.

Step 1. Simple invagination of the region of the cell membrane bearing the disc-like structures (described above in *Mycoplasma gallisepticum*), followed by subsequent lateral outgrowth and ultimate fusion of this invagination, could easily have generated a typical eukaryotic nucleus with intranuclear discs that would be morphologically equivalent to intranuclear NAOs. Initially the nucleus may have remained attached to the cell membrane, but probably would have soon become detached. During this sequence of events, genome segregation could continue by the same mechanism used by *M. gallisepticum;* the presence of the extra, probably fluid, membrane would be a very minor extra load. In this

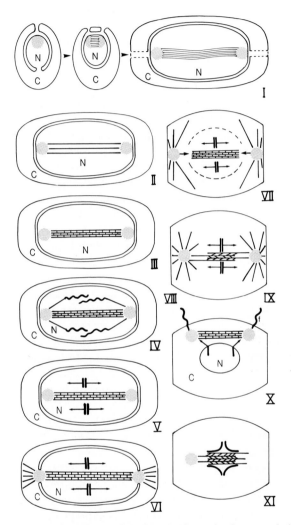

FIG. 9.   Diagrammatic representation of the Steps 1–10 postulated to occur during the evolution of mitosis. For a full description see the text, Section III, B. C, Cytoplasm; N, nucleus; f, flagellum. Bridges at right angles to the spindle microtubules denote static cross links whereas angled ones (VII, IX, and XI) denote active sliding linkages. Stippled circles are polar structures and the thickest lines are the chromosomes. The thin wavy lines in I denote actin rather than microtubules.

system, the actin spindle could continue functioning as it did prior to any membrane elaboration. There is no need to postulate the sudden appearance of a functionally intermediate system nor a radically different mechanism that would only function when mature. Selection for the formation of the nuclear envelope was presumably driven by the increased accuracy and efficiency that could be obtained by segregating chromosome-related activity from general cellular metabolism. The nuclear pore apparatus subsequently evolved to facilitate and control nuclear–cytoplasmic communication.

Step 2. Based on intuition and current usage in eukaryotic cells, we can conclude that microtubules are mechanically more rigid and stronger than actin filaments. Thus, as the early cells enlarged from the 0.3 to 1.0 μm range of extant mycoplasmas to the 10 μm typical for the diameter of a eukaryotic cell, we could predict a strong selection pressure for a stronger material to separate the intranuclear discs. This pressure may have been responsible for the evolution of the microtubule. It is likely that, for a hypothetical proto-eukaryotic cell, it was more crucial to survival to consistently maintain a large distance between separated genomes than to develop anisotropy in cell shape. Thus, one would predict the evolution of microtubules in this role prior to their development in the cytoplasm. Because the genomes were presumably still attached to their discs, it would have been most reliable to use these discs as spindle microtubule nucleating sites, thereby ensuring spindle development between discs, and thus between replicated genomes. Cells would then have an intranuclear spindle composed solely of a number of continuous microtubules running between disc-like polar structures to which the chromatin was directly attached. Probably the first formed spindles contained very few microtubules. The number might be expected to increase as the genome size and chromosome control-associated proteins increased in size and abundance. The actin formerly used in the ''spindle'' was presumably made redundant and/or transferred for use in cytoplasmic cleavage.

Step 3. Rigid inter-microtubule cross-linkages may have evolved to impart greater strength to the pole-to-pole spindle. Spindle elongation would have been simply achieved by microtubule polymerization. At either Step 2 or 3, polyspindles would have been viable.

Step 4. With increasing genome size and spindle size, it may have become increasingly inefficient (genetically inefficient, i.e., error prone) to synthesize sufficiently large discs to accommodate the necessary attachment points and to separate these discs and attachment points during prophase prior to spindle formation. One answer to this potential problem would be to evolve the use of a radial array of microtubules between the disc and the chromosomes. These kinetochore microtubules may have a smaller attachment area at the disc. (Data from *M. gallisepticum* suggests that in that system one chromosome requires on attachment disc of 55 nm diameter. A microtubule has a diameter of only 25 nm; thus, the area saving on this simplistic basis is approximately a 5-fold reduction.) The smaller attachment area would make available to the chromosomes a larger volume (especially useful when the chromosomes are condensed) if arranged in a hemispherical array at some distance remote from the disc. Thus, one would have smaller discs (equivalent to polar structures or NAOs) that could be structurally simpler and easier to duplicate and separate at prophase. Continuous association between the chromosomes and the discs could be maintained by having the kinetochore microtubule elongate as chromosome condensation necessitated more room for clustered chromosomes during metaphase. Kinetochore shortening, down to a single ring coincident with chromosome decondensation, would result in maintenance of the connection during interphase, with no visible kinetochore microtubules. Synthesis of these kinetochore microtubules would probably be nucleated by the NAO, as were the framework microtubules. We might initially expect that only one microtubule per chromosome would be sufficient for small chromosomes.

Step 5. Increasing genome complexity would entail increasing chromosome size. This, in turn, would generate evolutionary pressure for increasing complexity in chromosome structure to expedite chromosome movements during division and to aid in transcriptional regulation. With such complex chromosomes, it would probably be more difficult to separate the chromatids during mitosis. Thus, it may have been increasingly difficult to coordinate the replication and separation of NAOs with that of the chromosomes. This might lead to the evolution of substantial independence of the two phenomena. Back-to-back kinetochores, to which kinetochore microtubule nucleation activity was transferred, may have been the solution to the need to ensure equipartitioning of these complex and independent chromosomes. Retention of kinetochore microtubule binding sites at the poles would be expected. Metaphase may then have evolved as simply a time needed to permit separation of chromatids prior to anaphase movement. Furthermore, as chromosome size increased, so kinetochore size and number of attached kinetochore microtubules would probably also increase.

Step 6. Increasing competition lead to diversification of cytoplasmic features to permit increasing adaptation to diverse niches. Part of this process would have led to the evolution of systems capable of generating cellular projections, motile organs (e.g., flagella), means of generating water currents (e.g., cilia), and a means of ordering and moving cytoplasmic organelles and attaching flagella, etc., to the cell. Such factors seem to have led to the evolution of cytoplasmic microtubules that are now used in all of these roles. This pressure may then have led to the insertion of the formerly entirely intranuclear NAO *into* the nuclear envelope, so that it could control the nucleation of cytoplasmic microtubules as well as mitotic ones.

Step 7. As the control of cytoplasmic microtubules became temporally, and perhaps numerically, dominant to the formation of mitotic microtubules, so there might be pressure for the NAO to become increasingly external to, and ultimately largely remote from, the nuclear envelope. We could imagine the evolution of a system enabling transient migration of a cytoplasmic NAO back to the nucleus solely for mitosis. Polar fenestrae opening and resealing, such as appears to happen in *Coprinus* and the basidiomycetous yeasts, or more extensive nuclear envelope dispersal would be needed to allow the spindle microtubules to enter the nucleus from exclusively cytoplasmic NAOs. Transfer from the retention of both entirely intranuclear spindles and cytoplasmic microtubule systems to the use of cytoplasmically polymerized microtubules for mitosis may also have been favored by a possible greater efficiency achieved by forming microtubules in the cytoplasm near the site of synthesis of their subunits, rather than having to transport the individual subunits into the nucleus prior to their polymerization.

Step 8. Utilization of microtubules in various cytoplasmic processes led to the evolution of mechanochemical cross-bridges capable of generating sliding forces, either between adjacent microtubules (essential for flagellar activity) or between microtubules and adjacent organelles.

Step 9. Intermicrotubule cross-bridges were transferred back into the spindle for use in spindle elongation and possibly chromosome-to-pole movement. This step would have produced interdigitating framework microtubules whose sliding would be mediated by the mechanochemical cross-bridges. It should be noted that, because there seems to be greater energetic efficiency in the use of cross-bridges rather than microtubule polymerization to generate spindle elongation (see Section II, G, 4), these bridges may have evolved inside the nucleus for mitosis prior to their use in the cytoplasm. Thus, this step may well not have been needed. Initially, one might have expected the framework microtubules to form a coherent central bundle, rather than being interspersed among the chromosomes, on the grounds that development and control of movement must be easier to achieve if all elements are grouped together, rather than being dispersed. Dispersal of the framework would necessitate a more complex and subtle control system to activate all of the elements independently. Subsequent evolution would be expected to have led to subtle differences in the cross-bridges used in the various cellular locations due to increased specialization for specific functions.

These nine steps are seen as major ones that occurred in the evolution of the mitotic systems characteristic of higher organisms. In addition two specialized side steps also occurred in some groups of organisms.

Step 10. In some organisms the cytoplasmic-based motility systems, i.e., centrioles, flagella, and their associated root systems, became so large that a cytoplasmic spindle-like structure evolved to ensure their segregation. It was then more efficient to use this structure to also divide the nucleus than to maintain the organizational capacity to form two systems. Alternatively, with the development of cytoplasmic microtubule systems, it may have been equally efficient to use specially made cytoplasmic microtubules for mitosis (i.e., an extranuclear spindle) rather than maintain an additional, entirely intranuclear, spindle system or to thrust the cytoplasmic microtubules into the nucleus for mitosis. Either of these steps would necessitate the evolution of nuclear envelope-located kinetochores by which the chromosomes could interact with the extranuclear spindle. It may be more likely that this stage occurred soon after Step 3, when the chromosomes were still attached to the nuclear envelope, rather than after Step 4, when nucleoplasm-located kinetochores had evolved. However, such is by no means necessary. If intermediate stages in separation of kinetochore and membrane could evolve and function (as must have occurred in all theories), so the reverse procedure is equally likely, as the extranuclear spindle ascended over the intranuclear spindle.

Step 11. An alternate solution to the problems of increased chromosome complexity, representing a deviation from the main line of mitotic evolution at about Step 5, would have been to evolve chromatin independence from the NAOs and to utilize lateral sliding interactions between chromatin and framework microtubules, rather than increased development of kinetochore microtubules. Such a step would probably have had to evolve after the evolution of active microtubule-based sliding systems, as suggested in Steps 8 and 9.

What will now be attempted is to see at what steps the various extant mitoses ceased development and persisted in "relic" form to the present day.

C. The Phylogenetic Position of Extant Mitoses

A eukaryote that had not evolved past the hypothetical Step 1 described in the preceding section would have an intranuclear spindle containing no microtubules and only an actin framework. No eukaryotes with such a mitotic system are known at present. Probably such a system could not have satisfactorily handled the large genomes and nuclei of extant eukaryotes, and it perished in the small plastic ancestral eukaryotes.

After Step 2 we should have an intranuclear spindle with pole-to-pole microtubules and no kinetochore microtubules. The only organisms close to this condition are the zygomycetes such as *Mucor, Pilobolus,* and *Phycomyces,* but it should be remembered that these features are far from proven in these organisms. However, the existence of a spindle composed entirely of a few truly pole-to-pole microtubules throughout mitosis is highly probable. Since they are close together, they have probably progressed through Step 3 also, and are rigidly cross-linked to one another. None of the other organisms so far studied even come close to having evidence for both pole-to-pole microtubules and chromatin linked directly to the spindle poles without kinetochore microtubules. This suggests that, *if* (and this is still very much in doubt) kinetochores are ultimately proven absent from the mucoralean Zygomycetes, these organisms are the only ones that have retained the most primitive type of mitosis known.

Judging from the range of extant mitoses, Step 4 represents a major hurdle over which most eukaryotes leapt, but Step 5 was less crucial, because one has more organisms that appear to have their chromosomes linked to the spindle poles throughout the nuclear cycle, either via kinetochore microtubules of various lengths, or by direct connection during interphase. These organisms typically have the primitive feature of only one kinetochore microtubule per chromosome. Examples of this type of organization, with various framework arrangements, appear to occur among the Oomycetes, *Conidiobolus, Saccharomyces,* and the sporozoans, where the morphology of spindle development and the possession of polyspindles is at least consistent with the possibility of a continuous interaction between the chromatin and the spindle pole structures. This interaction utilizes a direct attachment at interphase and is mediated by kinetochore microtubules during mitosis. *Dictyostelium* may also come close to this arrangement because, while the NAO is extranuclear during interphase, the kinetochores seem to remain in the vicinity of the site of NAO reentry at the next mitosis, thus indicating some interphase attachment to a possible residual part of the NAO on the nuclear envelope. Other organisms may show similar features, but unfortunately, data on prophase events is largely absent from most reports of mitosis.

The vast majority of organisms have evolved past Step 5, so that as far as is known, most species listed in Table II have chromosomes that lie free in the nucleus and connect up to the spindle at prophase via various numbers of kinetochore microtubules. The degree of advancement among these would then be determined by the other parameters mentioned in the text, such as degree of chromosome condensation, size of chromosomes, complexity of kinetochores, number of kinetochore microtubules, behavior of the nucleolus and nuclear envelope, and location of the NAO.

The first part of Step 6 seems to have been a step that all extant eukaryotes took. There are no thoroughly studied, living eukaryotes that lack cytoplasmic microtubules at some stage in their life cycle; although, of course, there is tremendous diversity in the quantity and complexity of organization of such microtubules. However, this diversity is beyond the scope of this article.

The externalization of the NAO from the nucleus seems to be a step that many organisms did not take. All of the cells listed in Table II as having intact nuclear envelopes have retained at least some part of an intranuclear NAO. However, in many caes (e.g., *Plasmodium, Uromyces, Saccharomyces, Schizosacharomyces*), the structure is clearly set into a pore in the nuclear envelope and is bifacial, apparently nucleating both nuclear and cytoplasmic microtubules (equivalent to part two of Step 6). The behavior of *Physarum* plasmodial mitoses and *Oedogonium* is perhaps one of the most puzzling patterns. Here the NAO is intranuclear, but it is detached from the nuclear envelope and seems to function entirely free of the envelope. This behavior could be read either as an unusual case in which the formerly membrane-attached NAO has secondarily left the membrane for some reason, or as an indication that the hypothesized Step 1 is wrong. In the latter case, the disc detached itself from the membrane prior to nuclear envelope formation and subsequently reattached itself in most organisms. Which of these alternatives is correct is unclear at present.

Step 7 is clearly a major fence over which many organisms climbed. Essentially all organisms in Table II with polar fenestrae, dispersive nuclear envelopes, extranuclear spindles, or an extranuclear NAO entering the nucleus at prophase (e.g., *Rhodosporidium, Coprinus*) seem to have their spindle nucleating structures located entirely in the nucleoplasm and appropriate behavior of the envelope to permit spindle formation. These organisms would, in this respect, all be seen as advanced relative to those with entirely intranuclear or nuclear envelope-based NAOs.

The evolution of a spindle framework of primarily interdigitating microtubules is suggested to be an advanced state resulting from Steps 8 and 9. On this basis probably most spindles are of the advanced type, but conclusive data on this point are not widespread. It seems probable that the mucoralean Zygomycetes, *Uromyces, Saccharomyces,* and *Barbulanympha* all possess predominantly genuine pole-to-pole microtubules, and thus, are primitive in this respect. Conversely, it is clear that the reported diatoms, *Thraustotheca, Dictyostelium,* and *Cryptomonas,* all have interdigitating spindles, as probably do the Acantharia and many others. Unfortunately, the detailed type of analysis needed to determine the arrangement of the framework microtubules has rarely been achieved, so that the published data do not permit an adequate distinction between the two types of system. Organisms, presumably advanced on the basis of their framework microtubules being dispersed among the chromosomal microtubules rather than being formed into a tight bundle, are clearly indicated in the table and are not discussed further.

Organisms, located in the sidestream of the main current of mitotic evolution because they took Step 10, are, of course, the dinoflagellates and the zooflagellates, plus those few other species that have utilized extranuclear spindles (Table II). A further piece of evidence to support the hypothesis that these organisms contain a sidestream type of mitosis rather than an ancestral form is that, in general terms, they are complex organisms that appear to be some of the most highly evolved protists. These organisms are not a group among which we could expect to find relic mitoses. On the other sidestream, with respect to chromatin behavior only, are the diatoms, and possibly *Prymnesium,* which took Step 11 in order to handle their chromatin during mitosis.

IV. Mitosis as an Indicator of Phylogeny

The use of variations in the construction of the mitotic apparatus as indicators of phylogenetic relationships among protists in particular has been discussed by a number of authors (e.g., Pickett-Heaps, 1969, 1972e, 1975a,b; Pickett-Heaps and Marchant, 1972; Kubai, 1975, 1978; Fuller, 1976; Heath, 1974a, 1975a). A number of the advantages and limitations of their use for this purpose have been ably reviewed by Oakley (1978). Some of these points should be reiterated and others added.

A. ADVANTAGES

With respect to other more commonly used characters, the following points are advantages to the use of mitosis as a phylogenetic marker.

1. It is applicable to all eukaryotic cells and thus is valuable across boundaries where other structures are absent on one side and present in various forms on the other side.

2. It is part of the basic cell. A good case can be made for the evolution of chloroplasts and mitochondria via endosymbiosis (e.g., Margulis, 1970; Taylor, 1974) so that with the possibility of multiple endosymbioses of different symbionts by one host type, or multiple hosts for one symbiont, it becomes clear that these organelles are not ideal indicators of phylogenetic relationships.

3. It is composed of a large number of components, so that the chances of convergent evolution of the total apparatus are remote. Similarly, the diversity of components yields a substantial amount of information that can contain variability and thus provide good raw material with which to work.

B. DISADVANTAGES

Some of the problems that are associated with the use of mitosis as a phylogenetic indicator, but are by no means unique to mitosis, are as follows:

1. Many of its components are "either/or" features so that in any one feature, as mentioned earlier, convergent evolution is highly likely.

2. Unambiguous data on a number of features of potential value is extremely laborious to obtain and prone to artifact (e.g., microtubule lengths and types).

3. There may well be strong selection pressures working on mitosis and leading to convergent evolution (e.g., rapid mitoses may be advantageous for small motile cells; thus, time is of limited value).

4. The functional significance of many characteristics is unknown. Thus, there may be unexpected selection pressures driving toward convergent evolution; and yet, because we do not understand the role of the structure, we may assume an absence of such pressure.

5. A current disadvantage, but not an intrinsic one, is the highly variable quality and quantity of information on different organisms. This variability makes comparisons difficult.

6. There is no a priori reason to believe that the rate of evolution of the mitotic apparatus will necessarily correlate with the rate of evolution of the entire organism. Equally, individual components of the system may evolve at different rates to one another. Thus, one can potentially envisage two organisms that have retained a very primitive mitotic system from a very remote common ancestor, yet have diverged dramatically in other features. In other words, it is highly likely that the selection pressures operating in the mitotic system will differ substantially from those operating on the entire organism.

C. Tentative Relationships

Providing we bear in mind the above points, what relationship can we discern among organisms, based upon their mitotic characteristics? The following comments are based entirely on mitotic characteristics as contained in Table II and the text of this paper. In the first place, we can see that a number of groups of organisms clearly stand in isolation from other taxa. Thus, based on spindle structure, polar structure morphology, and type of chromatin interaction with the spindle, it seems likely that the diatoms represent a cohesive group of relatively advanced organisms that are rather distantly related to any other protists. A similar situation probably applies to the dinoflagellates, which seem to be even more remote on the basis of both their chromosome morphology and spindle structure. There may well be some relationship between the dinoflagellates and the Trichomonadida, Polymastigida, and Hypermastigida zooflagellates. Some of the dinoflagellates may represent the simplest examples of extranuclear spindles. It is also very clear that the Protomonadida are very distantly related to the rest of the zooflagellates; thus, one might seriously question their inclusion in the same phylum. The possession of the micronucleus and macronucleus must serve to place the ciliates in a rather isolated position also. However, on the basis of micronuclear structure, which is the nucleus most closely similar to that of other organisms in its functions, we find the ciliates showing closer affinities with the oomycetes and sporozoans. Another somewhat isolated, and apparently primitive, group is the Englenophyta, which shows a mixture of primitive (e.g., nucleolus, nuclear envelope) and advanced (e.g., condensed chromosomes, kinetochores) characteristics. There is no obvious way of relating these organisms to any of the other groups on the basis of their mitotic system.

The second conclusion that can be drawn is that a number of phyla and lower taxa seem to be rather unacceptably diverse in the mitotic systems of their members. The case of the Zoomastigina has already been mentioned. Other groups that seem to be rather heterogeneous are the Rhodophyta, Sarcodina, Foraminifera, Myxomycota, Zygomycota, and Basidiomycota. Much of this heterogeneity is, of course, reduced if comparisons are made within taxa below the phylum level. However, the important point is that when some phyla are very

68   I. BRENT HEATH

homogeneous, and others very heterogeneous, we must question the wisdom of drawing the phyla lines where they are drawn here. In the more heterogeneous phyla, it might be wiser to elevate some of the lower taxa to phylum status to achieve a more even level of homogeneity in each phylum. Extending the considerations of intrataxa heterogeneity to a lower level, we can certainly pick out genera or species that seem to be out of place in otherwise reasonably homogeneous taxa. Examples of such cases are *Hydrodictyon, Chlamydomonas moewusii,* and *Microspora* among the green algae.

A third result of our analysis is the questioning of common concepts of the state of evolutionary advancement of certain organisms. Thus, the Myxomycota generally seem to be rather highly advanced organisms, with the exception of the Dictyostelia, which should either be removed from this phylum entirely or at least be considered the most primitive group within the phylum. Among the algae, there is very little reason to believe that the Cryptophyta, Chrysophyta, Chloromonodophyta, or Haptophyta are particularly primitive relative to other

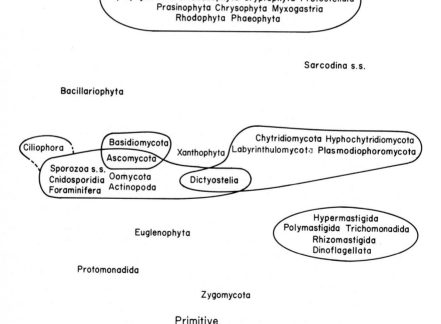

Fig. 10. Groupings of organisms based on overall similarities in mitotic systems. Height on the page denotes a rough indication of evolutionary advancement.

algae; their mitotic systems are of the most advanced type in the majority of characteristics. While the red and brown algae are undoubtedly somewhat primitive relative to many of the greens, the differences are not marked, thus suggesting a rather closer relationship than is often accorded by other parameters.

Among the considered taxa, there are a number of relationships that can be discerned, and others that have been suggested on other grounds, but seem particularly questionable on the basis of mitotic features. Such questionable relationships might include the Dictyostelia and Myxomycota referred to above and, on a larger scale, the heterokonta of Cavalier-Smith (1978). It is hard to envisage a more diverse array of mitotic systems than those grouped into this proposed kingdom. On the other hand, we can arrive at some reasonably homogeneous groupings of mitotic systems as outlined in Fig. 10. Relationships within these groupings are generally uncertain given the currently available data and comprehension of its value. Similarly, it is not possible to reliably trace connections between these groups. However, they may prove useful in comparison with other groupings arrived at on other phylogenetic parameters.

From the above considerations, it is evident that there is much potential phylogenetic value to be obtained from a consideration of the variations on a theme of mitosis. What is required at present is more complete data on more representatives of the various phyla and a less biased approach to the analysis of such data, perhaps by use of the techniques of numerical taxonomy. Unfortunately, what still seems to be lacking from the mitotic data is any clear indication of the relationship of some of the more remote taxa such as the Euglenophyta and dinoflagellates. Mitotic data merely reinforce previous conclusions that these groups have indeed diverged from other organisms.

V. Conclusions

From the preceding work, it is clear that there is a wealth of diversity among the characteristics of mitosis in the protists and fungi. Some of this diversity may prove to be artifact in the light of future work. Some may prove to be largely illusory in that we are currently looking at extremes of a spectrum of variation in which we cannot define unambiguous discontinuities. Such features will have to be discarded. Conversely, further work will undoubtedly reveal new features that will prove useful. However, there is at least some justification for the following positive statements, made in order of sharply decreasing certainty. The process of mitosis has undergone evolution. More data are becoming available to help explain possible universal, or variant, mitotic mechanisms. These data are giving a better understanding of these mechanisms and are explaining how mitosis evolved. The phylogeny of extant organisms is becoming clearer as a result of our analysis of mitosis. Unfortunately, what is very clear is that we need much more data with which to test some of the above hypotheses. Hopefully, the publication of this article will alert those generating these data to some of the facets of their work that may have value that they would not otherwise have perceived and would have thus overlooked or omitted.

ACKNOWLEDGMENTS

The preparation of this review was aided by an operating grant from the Natural Sciences and Engineering Research Council of Canada and by the provision of sabbatical leave from York University. The critical reviews of the manuscript and numerous discussions with Drs. M. C. Heath and P. B. Moens were very helpful, as was the excellent secretarial work of Dorothy Gunning.

REFERENCES

1. Aikawa, M., and Beaudoin, R. L. (1968). *J. Cell Biol.* **39,** 749.
2. Aikawa, M., Huff, C. G., and Sprinz, H. (1967). *J. Cell Biol.* **34,** 229.
3. Aikawa, M., and Jordan, H. B. (1968). *J. Parasitol.* **54,** 1023.
4. Aikawa, M., Sterling, C. R., and Rabbege, J. (1972). *Proc. Helminth. Soc. (Washington)* **39** (Special Issue), 174.
5. Aist, J. R. (1969). *J. Cell Biol.* **40,** 120.
6. Aist, J. R., and Williams, P. H. (1972). *J. Cell Biol.* **55,** 368.
7. Aldrich, H. C. (1967). *Mycologia* **59,** 127.
8. Aldrich, H. C. (1969). *Am. J. Bot.* **56,** 290.
9. Aldrich, H. C., and Carroll, G. C. (1971). *Mycologia* **63,** 308.
10. Alov, I. A., and Lyubskii, S. C. (1977). *Int. Rev. Cytol. Suppl.* **6,** 59.
11. Altman, P. L., and Dittmer, D. S. (1972). "Biology Data Book" (2nd ed.), Vol. 1, pp. 1–136. Federation of American Societies for Experimental Biology, Bethesda, Maryland.
12. Amoore, J. E. (1963). *J. Cell Biol.* **18,** 555.
13. Ashton, M. L. (1978). *In* "Nuclear Division in the Fungi" (I. B. Heath, ed.), p. 100. Academic Press, New York.
14. Atkinson, A. W., Gunning, B. E. S., John, P. C. L., and McCullough, W. (1971). *Nature (London) New Biol.* **234,** 24.
15. Bajer, A., and Molè-Bajer, J. (1972). *Int. Rev. Cytol. Suppl.* **3,** 1.
16. Barber, H. N. (1939). *Chromosoma* **1,** 33.
17. Bardele, C. F. (1969). *Z. Zellforsch. Mikrosc. Anat.* **93,** 93.
18. Beakes, G. W., and Gay, J. L. (1977). *Trans. Br. Mycol. Soc.* **69,** 459.
19. Bech-Hansen, C. W., and Fowke, L. C. (1972). *Can. J. Bot.* **50,** 1811.
20. Beckett, A., and Crawford, R. M. (1970). *J. Gen. Microbiol.* **63,** 269.
21. Belar, K. (1921). *Arch. Protistenk.* **43,** 287.
22. Belar, K. (1922). *Arch. Protistenk.* **46,** 1.
23. Berezney, R., and Coffey, D. S. (1977). *J. Cell Biol.* **73,** 616.
24. Bianchi, L., Rondanelli, E. G., Carosi, G., and Gerna, G. (1969). *J. Parasitol.* **55,** 1091.
25. Bibring, T., Baxandall, J., Denslow, S., and Walker, B. (1976). *J. Cell Biol.* **69,** 301.
26. Bland, C. E., and Lunney, C. Z. (1975). *Cytobiologie* **11,** 382.
27. Blum, J. J., Sommer, J. R., and Kahn, V. (1965). *J. Protozool.* **12,** 202.

28. Borisy, G. G., Peterson, J. B., Hyams, J. S., and Ris, H. (1975). *J. Cell Biol.* **67**, 38a.
29. Bouck, G. B., and Brown, D. L. (1973). *J. Cell Biol.* **56**, 340.
30. Bradbury, P. C., and Trager, W. (1968). *J. Protozool.* **15**, 700.
31. Braselton, J. P., and Miller, C. E. (1973), *Mycologia* **65**, 220.
32. Braselton, J. P., Miller, C. E., and Pechak, D. G. (1975). *Am. J. Bot.* **62**, 349.
33. Brinkley, B. R., and Cartwright, J. (1971). *J. Cell Biol.* **50**, 416.
34. Brinkley, B. R., Mace, M. L., and McGill, M. (1974). *In* "Electron Microscopy" (J. V. Sanders and D. J. Goodchild, eds.), Vol. 2, pp. 248–249. Australian Academy of Sciences, Canberra.
35. Brockaw, C. J. (1975). *In* "Molecules and Cell Movement" (S. Inoué and R. E. Stephens, eds.), pp. 165–179. Raven, New York.
36. Bronchart, R., and Demoulin, V. (1977). *Nature (London)* **268**, 80.
37. Brugerolle, G. (1975). *Protistologica* **11**, 457.
38. Burr, F. A., and West, J. A. (1970). *Phycologia* **9**, 17.
39. Byers, B., and Goetsch, L. (1975a). *Proc. Natl. Acad. Sci. U.S.A.* **72**, 5056.
40. Byers, B., and Goetsch, L. (1975b). *J. Bacteriol.* **124**, 511.
41. Byers, B., Shriver, K., and Goetsch, L. (1978). *J. Cell Sci.* **30**, 331.
42. Cachon, J., and Cachon, M. (1970). *Protistologica* **6**, 57.
43. Cachon, J., and Cachon, M. (1974). *C.R. Acad. Sci., Ser. D.* **278**, 1735.
44. Cachon, J., and Cachon, M. (1977). *Chromosoma* **60**, 237.
45. Camp, R. R., Mattern, C. F. T., and Honigberg, B. M. (1974). *J. Protozool.* **21**, 69.
46. Cande, W. Z., Snyder, J., Smith, D., Summers, K., and McIntosh, J. R. (1974). *Proc. Natl. Acad. Sci. U.S.A.* **71**, 1559.
47. Canning, E., and Sinden, R. E. (1973). *Parasitology* **67**, 29.
48. Cappuccinelli, P., and Ashworth, J. M. (1976). *Exp. Cell Res.* **103**, 387.
49. Carasso, N., and Favard, P. (1965). *J. Microscopie* **4**, 395.
50. Cavalier-Smith, T. (1978). *BioSystems,* **10**, 93.
51. Cesana, D. (1971). *C. R. Acad. Sci., Ser. D.* **272**, 3057.
52. Chaley, N., Lord, A., and Lafontaine, J. G. (1977). *J. Cell Sci.* **27**, 23.
53. Chapman, V. J., and Chapman, D. J. (1973). "The Algae" (2nd ed.). Mcmillan, London.
54. Comings, D. E., and Okada, T. A. (1972). *Chromosoma* **37**, 177.
55. Comings, D. E., and Okada, T. A. (1976). *Exp. Cell Res.* **103**, 341.
56. Coss, R. A. (1974). *J. Cell Biol.* **63**, 325.
56a. Coss, R. A., and Pickett-Heaps, J. D. (1973). *Protoplasma* **78**, 21.
56b. Coss, R. A., and Pickett-Heaps, J. D. (1974). *J. Cell Biol.* **63**, 84.
57. Danforth, H. D., and Hammond, D. M. (1972). *J. Protozool.* **19**, 454.
58. Daniels, E. W., and Roth, L. E. (1964). *J. Cell Biol.* **20**, 75.
59. Davidse, L. C., and Flach, W. (1977). *J. Cell Biol.* **72**, 174.
60. Davidson, L., and La Fountain, J. R. (1975). *Biosystems* **7**, 326.
61. Day, A. W. (1972). *Can J. Bot.* **50**, 1337.
62. Deacon, T. R., and Darden, W. H. (1971). *In* "Contributions in Phycology" (B. C. Parker and R. M. Brown, eds.), pp. 67–90. Allen, Lawrence, Kansas.
63. Demoulin, V. (1974). *Bot. Rev.* **40**, 315.
64. de Souza, W., and Meyer, H. (1974). *J. Protozool.* **21**, 48.
65. Desportes, I. (1970). *Ann. Sci. Naturelles Zool. Paris, Ser. 12* **12**, 73.
66. Desportes, I. (1976). *Protistologica* **12**, 121.
67. Desser, S. S. (1972a). *Can. J. Zool.* **50**, 477.
68. Desser, S. S. (1972b). *Can. J. Zool.* **50**, 707.
69. Dick, C., and Johns, E. W. (1967). *Biochem. J.* **105**, 46p.
70. Dodge, J. D. (1971). *Protoplasma* **73**, 145.

71. Dodge, J. D., and Bibby, B. T. (1973). *Bot. J. Linn. Soc.* **67,** 175.
72. Dodge, J. D., and Crawford, R. M. (1968). *Protistologica* **4,** 231.
73. Dubremetz, J.-F. (1971). *J. Microscopie* **12,** 453.
74. Dubremetz, J.-F. (1973). *J. Ultrastruct. Res.* **42,** 354.
75. Dusenbery, R. L. (1975). *Biochim. Biophys. Acta* **378,** 363.
76. Dutta, S. K., and Mandel, M. (1972). *J. Protozool.* **19,** 538.
77. Dykstra, M. J. (1976). *Protoplasma* **87,** 347.
78. Dylewski, D. P.. Braselton, J. P., and Miller, C. E. (1978). *Am. J. Bot.* **65,** 258.
79. Edds, K. T. (1975). *J. Cell Biol.* **66,** 145.
80. Eigsti, O. J., and Dustin, P. (1955). "Colchicine in Agriculture, Medicine, Biology and Chemistry." Iowa State College Press, Ames, Iowa.
81. Elsner, P. R., Van der Molen, G. E., Horton, J. C., and Bowen, C. C. (1970). *Phytopathology* **60,** 1765.
82. Ellzey, J. T. (1974). *Mycologia* **66,** 32.
83. Falk, H., Wunderlich, F., and Franke, W. W. (1968). *J. Protozool.* **15,** 776.
84. Febvre, J. (1977). *J. Ultrastruct. Res.* **60,** 279.
85. Firtel, R. A., and Bonner, J. (1972). *J. Mol. Biol.* **66,** 339.
86. Fitch, W. M. (1976). *J. Mol. Evol.* **8,** 13.
87. Floyd, G. L., Stewart, K. D., and Mattox, K. R. (1972a). *J. Phycol.* **8,** 68.
88. Floyd, G. L., Stewart, K. D., and Mattox, K. R. (1972b). *J. Phycol.* **8,** 176.
89. Forer, A. (1966). *Chromosoma* **19,** 44.
90. Forer, A. (1969). *In* "Handbook of Molecular Cytology" (A. Lima-de-Faria, ed.), pp. 553–601. North-Holland Publ., Amsterdam.
91. Forer, A. (1978). *In* "Nuclear Division in the Fungi" (I. B. Heath, ed.), pp. 21–88. Academic Press, New York.
92. Fowke, L. C., and Pickett-Heaps, J. D. (1969). *J. Phycol.* **5,** 240.
93. Franke, W. W., and Reau, P. (1973). *Arch. Mikrobiol.* **90,** 121.
94. Frenster, J. H. (1974). *In* "The Cell Nucleus" (H. Busch, ed.), Vol. 1, pp. 565–580. Academic Press, New York.
95. Friedländer, M., and Wahrman, J. (1970). *J. Cell Sci.* **7,** 65.
96. Fuge, H. (1977). *Int. Rev. Cytol. Suppl.* **6,** 1.
97. Fuller, M. S. (1976). *Int. Rev. Cytol.* **45,** 113.
98. Fuller, M. S., and Reichle, R. E. (1965). *Mycologia* **57,** 946.
99. Fulton, G. C., and Dingle, A. D. (1971). *J. Cell Biol.* **51,** 826.
100. Furtado, J. S., and Olive, L. S. (1970). *Cytobiologie* **2,** 200.
101. Furtado, J. S., and Olive, L. S. (1971). *Nova Hedwigia* **21,** 537.
102. Garber, R. C., and Aist, J. R. (1977). *In* "Abstracts, 2nd International Mycological Congress" (H. E. Bigelow and E. G. Simmons, eds.), p. 217.
103. Ghosh, A., Maniloff, J., and Gerling, D. A. (1978). *Cell* **13,** 57.
104. Gillott, M. A., and Triemer, R. E. (1978). *J. Cell Sci.* **31,** 25.
105. Girbardt, M. (1968). *In* "Aspects of Cell Motility" (P. L. Miller, ed.), Vol. 22, pp. 249–259. Cambridge Univ. Press, London.
106. Girbardt, M. (1978). *In* "Nuclear Division in the Fungi" (I. B. Heath, ed.), pp. 1–20. Academic Press, New York.
107. Girbardt, M., and Hädrich, H. (1975). *Z. Allg. Mikrobiol.* **15,** 157.
108. Godward, M. B. E. (1966). "The Chromosomes of the Algae." St. Martins, New York.
109. Goldstein, L., Rubin, R., and Ko, C. (1977). *Cell* **12,** 601.
110. Goodman, E. M., and Ritter, H. (1969). *Arch. Protistenk.* **111,** 161.
111. Gould, R. R., and Borisy, G. G. (1977). *J. Cell Biol.* **73,** 601.
112. Graham, L. E., and McBride, G. E. (1978). *J. Phycol.* **14,** 132.

113. Grain, J. (1966). *Protistologica* 2, (Part 2), 5.
114. Grell, K. G. (1973). "Protozoology" (2nd ed.). Springer-Verlag, Berlin and New York.
115. Griffin, D. H., Timberlake, W. E., and Cheney, J. C. (1974). *J. Gen. Microbiol.* 80, 381.
116. Gull, K., and Newsam, R. J. (1976). *Protoplasma* 90, 343.
117. Gull, K., and Trinci, A. P. J. (1974). *Trans. Br. Mycol. Soc.* 63, 457.
118. Guth, E., Hashimoto, T., and Conti, S. F. (1972). *J. Bacteriol.* 109, 869.
119. Guttes, S., Guttes, E., and Ellis, R. A. (1968). *J. Ultrastruct. Res.* 22, 508.
120. Haber, J. E., Peloquin, J. G., Halvorson, H. O., and Borisy, G. G. (1972). *J. Cell Biol.* 55, 355.
121. Hammond, D. M., Roberts, W. L., Youssef, N. N., and Danforth, H. D. (1973). *J. Parsitol.* 59, 581.
122. Hammond, D. M., Scholtyseck, E., and Chobotar, B. (1969). *Z. Parasitenk.* 33, 65.
123. Harder, D. E. (1976a). *Can. J. Bot.* 54, 981.
124. Harder, D. E. (1976b). *Can. J. Bot.* 54, 995.
125. Hardham, A. R., and Gunning, B. E. S. (1978). *J. Cell Biol.* 77, 14.
126. Harris, P. (1975). *Exp. Cell Res.* 94, 409.
127. Hauser, M. (1972). *Chromosoma* 36, 158.
128. Hauser, M. (1973). *Chromosoma* 44, 49.
129. Heath, I. B. (1974a). *In* "The Cell Nucleus" (H. Busch, ed.), Vol. 2, pp. 487–515. Academic Press, New York.
130. Heath, I. B. (1974b). *J. Cell Biol.* 60, 204.
131. Heath, I. B. (1974c). *Mycologia* 66, 354.
132. Heath, I. B. (1975a). *BioSystems* 7, 351.
133. Heath, I. B. (1975b). *Protoplasma* 85, 147.
134. Heath, I. B. (1975c). *Protoplasma* 85, 177.
135. Heath, I. B. (1978a). *In* "Nuclear Division in the Fungi" (I. B. Heath, ed.), pp. 89–176. Academic Press, New York.
136. Heath, I. B. (1978b). *J. Cell Biol.* 79, 122a.
137. Heath, I. B., and Greenwood, A. D. (1968). *J. Gen Microbiol.* 53, 287.
138. Heath, I. B., and Greenwood, A. D. (1970). *J. Gen. Microbiol.* 62, 139.
139. Heath, I. B., and Heath, M. C. (1976). *J. Cell Biol.* 70, 592.
140. Heath, M. C., and Heath, I. B. (1978). *Can. J. Bot.* 56, 648.
141. Heckmann, K. (1966). Institut f. Wissenschaftlichen Film E913. Göttingen.
142. Hemmes, D. E., and Hohl, H. R. (1973). *Can. J. Bot.* 51, 1673.
143. Hepler, P. K. (1977). *In* "Mechanisms and Control of Cell Division" (T. L. Rost and E. M. Gifford, eds.), pp. 212–232. Dowden, Hutchinson & Ross, Stroudsburg, Pennsylvania.
144. Herlan, G., Quevedo, R., and Wunderlich, F. (1978). *Exp. Cell Res.* 115, 103.
145. Heywood, P. (1978). *J. Cell Sci.* 31, 37.
146. Heywood, P., and Godward, M. B. E. (1972). *Chromosoma* 39, 333.
147. Heywood, P., and Godward, M. B. E. (1973). *Ann. Bot.* 37, 423.
148. Heywood, P., and Godward, M. B. E. (1974). *Am. J. Bot.* 61, 311.
149. Heywood, P., and Weinman, D. (1978). *J. Protozool.* 25, 287.
150. Hinchee, A. A. (1976). Ph.D. Thesis, University of Washington, Seattle.
151. Hoch, H. C., and Mitchell, J. E. (1972). *Protoplasma* 75, 113.
152. Hoffman, L. R. (1976). *Protoplasma* 87, 191.
153. Hollande, A. (1974). *Protistologica* 10, 413.
154. Hollande, A., Cachon, J., and Cachon, M. (1969). *C. R. Acad. Sci. Ser. D* 269, 179.
155. Hollande, A., and Carruette-Valentin, J. (1970). *C. R. Acad. Sci. Ser. D* 270, 1476.
156. Hollande, A., and Carruette-Valentin, J. (1971). *Protistologica* 7, 5.
157. Hollande, A., and Carruette-Valentin, J. (1972). *Protistologica* 8, 267.

158. Hollande, A., and Valentin, J. (1968). *C. R. Acad. Sci. Ser. D* **266,** 367.
159. Hoppe, G. (1976). *Protistologica* **12,** 169.
160. Horgen, P. A., and Silver, J. C. (1978). *Annu. Rev. Microbiol.* **32,** 249.
161. Howard, K. L., and Moore, R. T. (1970). *Bot. Gaz.* **131,** 311.
162. Howells, R. E., and Davies, E. E. (1971). *Ann. Trop. Med. Parasitol.* **65,** 451.
163. Huang, H. C., Tinline, R. D., and Fowke, L. C. (1975). *Can. J. Bot.* **53,** 403.
164. Hudson, P. R., and Waaland, J. R. (1974). *J. Cell Biol.* **62,** 274.
165. Hudspeth, M. E. S., Timberlake, W. E., and Goldberg, R. B. (1977). *Proc. Natl. Acad. Sci. U.S.A.* **74,** 4332.
166. Huettner, A. F. (1933). *Z. Zellforsch. Mikrosk. Anat.* **19,** 119.
167. Hughes, A. (1952). "The Mitotic Cycle." Butterworths, London.
168. Humber, R. A. (1977). Personal communication.
169. Hung, C.-Y., and Wells, K. (1971). *J. Gen. Microbiol.* **66,** 15.
170. Hung, C.-Y., and Wells, K. (1977). *Mycologia* **69,** 685.
171. Ichida, A. A., and Fuller, M. S. (1968). *Mycologia* **60,** 141.
172. Inaba, F., and Sotokawa, Y. (1968). *J. Protozool.* **15,** Suppl. 28.
173. Inoki, S., and Ozeki, Y. (1969). *Biken J.* **12,** 31.
174. Inoué, S., and Ritter, H. (1975). *In* "Molecules and Cell Movement" (S. Inoué and R. E. Stephens, eds.), pp. 3–30. Raven, New York.
175. Inoué, S., and Ritter, H. (1978). *J. Cell Biol.* **77,** 655.
176. Inoué, S., and Sato, H. (1967). *J. Gen. Physiol.* **50,** 259.
177. Jaeckel-Williams, R. (1978). *J. Cell Sci.* **34,** 303.
178. Jenkins, R. A. (1967). *J. Cell Biol.* **34,** 463.
179. Jenkins, R. A. (1977). *J. Protozool.* **24,** 264.
180. Johnson, U. G., and Porter, K. R. (1968). *J. Cell Biol.* **38,** 403.
181. Jordan, E. G., and Godward, M. B. E. (1969). *J. Cell Sci.* **4,** 3.
182. Jurand, A., and Selman, G. G. (1970). *J. Gen. Microbiol.* **60,** 357.
183. Karadzhan, B. P., and Raikov, I. B. (1977). *Protistologica* **13,** 15.
184. Kazama, F. Y. (1974). *Protoplasma* **82,** 155.
185. Kelley, G. L., and Hammond, G. M. (1972). *Z. Parasitenk.* **38,** 271.
186. Kelley, G. L., and Hammond, G. M. (1973). *J. Parasitol.* **59,** 1071.
187. Kerr, S. J. (1967). *J. Protozool.* **14,** 439.
188. Keskin, B. (1971). *Arch. Mikrobiol.* **77,** 344.
189. Khan, S. R. (1976). *Can. J. Bot.* **54,** 168.
190. Koevenig, J. L. (1964). *Mycologia* **56,** 170.
191. Kohlmeyer, J. (1975). *Bioscience* **25,** 86.
192. Kubai, D. F. (1973). *J. Cell Sci.* **13,** 511.
193. Kubai, D. F. (1975). *Int. Rev. Cytol.* **43,** 167.
194. Kubai, D. F. (1978). *In* "Nuclear Division in the Fungi" (I. B. Heath, ed.), pp. 177–229. Academic Press, New York.
195. Kubai, D. F., and Ris, H. (1969). *J. Cell Biol.* **40,** 508.
196. Kudo, R. R. (1966). "Protozoology" (5th ed.). Thomas, Springfield, Massachusetts.
197. Künkel, W., and Hädrich, H. (1977). *Protoplasma* **92,** 311.
198. Kuzmich, M. J., and Zimmerman, A. M. (1972). *Exp. Cell Res.* **72,** 441.
199. Laane, M. M. (1974). *Norwegian J. Bot.* **21,** 125.
200. Laane, M. M., and Haugli, F. B. (1974). *Norwegian J. Bot.* **21,** 309.
201. La Fountain, J. R., and Davidson, L. A. (1978). *J. Cell Biol.* **79,** 284a.
202. Lauer, G. D., Roberts, T. M., and Klotz, L. C. (1977). *J. Mol. Biol.* **114, 507.**
203. Leadbeater, B., and Dodge, J. D. (1967). *Arch. Mikrobiol.* **57,** 239.

204. Lecher, P. (1973). *In* "Chromosomes Today" (J. Wahrman and K. R. Lewis, eds.), Vol. 4, pp. 225–234. Halsted, New York.
205. Leedale, G. F. (1968). *In* "The Biology of Euglena" (D. E. Beutow, ed.), Vol. 1, pp. 185–242. Academic Press, New York.
206. Leibowitz, P. J., and Schaechter, M. (1975). *Int. Rev. Cytol.* **41**, 1.
207. Lerbs, V. (1971). *Arch. Mikrobiol.* **77**, 308.
208. Lerbs, V., and Thielke, C. (1969). *Arch. Mikrobiol.* **68**, 95.
209. Lessie, P. E., and Lovett, J. S. (1968). *Am. J. Bot.* **55**, 220.
210. Lewis, L. M., Witkus, E. R., and Vernon, G. M. (1976). *Protoplasma* **89**, 203.
211. Lewis, R., and Scott, J. L. (1979). In preparation.
212. Liddle, L., Berger, S., and Schweiger, H.-G. (1976). *J. Phycol.* **12**, 261.
213. Loubes, C., and Maurand, J. (1975). *Protistologica* **11**, 233.
214. Løvlie, A., and Bråten, T. (1970). *J. Cell. Sci.* **6** 109.
215. Lu, B. C. (1967). *J. Cell Sci.* **2**, 529.
216. Lu, B. C. (1978). *J. Cell Biol.* **76**, 761.
217. Luftig, R. B., McMillan, P. N., Weatherbee, J. A., and Weihing, R. R. (1977). *J. Histochem. Cytochem.* **25**, 175.
218. Luykx, P. (1970). *Int. Rev. Cytol. Suppl.* **2**, 1.
219. McCully, E. K., and Robinow, C. F. (1971). *J. Cell Sci.* **9**, 475.
220. McCully, E. K., and Robinow, C. F. (1972a). *J. Cell Sci.* **10**, 857.
221. McCully, E. K., and Robinow, C. F. (1972b). *J. Cell Sci.* **11**, 1.
222. McCully, E. K., and Robinow, C. F. (1973). *Arch. Mikrobiol.* **94**, 133.
223. McDonald, K. (1972). *J. Phycol.* **8**, 156.
224. McDonald, K., and Pickett-Heaps, J. D. (1976). *Am. J. Bot.* **63**, 592.
225. McDonald, K., Pickett-Heaps, J. D., McIntosh, J. R., and Tippit, D. H. (1977). *J. Cell Biol.* **74**, 377.
226. McKeen, W. E. (1972). *Can. J. Microbiol.* **18**, 1915.
227. McLaughlin, D. J. (1971). *J. Cell Biol.* **50**, 737.
228. McManus, S. M. A., and Roth, L. E. (1968). *Mycologia* **60**, 426.
229. McNitt, R. (1973). *Can. J. Bot.* **51**, 2065.
230. Madelin, M. F., and Beckett, A. (1972). *J. Gen. Microbiol.* **72**, 185.
231. Maltaux, M., and Massart, J. (1906). *Rec. Inst. bot. Léo Errera, Bruxelle* **6**, 369.
232. Maniloff, J., and Quinlan, D. C. (1973). *Ann. N.Y. Acad. Sci.* **225**, 181.
233. Maniloff, J., and Quinlan, D. C. (1974). *J. Bacteriol.* **120**, 495.
234. Maniloff, J., and Morowitz, H. J. (1972). *Bacteriol Rev.* **36**, 263.
235. Manton, I. (1964). *J. R. Microsc. Soc.* **83**, 317.
236. Manton, I., Kowallik, K., and von Stosch, H. A. (1969a). *J. Microsc.* **89**, 295.
237. Manton, I., Kowallik, K., and von Stosch, H. A. (1969b). *J. Cell Sci.* **5**, 271.
238. Manton, I., Kowallik, K., and von Stosch, H. A. (1970a). *J. Cell Sci.* **6**, 131.
239. Manton, I., Kowallik, K., and von Stosch, H. A. (1970b). *J. Cell Sci.* **7**, 407.
240. Marchant, H. J. (1974). *J. Phycol.* **10**, 107.
241. Marchant, H. J., and Pickett-Heaps, J. D. (1970). *Aust. J. Biol. Sci.* **23**, 1173.
242. Marchant, H. J., and Pickett-Heaps, J. D. (1973). *J. Phycol.* **9**, 461.
243. Margulis, L. (1970). "Origin of Eukaryotic Cells." Yale Univ. Press, New Haven, Connecticut.
244. Margulis, L., To, L., and Chase, D. (1978). *Science* **200**, 1118.
245. Markey, D. R., and Wilce, R. T. (1975). *Protoplasma* **85**, 219.
246. Matile, P. H., Moor, H., and Robinow, C. F. (1969). *In* "The Yeasts" (A. H. Rose and J. S. Harrison, eds.), Vol. 1, pp. 219–302. Academic Press, New York.

247. Mattox, K. R., and Stewart, K. D. (1977). *Am. J. Bot.* **64**, 931.
248. Mazia, D. (1961). *In* "The Cell" (J. Brachet and A. E. Mirskey, eds.), Vol. III, pp. 77–412. Academic Press, New York.
249. Mehlhorn, H. (1972). *Z. Parasitenk.* **40**, 243.
250. Mehlhorn, H., Senaud, J., and Scholtyseck, E. (1972). *C. R. Acad. Sci., Ser. D* **275**, 835.
251. Mignot, J.-P. (1966). *Protistologica* **2**, (Part 3), 51.
252. Millecchia, L. L., and Rudzinska, M. A. (1971). *Z. Zellforsch. Mikrosk. Anat.* **115**, 149–164.
253. Milovidov, P. F. (1949). "Physik und Chemie des Zellkernes." Protoplasma-Monographien 20, pt. 1. Bornträger, Berlin.
254. Mims, C. W. (1972). *J. Gen. Microbiol.* **71**, 53.
255. Mims, C. W. (1973). *Protoplasma* **77**, 35.
256. Mims, C. W. (1977). *Can. J. Bot.* **55**, 2319.
257. Moens, P. B. (1976). *J. Cell Biol.* **68**, 113.
258. Moens, P. B. (1978a). *J. Cell Biol.* **79**, 292a.
259. Moens, P. B. (1978b). *Chromosoma* **67**, 41.
260. Moens, P. B., Mowat, M., Esposito, M. S., and Esposito, R. E. (1977). *Trans. R. Soc. London, Ser. B.* **277** 351.
261. Moens, P. B., and Rapport, E. (1971). *J. Cell Biol.* **50**, 344.
262. Mohberg, J. (1977). *J. Cell Sci.* **24**, 95.
263. Molnar, K. E., Stewart, K. D., and Mattox, K. R. (1975). *J. Phycol.* **11**, 287.
264. Moor, H. (1966). *J. Cell Biol.* **29**, 153.
265. Moorman, G. W. (1976). *Mycologia* **68**, 902.
266. Motta, J. (1967). *Mycologia* **59**, 370.
267. Motta, J. J. (1969). *Mycologia* **61**, 873.
268. Mughal, S., and Godward, M. B. E. (1973). *Chromosoma* **44**, 213.
269. Neimark, H. C. (1977). *Proc. Nat. Acad. Sci. U.S.A.* **74**, 4041.
270. Neushul, M., and Dahl, A. L. (1972). *Am. J. Bot.* **59**, 401.
271. Nicklas, R. B. (1971). *Adv. Cell Biol.* **2**, 225.
272. Nicolaides, A. A., and Holland, I. B. (1978). *J. Bacteriol.* **135**, 178.
273. Nilshammar, M., and Walles, B. (1974). *Protoplasma* **79**, 317.
274. Nilsson, J. R. (1976). *C.R. Trav. Lab. Carlsberg* **40**, 215.
275. Oakley, B. R. (1978). *BioSystems* **10**, 59.
276. Oakley, B. R., and Bisalputra, T. (1977). *Can. J. Bot.* **55**, 2789.
277. Oakley, B. R., and Dodge, J. D. (1973). *Nature (London)* **244**, 521.
278. Oakley, B. R., and Dodge, J. D. (1974). *J. Cell Biol.* **63**, 322.
279. Oakley, B. R., and Dodge, J. D. (1976). *Protoplasma* **88**, 241.
280. Oakley, B. R., and Heath, I. B. (1978). *J. Cell Sci.* **31**, 53.
281. Oakley, B. R., and Morris, N. R. (1978). *J. Cell Biol.* **79**, 300a.
282. Ochs, S. (1972). *Science* **176**, 252.
283. Olive, L. S. (1953). *Botan. Rev.* **19**, 439.
284. Olive, L. S. (1975). "The Mycetozoans." Academic Press, New York.
285. Olmsted, J. B., and Borisy, G. G. (1975). *Biochemistry* **14**, 2996.
286. Olson, L. W. (1974a). *C.R. Trav. Lab. Carlsberg* **40**, 113.
287. Olson, L. W. (1974b). *C.R. Trav. Lab. Carlsberg* **40**, 125.
288. Oppenheim, D. S., Hauschka, B. T., and McIntosh, J. R. (1973). *Exp. Cell Res.* **79**, 95.
289. Oppenheim, A., and Katzir, N. (1971). *Exp. Cell Res.* **68**, 224.
290. Ott, D. W., and Brown, R. M. (1972). *Br. Phycol J.* **7**, 361.
291. Pearson, B. R., and Norris, R. E. (1975). *J. Phycol.* **11**, 113.
292. Perkins, F. O. (1968). *J. Invert. Pathol.* **10**, 287.

293. Perkins, F. O. (1969). *J. Parasit.* **55**, 897.
294. Perkins, F. O. (1970). *J. Cell Sci.* **6**, 629.
295. Perkins, F. O. (1975). *J. Cell Sci.* **18**, 327.
296. Perkins, F. O., and Amon, J. P. (1969). *J. Protozool.* **16**, 235.
297. Peterson, J. B., Gray, R. H., and Ris, H. (1972). *J. Cell Biol.* **53**, 837.
298. Peterson, J. B., and Ris, H. (1976). *J. Cell Sci.* **22**, 219.
299. Peyrière, M. (1971). *C.R. Acad. Sci., Ser. D* **273**, 2071.
300. Pickett-Heaps, J. D. (1967). *Aust. J. Biol. Sci.* **20**, 883.
301. Pickett-Heaps, J. D. (1969). *Cytobios* **1**, 257.
302. Pickett-Heaps, J. D. (1970a). *Protoplasma* **70**, 325.
303. Pickett-Heaps, J. D. (1970b). *Cytobios* **2**, 69.
304. Pickett-Heaps, J. D. (1972a). *J. Phycol.* **8**, 343.
305. Pickett-Heaps, J. D. (1972b). *Cytobios* **6**, 167.
306. Pickett-Heaps, J. D. (1972c). *Ann. Bot.* **36**, 693.
307. Pickett-Heaps, J. D. (1972d). *New Phytol.* **72**, 347.
308. Pickett-Heaps, J. D. (1972e). *Cytobios* **5**, 59.
309. Pickett-Heaps, J. D. (1973a). *Ann. Bot.* **37**, 1017.
310. Pickett-Heaps, J. D. (1973b). *J. Phycol.* **9**, 408.
311. Pickett-Heaps, J. D. (1974a). *Br. Phycol. J.* **9**, 63.
312. Pickett-Heaps, J. D. (1974b). *BioSystems* **6**, 37.
313. Pickett-Heaps, J. D. (1975a). *Ann. N.Y. Acad. Sci.* **253**, 352.
314. Pickett-Heaps, J. D. (1975b). "Green Algae." Sinauer, Sunderland, Massachusetts.
315. Pickett-Heaps, J. D., and Fowke, L. C. (1969). *Aust. J. Biol. Sci.* **22**, 857.
316. Pickett-Heaps, J. D., and Fowke, L. C. (1970a). *Aust. J. Biol. Sci.* **23**, 71.
317. Pickett-Heaps, J. D., and Fowke, L. C. (1970b). *J. Phycol.* **6**, 189.
318. Pickett-Heaps, J. D., and Marchant, H. J. (1972). *Cytobios* **6**, 255.
319. Pickett-Heaps, J. D., and McDonald, K. (1975). *New Phytol.* **74**, 235.
320. Pickett-Heaps, J. D., McDonald, K., and Tippit, D. H. (1975). *Protoplasma* **86**, 205.
321. Pickett-Heaps, J. D., and Ott, D. W. (1974). *Cytobios* **11**, 41.
322. Pickett-Heaps, J. D., and Staehelin, L. A. (1975). *J. Phycol.* **11**, 186.
323. Pickett-Heaps, J. D., and Tippit, D. H. (1978). *Cell* **14**, 455.
324. Pickett-Heaps, J. D., Tippit, D. H., and Andreozzi, J. A. (1978a). *Biol. Cell.* **33**, 71.
325. Pickett-Heaps, J. D., Tippit, D. H., and Andreozzi, J. A. (1978b). *Biol. Cell.* **33**, 79.
326. Pickett-Heaps, J. D., and Weik, K. L. (1977). *In* "Mechanisms and Control of Cell Division" (T. L. Rost and E. M. Gifford, eds.), pp. 308–336. Dowden, Hutchinson & Ross, Stroudsburg, Pennsylvania.
327. Pollard, T. D. (1976). *J. Supramol. Struct.* **5**, 317.
328. Poon, N. H., and Day, A. W. (1976a). *Can. J. Microbiol.* **22**, 507.
329. Poon, N. H., and Day, A. W. (1976b). *Can. J. Microbiol.* **22**, 495.
330. Porchet-Hennere, E. (1977). *Protistologica* **13**, 31.
331. Porter, D. (1972). *Protoplasm* **74**, 427.
332. Powell, M. J. (1975). *Can. J. Bot.* **53**, 627.
333. Prescott, D. M. (1964). *Prog. Nucleic Acid Res. Mol. Biol.* **3**, 33.
334. Prescott, D. M. (1976). "Reproduction of Eukaryotic Cells." Academic Press, New York.
335. Raikov, I. B. (1966). *Arch. Protistenk.* **109**, 71.
336. Raikov, I. B. (1973). *C.R. Acad. Sci. Ser. D.* **276**, 2385.
337. Raju, N. B., and Lu, B. C. (1973). *J. Cell Sci.* **12**, 131.
338. Ris, H. (1975). *BioSystems* **7**, 298.
339. Ris, H., and Kubai, D. F. (1974). *J. Cell Biol.* **60**, 702.

340. Ritter, H., Inoué, S., and Kubai, D. F. (1978). *J. Cell Biol.* **77**, 638.
341. Roberts, W. L., Hammond, D. M., Anderson, L. C., and Speer, C. A. (1970). *J. Protozool.* **17**, 584.
342. Robinow, C. F. (1978). *In* "Nuclear Division in the Fungi" (I. B. Heath, ed.), p. 128. Academic Press, New York.
343. Robinow, C. F., and Bakerspigel, A. (1965). *In* "The Fungi, an Advanced Treatise" (G. C. Ainsworth and A. S. Sussman, eds.), Vol. 1, pp. 119-142. Academic Press, New York.
344. Robinow, C. F., and Caten, C. E. (1969). *J. Cell Sci.* **5**, 403.
345. Robinow, C. F., and Marak, J. (1966). *J. Cell Biol.* **29**, 129.
346. Rooney, L., and Moens, P. B. (1973). *Can. J. Microbiol.* **19**, 1383.
347. Roos, U.-P. (1975a). *J. Cell Biol.* **64**, 480.
348. Roos, U.-P. (1975b). *J. Cell Sci.* **18**, 315.
349. Ross, I. K. (1967). *Am. J. Bot.* **54**, 617.
350. Ross, I. K. (1968). *Protoplasma* **66**, 173.
351. Ross, I. K., Pommerville, J. C., and Damm, D. L. (1976). *J. Cell Sci.* **21**, 175.
352. Roth, L. E., and Daniels, E. W. (1962). *J. Cell Biol.* **12**, 57.
353. Roth, L. E., Pihlaja, D. J., and Sigenaka, Y. (1970). *J. Ultrastruct. Res.* **30**, 7.
354. Roth, L. E., and Shigenaka, Y. (1964). *J. Cell Biol.* **20**, 249.
355. Ruthman, A., and Hauser, M. (1974). *Chromosoma* **45**, 261.
356. Ruthman, A., and Permantier, Y. (1973). *Chromosoma* **41**, 271.
357. Ryser, U. (1970). *Z. Zellforsch. Mikrosk. Anat.* **110**, 108.
358. Sakai, A., and Shigenaga, M. (1972). *Chromosoma* **37**, 101.
359. Scheetz, R. W. (1972). *Mycologia* **64**, 38.
360. Schmitt, F. O. (1968). *Proc. Natl. Acad. Sci. U.S.A.* **60**, 1092.
361. Schnepf, E., and Deichgräber, G. (1978). *Arch. Microbiol.* **116**, 141.
362. Schornstein, K., and Scott, J. L. (1979). *Nature (London)*, in press.
363. Schrantz, J. P. (1967). *C.R. Acad. Sci., Ser. D.* **264**, 1274.
364. Schrével, J., Asfaux-Foucher, G., and Bafort, J. M. (1977). *J. Ultrastruct. Res.* **59**, 332.
365. Schuster, F. L. (1968). *J. Parasitol.* **54**, 725.
366. Schuster, F. L. (1975). *Tissue Cell* **7**, 1.
367. Schuster, F. L., Goldstein, S., and Hershenov, B. (1968). *Protistologica* **4**, 141.
368. Schwab, D. (1968). *Naturwissenschaften* **55**, 88.
369. Schwab, D. (1969). *Z. Zellforsch. Mikrosk. Anat.* **96**, 295.
370. Schwab, D. (1972). *Protoplasma* **75**, 79.
371. Schwab, D. (1973). *Protoplasma* **73**, 339.
372. Schwartz, R. M., and Dayhoff, M. O. (1978). *Science* **199**, 395.
373. Scott, J. L., and Thomas, J. P. (1975). *J. Phycol.* **11**, 474.
374. Scott, J. L., Bosco, C., Schornstein, K., and Thomas, J. (1979). In preparation.
375. Scott, J. L., and Bullock, K. W. (1976). *Can. J. Bot.* **54**, 1546.
376. Searcy, D. G., Stein, D. B., and Green, G. R. (1978). *BioSystems* **10**, 19.
377. Senand, J., and Černà, Z. (1978). *Protistologica* **14**, 155.
378. Setliff, E. C. (1977). *In* "Abstracts, 2nd International Mycological Congress" (H. E. Bigelow and E. G. Simmons, eds.), p. 607.
379. Setliff, E. C., Hoch, H. C., and Patton, R. F. (1974). *Can. J. Bot.* **52**, 2323.
380. Shapiro, H. S. (1976). *In* "FASEB Biological Handbooks, 1. Cell Biology" (P. L. Altman and D. D. Katz, eds.), pp. 367-378. Federation of American Societies for Experimental Biology, Bethesda, Maryland.
381. Sheffield, H. G., and Melton, M. L. (1968). *J. Parasitol.* **54**, 209.
382. Siebert, A. E., and West, J. A. (1974). *Protoplasma* **81**, 17.
383. Slankis, T., and Gibbs, S. P. (1972). *J. Phycol.* **8**, 243.

384. Smetana, K., and Busch, H. (1974). *In* "The Cell Nucleus" (H. Busch, ed.), Vol. 1, pp. 73–147. Academic Press, New York.
385. Snigirevskaya, E. S. (1969). *Acta Protozool.* **7,** 57.
386. Snyder, J. A., and McIntosh, J. R. (1976). *Annu. Rev. Biochem.* **45,** 699.
387. Soyer, M.-O. (1969). *C.R. Acad. Sci., Ser. D.* **268,** 2082.
388. Soyer, M.-O. (1971). *Chromosoma* **33,** 70.
389. Soyer, M.-O. (1977). *C.R. Acad. Sci., Ser. D.* **285,** 693.
390. Spiegel, F. W. (1978). Ph.D. Thesis, University of North Carolina, Chapel Hill.
391. Sprague, V., and Vernick, S. H. (1968). *J. Protozool.* **15,** 547.
392. Steffens, W. L., and Wille, J. J. (1977). *In* "Abstracts, 2nd International Mycological Congress" (H. E. Bigelow and E. G. Simmons, eds.), p. 635.
393. Stephens, R. E. (1978). *Biochemistry* **17,** 2882.
394. Sterling, C. R., Aikawa, M., and Nussenzweig, R. S. (1972). *Proc. Helminth Soc. Washington* **39** (Special Issue), 109.
395. Stevenson, I. (1972). *Aust. J. Biol. Sci.* **25,** 775.
396. Stevenson, I., and Lloyd, F. P. (1971a). *Aust. J. Biol. Sci.* **24,** 963.
397. Stevenson, I., and Lloyd, F. P. (1971b). *Aust. J. Biol. Sci.* **24,** 977.
398. Stewart, K. D., Mattox, K. R., and Chandler, C. D. (1974). *J. Phycol.* **10,** 65.
399. Stiers, D. L. (1976). *Can. J. Bot.* **54,** 1714.
400. Summer, J. R., and Blum, J. J. (1965). *Exp. Cell Res.* **39,** 504.
401. Sun, N.C., and Bowen, C. C. (1972). *Caryologia* **25,** 471.
402. Sundberg, W. J. (1977). *In* "Abstracts, 2nd International Mycological Congress" (H. E. Bigelow and E. G. Simmons, eds.), p. 644.
403. Švejda, J., and Hornak, O. (1969). *Neoplasma* **16,** 585.
404. Tanaka, K. (1970). *Protoplasma* **70,** 423.
405. Tanaka, K. (1973). *J. Cell Biol.* **57,** 220.
406. Tanaka, K. (1977). *In* "Abstracts, 2nd International Mycological Congress" (H. E. Bigelow and E. G. Simmons, eds.), p. 652.
407. Tanaka, K. (1977b). *In* "Growth and Differentiation in Microorganisms" (T. Ishikawa, Y. Maruyama, and H. Matsumiya, eds.), pp. 229–254. Tokyo Univ. Press, Tokyo.
408. Taylor, F. J. R. (1974). *Taxon* **23,** 229.
409. Terzakis, J. A. (1971). *J. Protozool.* **18,** 62.
410. Terzakis, J. A., Sprinz, H., and Ward, R. A. (1967). *J. Cell. Biol.* **34,** 311.
410a. Thielke, C. (1973). *Naturwissenschaften* **59,** 471.
411. Thielke, C. (1974). *Arch. Mikrobiol.* **98,** 225.
412. Thielke, C. (1975). Institut f. Wissenschaftlichen Film E 2179, Göttingen.
413. Thielke, C. (1976). *Z. Piltzkunde* **42,** 57.
414. Thielke, C. (1978). *Z. Mykol.* **44,** 71.
415. Tilney, L. G., and Porter, K. R. (1965). *Protoplasma* **50,** 317.
416. Tippit, D. H., McDonald, K. L., and Pickett-Heaps, J. D. (1975). *Cytobiologie* **12,** 52.
417. Tippit, D. H., and Pickett-Heaps, J. D. (1976). *J. Cell Sci.* **21,** 273.
418. Tippit, D. H., and Pickett-Heaps, J. D. (1977). *J. Cell Biol.* **73,** 705.
419. Tippit, D. H., Schulz, D., and Pickett-Heaps, J. D. (1978). *J. Cell Biol.* **79** 737.
420. Tobin, S. L., and Laird, C. D. (1977). *J. Cell Biol.* **75,** 150a.
421. Triemer, R. E., and Brown, R. M. (1974). *J. Phycol.* **10,** 419.
422. Tucker, J. B. (1967). *J. Cell Sci.* **2,** 481.
423. Tucker, J. B., Dunn, M., and Pattisson, J. B. (1975). *Dev. Biol.* **47,** 439.
424. Turian, G., and Oulevey, N. (1971). *Cytobiologie* **4,** 250.
425. Turner, F. R. (1968). *J. Cell Biol.* **37,** 370.
426. Turner, F. R. (1970). *J. Cell Biol.* **46,** 220.

427. Van Winkle, W. B., Biesele, J. J., and Wagner, R. P. (1971). *Can. J. Genet. Ctyol.* **13,** 873.
428. Vavra, J. (1965). *C.R. Acad. Sci., Ser. D.* **261,** 3467.
429. Vickerman, K., and Preston, T. M. (1970). *J. Cell Sci.* **6,** 365.
430. Vinckier, D., Devauchelle, G., and Prensier, G. (1971). *Protistologica* **7,** 273.
431. Vinnikova, N. V. (1976). *Protistologica* **12,** 7.
432. Vivier, E. (1965). *J. Microsc.* **4,** 559.
433. Vivier, E., Petitprez, A., and Landau, I. (1972). *Protistologica* **8,** 315.
434. Walker, G. K. (1976). *Protistologica* **12,** 271.
435. Walker, G. K., and Goode, D. (1976). *Cytobiologie* **14,** 18.
436. Wallace, D. C., and Morowitz, H. J. (1973). *Chromosoma* **40,** 121.
437. Weisenberg, R. C., and Deery, W. J. (1976). *Nature (London)* **263,** 792.
438. Wells, K. (1970). *Mycologia* **62,** 761.
439. Wells, K. (1978). *Protoplasma* **94,** 83.
440. Westergaard, M., and Von Wettstein, D. (1969–70). *C.R. Trav. Lab. Carlsberg* **37,** 195.
441. Wheatley, D. N. (1974). *J. Gen. Virol.* **24,** 395.
442. Whisler, H. C., and Travland, L. B. (1973). *Arch. Protistenk.* **115,** 69.
443. Whittaker, R. H., and Margulis, L. (1978). *BioSystems* **10,** 3.
444. Wick, S. M., and Hepler, P. K. (1976). *J. Cell Biol.* **70,** 209a.
445. Williams, N.E., and Williams, R. J. (1976). *J. Cell Sci.* **20,** 61.
446. Wilson, H. J., Wanaka, F., and Linskens, H. F. (1973). *Planta* **109,** 259.
447. Wolpert, L. (1965). *Symp. Soc. Gen. Microbiol.* **15,** 270.
448. Wong, S. T. C., and Desser, S. S. (1978). *J. Protozool.* **25,** 302.
449. Woodcock, C. L. F., and Miller, G. J. (1973). *Protoplasma* **77,** 313.
450. Wright, R. G., Lennard, J. H., and Denhan, D. (1978). *Trans. Br. Mycol. Soc.* **70,** 229.
451. Wunderlich, F., and Herlan, G. (1977). *J. Cell Biol.* **73,** 271.
452. Wunderlich, F., and Speth, V. (1970). *Protoplasma* **70,** 139.
453. Zickler, D. (1970). *Chromosoma* **30,** 287.
454. Zickler, D. (1971). *C.R. Acad. Sci., Ser. D.* **273,** 1687.
455. Zickler, D. (1973). *Histochemie* **34,** 227.
456. Zickler, D., and Olson, L. W. (1975). *Chromosoma* **50,** 1.

INTERNATIONAL REVIEW OF CYTOLOGY, VOL. 64

The Centriolar Complex

SCOTT P. PETERSON AND MICHAEL W. BERNS

Developmental and Cell Biology, University of California, Irvine, Irvine, California

I.	Introduction	81
II.	Organization of the Mitotic Centriolar Complex	82
III.	Centriolar Movements during Mitosis	84
IV.	Systems without Centrioles	85
V.	The Two Functional States of the Centriole	90
VI.	Indirect Immunofluorescence and Its Use in Mitotic Investigation	91
VII.	*In Vitro* Studies on Microtubular Nucleation	96
VIII.	The Role of Nucleic Acids in Centriolar Function	98
IX.	Multicentriolar Cells	100
X.	Conclusions	102
	References	103

I. Introduction

Cell division has attracted the attention of biologists since the first quarter of this century (see discussion by Wilson, 1925). Yet despite all the attention given to this problem, our understanding of the process is still remarkably incomplete. The field of cell division today could best be described as "voluminous" in terms of description and "lacking" in terms of understanding. The literature is populated with numerous "models" to explain the complex process of chromosome separation. Too frequently, the model itself becomes the center of attention with the ultimate goal, that of understanding the process, being lost. For example, the dogmatic adherence to any of the existing models can easily result in disregard of valid observations that just do not fit in with the model at hand. Only recently has the suggestion been made that perhaps there is not any one mechanism to explain mitosis but rather a combination of mechanisms that encompasses aspects of many of the currently popular (or unpopular) models (Pickett-Heaps and Bajer, 1977).

It is not the purpose of this article to discuss mitotic models. In fact, we will purposely avoid such a discussion. Rather, our purpose is to focus attention on one component of the cell division process: the centriolar complex. We feel that it is fitting to devote an article to this component because, like the entire process of cell division, the centriolar complex has enjoyed considerable controversy with respect to origin, function, and composition.

81

II. Organization of the Mitotic Centriolar Complex

Much of the detailed structural information concerning the centriolar complex has centered around the mitotic phase of the cell cycle. However, there is some evidence to suggest that the centriole and its associated material may change considerably during the cell cycle. For example, it is well known that in L cells a new daughter centriole "grows" out from the older centriole at the G_1–S border of the cell cycle (Phillips and Rattner, 1976). This process undoubtedly involves substantial biochemical activity and structural reorganization. Similarly, Snyder and McIntosh (1975) have suggested that the centriolar complex undergoes a maturation process that is cell cycle-dependent. They demonstrated that inter-phase complexes can only initiate a few microtubules but that prometaphase cen-triolar complexes can initiate large numbers of microtubules. More recently, Heidemann and Kirschner (1975, 1978) have emphasized that the "mature" mitotic center may be a result of the interaction of the centriolar region and the cytoplasm in preparation for mitosis. These suggestions are further strengthened by possible changes in the interphase microtubule network and cytoskeletal sys-tem during the cell cycle (and especially as mitosis approaches) that may be correlated with a chemical and functional change in the centriolar region (Lazarides and Weber, 1975).

As indicated in the previous discussion, it is likely that the centriolar complex is changing during the cell cycle. Therefore, the picture we present during cell division may not reflect the complete range of possibilities with respect to the organization of the centriolar region. In addition, the different cell types, cell lines, and preparative procedures used all contribute to potential variation. De-spite these pitfalls in describing the centriolar complex, it is necessary to struc-turally define the mitotic pole.

The centriolar complex may be divided into two general component regions: the centrioles and the cloud of "electron-dense" osmiophilic material surround-ing the centrioles. Collectively, the entire centriolar complex has often been referred to as "the centrosome." The "cloud" has been called the "peri-centriolar cloud," the "centrosphere," or just "pericentriolar material." The amount of cloud material appears to be quite variable from preparation to prepa-ration and among animal species. For example, the flat PTK_2(*Potorous*) cells have particularly abundant pericentriolar cloud material (Roos, 1973; Rattner and Berns, 1976b), whereas primary cultures of salamander lung epithelial cells, which also are relatively flat in mitosis (Berns *et al.,* 1969), appear to have only small amounts of cloud material (personal observations). In addition to varying amounts of cloud material, the organization of components within the pericen-triolar material also varies considerably. For example, within the cloud of PTK cells and Chinese hamster cells, in addition to the general diffuse osmiophilic material, it is often possible to detect darker, more compact electron-dense aggregates that have been referred to as "pericentriolar satellites." These struc-tures may be scattered throughout the pericentriolar cloud, or they may be clumped together. These satellites have been referred to as "virus-like" particles because of their morphological resemblance to viruses (Wheatley, 1974; Gould and Borisy, 1977). However, at times it is possible to detect even larger satellite-like material that may be different than the virus-like particles. In sum-mary, the pericentriolar cloud may vary considerably in both the quantity of material and the different components present.

The chemical composition of the pericentriolar material is rather controversial. As will be seen in Section VIII, the occurrence and location of nucleic acids in the centriolar complex is a hotly debated issue. The general feeling is that some nucleic acid (probably RNA) is located in the centriolar complex. Recent laser microbeam experiments employing photosensitization with acridine orange and psoralens seem to indicate the presence of nucleic acid in the pericentriolar cloud (Berns et al., 1977; Peterson and Berns, 1978). However, electron microscope studies using cytochemical stains and RNase digestion clearly suggest that RNA is located in close association with the triplet blades of the centriole proper (Dippell, 1976; Rieder, 1979). Of course, it is possible that nucleic acid is located in both regions. The possible function of this nucleic acid will be dis-cussed in Section VIII.

Another major component of the pericentriolar cloud is tubulin. Microtubules are usually abundant in the cloud material, and it often appears that it is the cloud that organizes the microtubules rather than the centrioles. Experiments have been performed in which the centrioles have been destroyed or partly removed from the cloud, yet spindle microtubules are still organized by the cloud (Berns et al., 1977; Berns and Richardson, 1978). In addition, Gould and Borisy (1977) have demonstrated that pericentriolar cloud material in vitro can nucleate mi-crotubules, and Brenner et al. (1977) have shown that unusual meiotic-like tissue culture cells contain functional spindle poles minus centrioles but with cloudlike material. Despite the above lines of evidence suggesting a major microtubule organizing function for the pericentriolar cloud, occasional micrographs also clearly show that microtubules do extend directly from the centriole itself.

The structural organization of the centriole is precise in certain respects and The structural organization of the centriole is precise in certain respects and quite variable in others. For example, it is by now classical to describe the centriole as a cylindrical structure that is between 0.20 and 0.25 μm in diameter and between 0.25 and several microns in length. With the electron microscope, the centriole can be seen to consist of a set of nine tubule-like triplet blades arranged at a precise angular pitch to each other. Each of these blades is com-posed of three microtubules. In particularly good cross sections, it is often possible to further resolve the internal structure of the centriole and even demon-strate connections between each of the triplet blades and between the triplets and the central region of the centriole where a small central vesicle 600 Å in diameter is often visible. The location of these internal "spokes," which create a cartwheel-like configuration, appears to be confined to one end of the centriole (Stubblefield and Brinkley, 1967). In fact, there is quite a bit of difference in centriole architecture depending upon which end of the centriole is being examined. In addition to the change in the internal "spokes," the angular pitch of the triplets may change considerably as one sections through the centriole. Stubblefield and Brinkley (1967) demonstrated a pitch change of 10° in three serial sections. The same authors also describe "fibrous" appendages that appear to radiate out from the centriole near one end. Similar structures are often seen in the plane that represents the junction between the basal body and the cilium (Brinkley and Stubblefield, 1970).

The chemical composition of the centriole is unclear. The triplets are undoubtedly composed of tubulin; however, other components, such as RNA (Rieder, 1979; Brinkley and Stubblefield, 1970; Zackroff et al., 1977; McGill et al., 1976) and even DNA (Went, 1977a,b) may also be present. There also appears to be a definite polarity to the centriole with the proximal end more electron dense and thicker than the distal end (Fig. 2). In fact, cross sections through the distal end do not reveal the "cartwheel" configuration, whereas sections through the proximal end do (Fig. 3).

The method of centriole reproduction is relatively unknown. In mitotic cells, the centrioles always exist in pairs (called a diplosome or duplex) with one at a precise right angle to the other. This association can be traced back to interphase when each centriole in the duplex appears to give rise to a new "pro-centriole" which grows at a right angle to the parent centriole in a true "mother-daughter" (father-son?) relationship (Fulton, 1971; Rattner and Phillips, 1973). However, very little is known about this process, and as mentioned earlier, DNA has not been clearly demonstrated in or around the centriole. Recent experiments by Went (1977a,b) using drugs, such as actinomycin D, BUdR, and chloramphenicol, on sand dollar eggs have resulted in the hypothesis that centrioles replicate by RNA-*directed* DNA synthesis.

III. Centriolar Movements during Mitosis

A considerable portion of this chapter will deal with the functional role of the centriolar complex in chromosome movement and, specifically, as a microtubule organizing center. However, as the cell enters mitosis, the two centriolar complexes, each containing a centriole duplex, can be seen in close proximity to each other in the cytoplasm near one end of the nucleus. As the nuclear envelope breaks down, one or both centriolar complexes migrate around, or through, the nucleus to eventually take up positions opposite each other. The result of this process is the establishment of opposite spindle poles. The controls for this migration and the forces involved in the movements are entirely unknown. However, detailed light and electron microscopic examination of this process has revealed that the migration of the two centriolar zones to opposite poles involves the formation of a large number of microtubules between the two regions (Molè-Bajer, 1975; Roos, 1973). In a clonal subline of PTK_2 cells, this process is particularly clear. In this subline of PTK_2, the two centriolar complexes are unusually visible and appear as two small dark dots in a "clear zone" when viewed with the phase contrast microscope (Figs. 1 and 4). As the cell proceeds further into prophase, the dots can be seen to separate and the "clear zone" elongates with the two dots at either end. Though the path of the centriolar complexes to establish the poles may be quite variable (Rattner and Berns, 1976a,b), the end result is always two spindle poles. When electron microscopic sections are used to reconstruct the process, numerous microtubules are seen extending between the two centriolar complexes (Fig. 1). Some of the microtubules appear to extend completely between the two centriolar complexes, but most do not. Large numbers of microtubules appear to cross each other at oblique angles, especially those along the outer margins of the microtubular bundles. This type of configuration suggests some sort of lateral interaction as a possible mechanism of force generation. However, just how the microtubules function in the separation of the centriolar zones is entirely unknown. It seems that understanding centriolar migration may be just as elusive as understanding chromosome movement, though a relationship between the two is almost certain.

IV. Systems without Centrioles

Are centrioles really necessary for division? To the uninitiated, such a question may seem foolish. From the time we were taught basic biology to modern day reviews in the journal *Science* (Marx, 1973), the centriole has always been ascribed a central, if not crucial, role as a spindle pole organizer. Yet it is a well known fact that most higher plants divide without centrioles, as do fungi (Byers *et al.*, 1978; Peterson *et al.*, 1972; Kubai, 1975) and diatoms (Pickett-Heaps and Tippit, 1978). However, these organisms often do not receive the attention of main line "animal" biologists, and it is therefore easy to explain away this lack of "centrioles" as an unusual evolutionary obscurity. It is more difficult, though, to explain why cells that normally do contain centrioles can be shown to undergo cell division without these organelles. In 1959, Dietz demonstrated that when the centriolar region was displaced off the spindle of the crane fly, the cells were still perfectly capable of undergoing anaphase and dividing. Also, it has been possible to destroy prophase centrioles with a laser microbeam and demonstrate that anaphase movements still continue (Berns and Richardson, 1977).

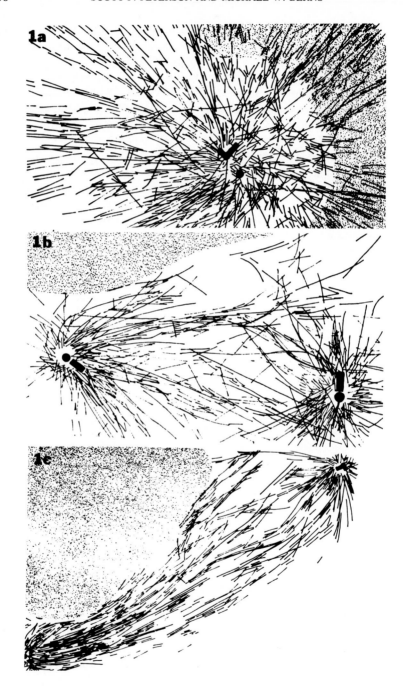

Furthermore, animal cells *in vitro* have been observed undergoing meiotic-like reduction division, and when examined ultrastructurally, some poles have been shown to be lacking centrioles (Brenner *et al.*, 1977).

These observations, on both naturally occurring non-centriole-requiring divisions and experimentally induced acentriolar divisions, raise the question of centriole function. It seems quite possible that in those cells where centrioles are normally present, they serve a function in the generation or organization of the pericentriolar cloud material. In Dietz' experiments (1959), no distinction was made between the centriole proper and the pericentriolar cloud. It is therefore not possible to tell whether both the cloud and the centrioles, or just the centrioles, were dislocated off the spindle. However, in the laser microbeam experiments on cultured PTK cells (Berns and Richardson, 1977), it was possible to damage the prophase centrioles and cause them to move considerably away from the spindle pole. When the cells were observed behaviorally, the chromosomes congressed towards the metaphase plate and underwent subsequent anaphase movements. Ultrastructural examination of similarly irradiated cells revealed that microtubules were focused into electron-dense, pericentriolar-like material even though the centrioles were considerably damaged and greatly removed from the pole region. This type of experiment strongly suggests that the centriole is not needed at the pole during mitosis and ascribes the role of microtubule organization to the pericentriolar material. The experiments of Gould and Borisy (1977) on isolated pericentriolar material and Berns *et al.* (1977) employing laser microirradiation indicated that the pericentriolar cloud was the major microtubule organizing center. In the latter experiments, acridine orange was used to selectively sensitize the pericentriolar material to visible laser light. Laser microirradiation appeared to damage this region without greatly disrupting the centrioles. Irradiation of centriolar regions in prophase PTK_2 cells resulted in cells undergoing cytokinesis without any anaphase chromosome movements. Ultrastructural studies revealed that the pericentriolar cloud was damaged, but the centrioles appeared normal. Microtubules could be seen emanating from the kinetochores of the chromosomes, but none were seen extending to opposite poles from the damaged pericentriolar material.

The conclusion from the above laser experiments should not be that centrioles are not needed for cell division since it may be possible that the centriole is involved in the production or organization of the pericentriolar cloud prior to prophase. This possibility has yet to be examined experimentally.

FIG. 1.   (a) Serial section reconstruction of 10 sections through centrosome region of PTK_2 cell prior to duplex separation. (b) Serial section reconstruction of 10 sections through centrosome regions of prophase PTK_2 cell showing duplex separation. (c) Serial section reconstruction of 8 sections of late prophase cell showing duplexes near completion of separation. (Courtesy of J. B. Rattner.)

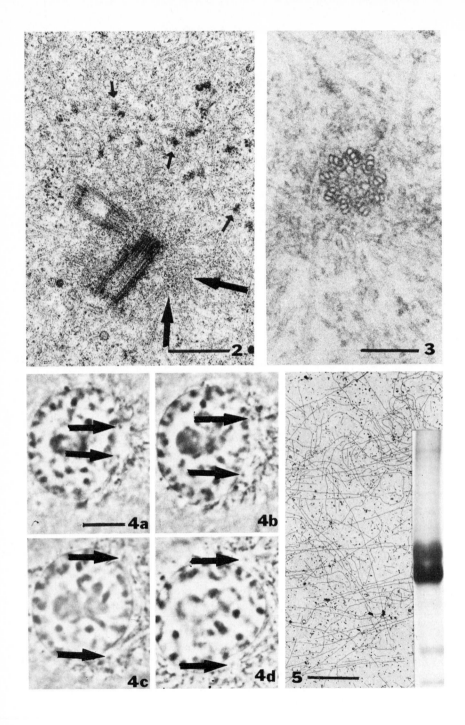

The central role of pericentriolar material as a major spindle pole organizer gets strong support from those unusual systems where centrioles are not found. In all of these cells, electron-dense material (perhaps analagous to pericentriolar cloud material) is found associated with the microtubules at the poles. For example, in the yeast *Saccharomyces cerevisiae,* electron-dense, amorphous, plaque-like bodies called "spindle pole bodies" (SPB) have been found at mitotic (Byers *et al.,* 1978; Peterson and Ris, 1976; Robinow and Marak, 1966) and meiotic spindle poles (Moens and Rapport, 1971). In fact, every microtubule appears to have at least one end associated with an SPB (Byers *et al.,* 1978). The chemical nature of the SPB and its functional capacity are currently under investigation. Presently, all that can be said is that the microtubule organizing capacity of this region is inhibited by trypsin treatment and not by DNase or RNase treatment.

Another group of organisms that undergoes mitosis without centrioles is the diatoms (Pickett-Heaps and Tippit, 1978). As pointed out in their review, Pickett-Heaps and Tippit suggest that most of our current ideas about the mechanisms of mitosis may not apply to the diatoms. In fact, the suggestion they are really making is that current mitotic models generally rely upon the biochemical and biophysical properties of microtubules *in vitro* whose relevance to mitosis *in vivo* is uncertain. These authors indicate that "our own understanding of how the diatom spindle functions now runs counter to virtually all the models of mitosis we know." Rather than suggesting that the diatoms have evolved an unusual way of undergoing mitosis, these authors feel that much of the basic "accepted" dogma about mitosis may indeed be incorrect. An example is that kinetochores act as microtubule organizing centers. There are several *in vitro* studies demonstrating microtubule growth from kinetochores of isolated chromosomes (Telzer *et al.,* 1975; Gould and Borisy, 1977). However, the only direct *in vivo* evidence of microtubule origin from the kinetochores is the early studies of Forer (1965) in which he used a UV microbeam to disrupt the birefringence in the kinetochore microtubules. The movement of the reduced birefringence towards the poles during metaphase and anaphase suggested that these fibers were emanating from the kinetochores and moving poleward as well. However, it has

---

FIG. 2.   Centriolar complex containing one centriolar duplex with the centrioles at right angles to each other. Note the denser end in one of the centrioles, the pericentriolar cloud (large arrows), and the pericentriolar satellites (small arrows). Bar = 0.2 μm.

FIG. 3.   Cross section through proximal end of centriole. Note the internal spokes and "cartwheel-like" appearance. Bar = 0.2 μm.

FIG. 4a–d.   Early prophase cell in which duplex separation and migration occur along the nuclear surface. Both duplexes are motile. Bar = 1.5 μm.

FIG. 5.   Bovine brain microtubules produced by polymerization *in vitro* by a modification of the procedures described by Berkowitz *et al.* (1977). Inset: 6–12% SDS-polyacrylamide gel of microtubule proteins. MAPs protein at the top of the gel with α and β tubulin, the two major bands, in the middle of the gel. Bar = 5 μm.

not been demonstrated that these areas of reduced birefringence are caused by microtubule disruption or that the decrease in birefringence results from the specific dissolution of the kinetochore microtubules (Pickett-Heaps and Tippit, 1978).

Whether or not the diatom spindle is an obscure evolutionary quirk or a related and relevant structure to understanding mitosis in a general sense may be debated for some time. Whichever is the case, the facts still remain that diatoms do have spindle poles, that chromosomes move towards them, that microtubules are involved in the process in some way, and that centrioles (or even structures closely resembling centrioles) are not involved. The structures that diatoms have at their spindle poles are unique, not well understood, and changing during division. At the beginning of division, an electron-dense structure called a microtubule center (MC) rests against the nucleus, and the prophase spindle appears to form close to it. However, as division proceeds, the MC disappears and a new structure called the polar complex (PC) appears. This structure is a plate- or rod-like structure consisting of one or more layers into which spindle microtubules are inserted. The plate-like structures often appear to be surmounted by a granular material that appears to be closely associated with Golgi and other membranous inclusions.

The lack of centrioles is intriguing because Manton *et al.* (1970a,b) have shown that in the marine diatom *Lithodesmium* centrioles are formed *de novo* between the first and second meiotic division and apparently are required to form the axonemal microtubules in the flagellum of the male gametes. One wonders why a cell that can make centrioles does not make them for spindle poles. Perhaps the centrioles are simply not needed for division. It is also interesting to observe that in animal tissue culture cells in G_0 (R. Tucker, personal communication) and in growing PTK_2 cells, cilia frequently can be seen emanating from one of the centrioles in interphase. Cells have even been observed entering mitosis with the remnants of one of these cilia still associated with the centriole. Is it possible that the role of the centriole is really to form cilia (or flagella) when the particular differentiated cell function needs these organelles? In these tissue culture cells, the normally repressed cilia differentiative reactions have been activated by some abnormal set of conditions.

V. The Two Functional States of the Centriole

In the preceding section, we saw that many cells do not even have centrioles; however, when centrioles are present, they may fulfill two functions in the cell. First, centrioles are found in the mitotic poles of dividing cells where they may function in the organization of the microtubules into a spindle during mitosis. Second, centrioles may be found as the basal bodies of cilia where microtubules are nucleated onto the centriole as a highly organized, fibrous array. It is not known how the centriole can fulfill this dual role in cells. Yet it is clear that the centriole can serve both functions at the same time. Jensen and Rieder (personal communication) have demonstrated that some PTK cells in culture may have a cilium growing out of one of the centrioles at the pole while the cell is in mitosis.

Examination of the basal body of cilia or the centrioles of cells shows them to be structurally identical and probably capable of participating in the same functional roles (Heidemann *et al.,* 1977; Anderson and Brenner, 1971; Fulton, 1971). In cilia or flagella, microtubular components may be seen actually growing out of the basal body (Wolfe, 1972), whereas in a mitotic spindle, the pericentriolar cloud material seems to be the nucleating material of microtubules (Pickett-Heaps, 1969). There is no explanation of this diversity of function for the centrioles of cells except the fact that both functions involve a form of cellular motility.

Both cilia and normal mitotic spindles may be observed in tissue culture cells. Stubblefield and Brinkley (1966) have found that treatment of tissue culture cells with low concentrations of colcemid will result in the formation of cilia from the centrioles of interphase cells. In our laboratory, we sometimes see cilia growing from the centrioles of interphase cells. Figure 9 shows one of these cilia. The apparent attachment of the microtubules in the cilia to the centriole (basal body) may be easily seen, although there seems to be no other change in the morphology of the centriole.

Immunological evidence for the identical nature of the centriole and the basal body also exists. Connolly and Kalnins (1978) have found a rabbit serum that is specific for the centrioles of tissue culture cells and will also cross-react with the basal bodies of epithelial cells. Such data provide additional evidence to the morphological and biochemical data (Fulton, 1971) for the similarity of the two functional states of the centriole.

Interconversion of the centriole between its two functional states of mitosis and cilia may be a normal occurrence in tissue culture cells. In senescent cells, cilia may be observed in tissue culture with a high frequency (Tucker *et al.,* 1978b; C. Jensen, personal communication). When the cells enter mitosis, the cilia usually disappear and the mitotic spindle with its fusiform array of microtubules appears. Since there is no apparent morphological change in the centriole between its two functional states, there may be a biochemical difference between the basal body of cilia and the centriole found in the poles of mitotic cells.

VI. Indirect Immunofluorescence and Its Use in Mitotic Investigation

It has been difficult to study the organization and orientation of the forming spindle in dividing cells because this structure is not normally visible by light microscopy. Polarizing microscopy has been utilized by many researchers (Ritter

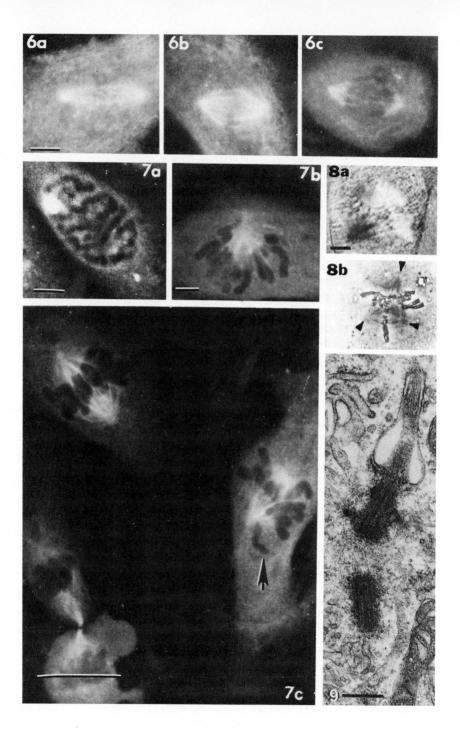

et al., 1978; Inoué and Ritter, 1978; Inoué, 1961) and allows visualization of the spindle, but only when the microtubules are highly organized into a parallel array (as in metaphase). The technique of indirect immunofluorescence has allowed observation of the forming spindle in the early stages of mitosis.

The technique of producing antibodies to cellular proteins and then reintroducing them to cells to label cellular organelles or areas has been used to examine the localization and function of proteins important in mitosis. Tubulin, the major component of the microtubules in the spindle, actin, and calcium-dependent regulatory protein (CDR) have all been localized to the spindle of dividing cells.

Commonly, indirect immunofluorescence is performed by first purifying the protein of interest by biochemical means. The protein is then injected into a rabbit to obtain specific antibodies. After fixation with a formaldehyde solution, cells are treated with an organic solvent and then the protein-specific antiserum that was produced in the rabbit. The rabbit antiserum is then rinsed from the cells and a fluorescent goat antibody to rabbit IgG is applied. After a second rinsing, the cells are viewed in a fluorescent microscope. The protein of interest will be fluorescently labeled (Fig. 10). Experiments such as this allow the localization of specific proteins to the mitotic apparatus.

Localization of tubulin by indirect immunofluorescence has been accomplished by many workers (De Brabander *et al.,* 1977; Tucker *et al.,* 1978a; Eckert and Snyder, 1978; Lazarides, 1976). In Fig. 7 is a series of photographs showing tubulin localization in mitotic PTK$_2$ cells. During prophase, there is a disappearance of the cytoplasmic microtubules that is correlated with a subsequent increase in fluorescence (tubulin) around the centrioles. As mitosis proceeds to prometaphase, the chromosomes are grouped around the centriolar region and appear to be connected to the mitotic poles by the increasing number of microtubules. Cells in metaphase normally have little tubulin in their general cytoplasm, whereas the spindle fluoresces brightly. Generally, through the early

---

FIG. 6. Anti-actin staining of the mitotic spindle of dividing cells. (a) Early metaphase, (b) metaphase, (c) anaphase. Bar is approximately 1 μm. (Courtesy of J. R. McIntosh, University of Colorado.)

FIG. 7. Anti-tubulin staining of the mitotic spindle of dividing cells. (a) Prophase cell. Centriolar regions are near nucleus and appear as a very bright region. Bar = 1 μm. (b) Prometaphase cell. Chromosomes are arranged around the centriolar regions connected by the many fluorescent microtubules. Bar = 1 μm. (c) Three mitotic cells. Late metaphase cell in upper left with most tubulin concentrated in the spindle. Early metaphase cell can also be seen, still with tubulin in the cytoplasm. A chromosome (arrow) has been irradiated in its kinetochores to block microtubule nucleation onto this chromosome. A late telophase cell (lower left) has many microtubules running between the two daughter cells and the amount of tubulin in the general cytoplasm is increasing. Bar = 5 μm.

FIG. 8. A tripolar hybrid PTK$_2$–human cell. (a) Polarizing micrograph showing three poles in this cell in division. (b) Coomassie Brilliant Blue–Safranin O stained tripolar cell; arrows indicate location of poles. Bar = 1 μm.

FIG. 9. Cilium growing from the centriole of an interphase PTK$_2$ cell. Bar = 0.2 μm.

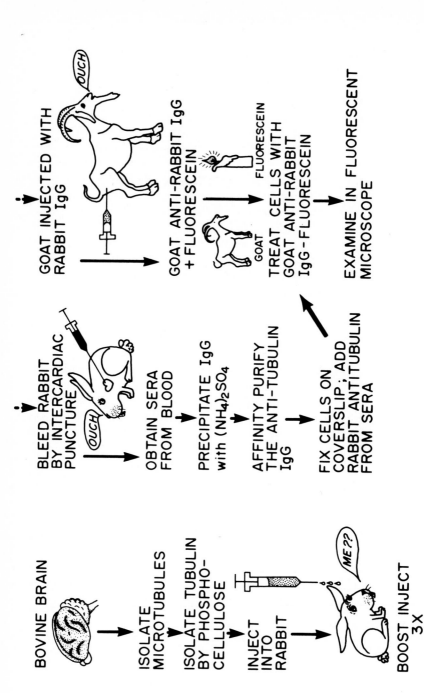

FIG. 10. Procedure outlining indirect immunofluorescence employing an antibody to tubulin.

spindle forming stages of mitosis, there is a decrease in the amount of fluorescence (tubulin) in the general cytoplasm coupled with an increase in the amount of fluorescence associated with the mitotic spindle. In the late stages of mitosis (anaphase, telophase), the amount of tubulin in the cytoplasm may be seen to increase as the spindle breaks down (Fig. 7).

Tubulin is not the only protein that is concentrated in the spindle of dividing cells. Cande *et al.* (1977) have shown that antibody to actin also binds in the mitotic spindle (Fig. 6). Again, as with tubulin, indirect immunofluorescence shows that as mitosis proceeds to metaphase, the amount of actin in the spindle increases. Electron microscopic evidence for the existence of actin in the spindle correlates well with the data obtained with direct immunofluorescence (Forer, 1974; Forer and Jackson, 1976). Actin may also be visualized in the spindle by treating mitotic cells with fluorescently labeled meromyosin (Herman and Pollard, 1978). In addition, antibody to myosin reveals the presence of this contractile protein in the spindle (Fujiwara and Pollard, 1976) as well as other areas of the cell. The function of actin–myosin in the spindle is at present unknown.

Structural proteins are not the only cellular components that are concentrated in the spindle of dividing cells. Calcium-dependent regulatory protein has also been shown to be compartmentalized in the spindle (Marcum *et al.*, 1978). This protein is believed to be important in controlling the assembly and disassembly of microtubules in dividing cells. Such a protein would be necessary in many of the present theories of spindle function and formation (Margolis *et al.*, 1978).

Through exposure of cells to a nonimmunized rabbit serum, a specific antibody to the centriolar region has been found (Connolly and Kalnin, 1978). Cells treated with the serum and subsequently labeled with fluorescent goat anti-rabbit IgG usually show one or two areas of bright fluorescence (centrioles), whereas the remainder of the cell appears dark. These results imply that there is an antigen in the centriolar region, which is at present unidentified, that is found only in close association with the centrioles. Basal bodies may also be labeled with the same nonimmune rabbit serum that binds to centrioles. Visualization of the centrioles in mitosis is possible with this antibody, and the results are similar to those seen with antitubulin antibody. During prophase, the centrioles are seen in close association with each other. As division proceeds, the centrioles are observed to separate and move to opposite sides of the cells. Results about the movement of centrioles obtained with either the nonimmune rabbit serum or with antitubulin correlate well with electron microscopic observation of mitosis (Rattner and Berns, 1976b).

Although most indirect immunofluorescence is performed with animal cells, antitubulin has also been employed with plant cells (Franke *et al.*, 1977). It has been shown that a protein with the same antigenic characteristics as tubulin may be localized to the mitotic spindle of *Leucojum aestivum,* a type of amaryllis.

Since indirect immunofluorescence normally requires extensive biochemical as well as cytological expertise, it has not been widely available to many laboratories. Wissinger and Wang (1978) employed a cytological staining technique to visualize the spindle of dividing cells in culture. This staining procedure employs Coomassie Brilliant Blue and Safranin 0 stains. Both the spindle and the chromosomes may be easily seen with this technique. Results using this method are the same as those obtained with antitubulin staining and polarizing microscopy. The advantages of this technique are that it may be accomplished with a simple light microscope and no biochemical isolation or antibody production is needed. Figure 8 shows a hybrid cell (Peterson and Berns, 1979) with several mitotic poles that has been stained by this technique to reveal the spindle. Unfortunately, this procedure may not be used to stain the microtubules of interphase cells.

VII. *In Vitro* Studies on Microtubular Nucleation

The previous section dealt with the combination of biochemical and cytological techniques to study the spindle. Biochemical *in vitro* experiments to examine the nucleation of microtubules onto the centriolar region have not yet been successful in producing a mitotic-like spindle of microtubules as is seen in a living cell. Although it has been possible to isolate basal bodies (Gould, 1975) and centrioles from tissue culture cells (Blackburn *et al.*, 1978), only asters of microtubules have been produced *in vitro* rather than the pole-to-pole growth observed in dividing cells (Snell *et al.*, 1974).

Classically, tubulin for use in *in vitro* experiments is isolated from the brains of freshly slaughtered animals like rats (Weisenberg, 1972), steers (Berkowitz *et al.*, 1977; Lee *et al.*, 1978), and pigs (Olmsted and Borisy, 1973). There is also evidence that neurotubulin from brains is not easily isolated as the organ ages because of the action of endogenous protease (Sloboda and Warbasse, 1978). The advent of tubulin isolation procedures from HeLa tissue culture cells (Weatherbee *et al.*, 1978) and from *Tetrahymena* (Maekawa and Sakai, 1978) makes *in vitro* experimentation with tubulin from a nonneural source possible.

In addition to tubulin, many proteins are known to purify as part of the microtubule. Among these are tau (Weingarten *et al.*, 1975; Cleveland *et al.*, 1977a,b), various high molecular weight proteins (Berkowitz *et al.*, 1977), and high molecular weight phosphoprotein (Klein *et al.*, 1978). Calcium-dependent regulatory protein can also promote microtubule assembly *in vitro* by chelating free calcium in solution (Marcum *et al.*, 1978). Other factors isolated from the cortex of sea urchin eggs have been found to inhibit the polymerization of microtubules *in vitro* (Naruse and Sakai, 1978). It is evident that many proteins may be required for microtubule nucleation onto the centriolar region, but which factors may or may not be needed in this process are unknown.

Microtubule-associated proteins are not necessary in the polymerization of tubulin *in vitro*. Zinc (Gaskin and Kress, 1977) and dimethyl sulfoxide (Himes *et al.*, 1977) have both been shown to promote tubulin polymerization and microtubule growth. It is not known how this polymerized tubulin relates to the microtubule structures seen in living cells.

Isolated basal bodies have been employed as *in vitro* nucleating sites for microtubules (Snell *et al.*, 1974). These authors have found that the part of the basal body that nucleates microtubules is related to the amount of tubulin subunit available in the surrounding medium. Growth from the ends of the basal bodies seems to be favored in solutions containing low tubulin concentration, whereas microtubules may be seen arising from the sides in higher tubulin concentrations. Isolated mitotic apparati from marine eggs (Rebhun *et al.*, 1974; Inoué *et al.*, 1973) and lysed mitotic HeLa cells (Cande *et al.*, 1974) have also been shown to be capable of the nucleation and polymerization of microtubules *in vitro*.. All of these authors find that tubulin from one species may be used by the centriolar region or basal bodies of another, nonrelated species to form or increase microtubules *in vitro*. Little is known about how nucleation and growth of microtubules is carried out by the centriolar region of cells.

Kinetochores have also been shown to be capable of microtubule nucleation *in vitro* (Telzer *et al.*, 1975). These workers have used isolated chromosomes as nucleating sites for microtubules. Electron microscopic examination of chromosomes after exposure to the correct microtubule nucleating conditions shows a specific association of tubules with only the kinetochore regions, showing that these regions are uniquely competent to nucleate microtubules.

Isolation of centrioles from tissue culture cells in reasonably pure preparations has been accomplished (Blackburn *et al.*, 1978). After cell lysis and centrifugation of the suspension, centrioles may be seen by electron microscopy. Single centrioles as well as duplex pairs of centrioles may be observed by this technique.

Recent experiments with [³H]GTP have shown that microtubules are assembled at one end and disassembled at the other (Margolis and Wilson, 1978). This data implies a directed assembly from the microtubule nucleating centers in the living cell (centrosome, kinetochores). These authors have proposed a mitotic mechanism that envisages assembly at the kinetochore for kinetochore microtubules and subsequent disassembly at the centrosdome, while the continuous microtubules grow from the two centrosomes to be associated in their ends (Margolis *et al.*, 1978). However, the role of microtubule-associated proteins and the mechanism of growth *in vivo* from centriolar regions are not explained by this model.

Lysed cells have been used to examine the growth of microtubules from the centrosomes and kinetochores of cells (Snyder and McIntosh, 1975). These authors find that the amount of tubulin available to a lysed mitotic cell will change the amount of birefringence as well as the number of microtubules in the spindle. Higher tubulin concentrations produce more birefringence, whereas low concentrations have the opposite effect. When prometaphase cells are treated with colcemid to break down the early spindle and then lysed into a medium containing tubulin, asters are seen to form around the centriolar region rather than the mitotic spindle of microtubules. Such astral growth is similar to that reported for the *in vitro* nucleation of microtubules onto basal bodies (Snell *et al.*, 1974).

The technique of cell lysis has also been employed by Cande and Wolniak (1978) to examine the effect of metabolic inhibitors on mitosis in cells. Vanadate, a potent ATPase-inhibiting drug, may be shown to reversibly inhibit the movement of chromosomes during mitosis. Because several groups have found that vanadate inhibits the dynein ATPase of flagella (Gibbons *et al.*, 1978; Kobayashi *et al.*, 1978), a spindle dynein is proposed to exist that is important in catalyzing the movement of chromosomes in mitotic cells.

Considerable debate about the role of the centriole in mitosis has been generated in recent years. Some believe that the centriole is not needed for nucleation (Pickett-Heaps, 1969), while others believe that it is important in this function (Heidemann and Kirschner, 1975). The only available evidence that directly examines this problem *in vitro* has been produced by Gould and Borisy (1977). These authors find that the centrosphere (pericentriolar cloud) is the nucleating material for the microtubules, and not the centrioles themselves. Interphase centriolar regions were found to nucleate only very few microtubules, while many arose from the pericentriolar material of mitotic cells. These observations imply a functional difference between the centrosome in different portions of the cell cycle. The only microtubules that could be observed actually "growing" from the centrioles themselves were formed into the cilia-like configuration rather than the spindle or astral arrays seen associated with mitosis.

VIII. The Role of Nucleic Acids in Centriolar Function

Renewed interest has arisen concerning the role that nucleic acids might play in the normal function of the centriolar region. Proposals of nucleic acid playing a part in the duplication of centrioles, as well as in the ability of centrioles to form a spindle, exist in the literature.

Staining various types of cells with nucleic acid-specific dyes has shown an apparent concentration of nucleic acids within the centriolar regions or basal bodies of cells. Randall and Disbrey (1965) have shown that the basal bodies of

Tetrahymena will attract and hold the nucleic acid dye, acridine orange. Other workers (Smith-Sonneborn) and Plaut, 1967; Dippel, 1976) have also obtained evidence for nucleic acid localization in the centriole or basal body region (see Fulton, 1971, for review). Long-term exposure of tissue culture cells to ethidium bromide (McGill *et al.*, 1976), a dye with a high affinity for nucleic acid phosphate groups, has produced ultrastructural changes in the centriole that can be observed in the electron microscope. Alteration of the normal nine triplet centriole, as well as changes in the internal centriolar architecture, have been reported by these authors. Autoradiographic evidence for nucleic acid presence in the centriolar region also exists (Brinkley and Stubblefield, 1970).

The exact location and function of the nucleic acids in the centriolar region is unknown, although work by Rieder (1979) seems to implicate a close association between the triplets of the centriole and ribonucleoprotein (RNP). By digesting fixed cells with ribonuclease (RNase), this author has shown the presence of RNP in both the centriolar region and the kinetochores of mitotic chromosomes in dividing cells. Other workers, using similar RNase digestions, have shown the presence of RNP in these two cellular organelles (Bielek, 1978; Brinkley and Stubblefield, 1966).

Laser microbeam experiments employing acridine orange to absorb laser light to heat-damage small areas of cells also implicate the presence of nucleic acid in the centriolar region. Dividing PTK_2 cells growing in medium containing acridine orange were irradiated in the centriolar region (Berns *et al.*, 1977). When examined in the electron microscope, ultra-structural damage was observed in the pericentriolar cloud, indicating that the dye had been held in this area by some cellular compound that binds acridine orange (nucleic acid).

Other laser microbeam experiments have used the laser in conjunction with the light-activated, nucleic acid-binding drugs, psoralens. By introducing these drugs to cells and subsequently exposing small cellular areas containing centrioles to laser light, it is possible to cause cells to abort mitosis (Peterson and Berns, 1978). Because of the chemical specificity of these drugs for nucleic acids, an RNA participating in the formation of the spindle is proposed to exist in the centriolar region.

Perhaps the best biochemical evidence that exists for the presence and functioning of centriolar region RNA has been produced by Heidemann *et al.* (1977). These authors have digested isolated basal bodies with RNase, as well as other enzymes, and then microinjected the treated basal bodies into *Xenopus* eggs. Electron microscopic examination revealed that the nucleation of microtubules and the formation of asters did not occur from RNase-digested basal bodies.

Many inhibitors of both protein and nucleic acid function may be shown to interfere with the duplication of the centriole. Phillips and Rattner (1976) have shown that protein inhibitors interfere with the replication of the centriole in cells. Nucleic acid-binding drugs, such as ethidium bromide (McGill *et al.*, 1976) also alter normal replication of the centriole. Many other chemicals, such as mercaptoethanol (Mazia, 1961) and chloramphenical (Deutch and Shumway, 1973) also change the normal duplication of centrioles. In any case, there seems to be both protein and nucleic acid components that are important in the replication and normal functioning of the centriole.

More evidence for the functioning of nucleic acid in the duplication of the centriole has been obtained by Went (1977a). Using various inhibitors of nucleic acid synthesis, this author has again shown interference with centriolar duplication caused by these compounds and has also proposed a theoretical model of centriolar duplication involving the production of a DNA molecule from an RNA template (Went, 1977b).

Evidence from most laboratories seems to support the presence of nucleic acids in the centriolar region as necessary to both the duplication of the centriole in interphase as well as its function in mitosis. Bryan *et al.* (1975) have found that RNA will complex with the microtubule-regulating protein tau. Although this is *in vitro* evidence that may not relate to the *in vivo* function of centriolar region RNA, it is the only biochemical explanation of how RNA may act in mitotic pole organization and control.

IX. Multicentriolar Cells

In normal mitosis, an animal cell forms a pair of mitotic poles that act to nucleate the microtubules of the spindle. However, cells often have more poles than the usual two. Multipolar spindles may be created naturally, as in random tissue culture events, or experimentally by fusion of cells. Treatment of eggs with ions can also be employed to produce multiple centrioles within cells.

Naturally occurring multipolar divisions have long been observed in tissue culture cells (Pera and Rainer, 1973; Pera and Schwarzacher, 1969). Ghosh *et al.* (1978) have examined multinucleate PTK$_2$ cells by light and electron microscopy and have observed the presence of multipolar–multidaughter mitoses. Commonly, the production of multiple daughters from a single mitosis is correlated with the presence of multiple centrioles in the parent cell. Brenner *et al.* (1977) have observed a unique cell division that resembles meiosis and has multiple mitotic poles, some of which lack centrioles but do appear to contain "pericentriolar-like" material. These reports all indicate that multipolar-multiple daughter mitoses are not unusual occurrences in tissue culture cells.

Extensive examination of PTK$_2$-human hybrid cells in mitosis has been accomplished by various microscopic methods (Peterson and Berns, 1979). Commonly, these cells have more than two pairs of centrioles and divide to produce three daughters. Examination of these hybrids by polarizing microscopy or staining techniques (Fig. 8) reveal very extensive and complex spindles. Electron microscopic examination of hybrid cells shows them to have many mitotic poles and often some centriolar regions that do not seem to nucleate spindle microtubules. The number of centriolar regions (mitotic poles) in a hybrid cell seems to have only a minor effect in determining the number of daughter cells produced in a single mitosis. There does not appear to be a one-to-one correspondence between the number of centriolar duplexes in a cell and the number of daughters produced at the completion of mitosis.

Sea urchin eggs, when treated with various ions, will produce many centriolar regions that form asters of microtubules rather than participating in mitosis directly (Loeb, 1913). Dozens of centriolar complexes may be produced within a single egg cell by this technique.

Work by Miki-Noumura (1977) has shown that treatment of eggs by Loeb's double method produces centriolar complexes *de novo*. In this case, the centrioles are fabricated directly from cytoplasmic material rather than growing from preexisting centrioles as in normal centriole duplication. Examination of these asters shows that they consist of a central centriole(s) with a small amount of material surrounding them and an extensive framework of microtubules. Actual spindle formation only seems to occur in those centriolar regions that are associated with chromosomes. This same author has found that inhibitors of mitosis (colchicine, colcemid) will interfere with the formation of asters in a manner similar to the way they block mitosis. Kato and Sugiyama (1971) have also examined the asters of activated sea urchin eggs by electron microscopy and have found similar results to those reported by the Miki-Noumura manuscript.

Kane (1962) has employed the activation of sea urchin eggs as a means of observing the microtubule nucleation formation of the centriolar region. Because of the large numbers of centrioles and asters in an activated sea urchin egg, they make an ideal source for the isolation of centrioles for electron microscopic observation.

Another method for increasing the number of centriolar complexes in a cell is by microinjection (Maller *et al.*, 1976). Partially purified, centriole-containing fractions from sperm are able to initiate spindle formation in *Xenopus* eggs. When many centriolar complexes were introduced into a single cell, it was noted that the egg would commonly cleave into multiple daughters as has been observed in tissue culture cells. Similar results have also been obtained by Heidemann and Kirschner (1975) where they have injected basal bodies from *Chlamydomonas* rather than the centrioles from sperm.

Variation in the normal pair of centriolar poles found in dividing cells may also be obtained by treating cells with mercaptoethanol. A recent report by Sluder and Begg (1978) has shown that embryo cells may be produced that only have a single mitotic pole. Polarizing microscopy reveals that these cells have what appears to be a half spindle with microtubules extending to the chromosomes in the center of the cell. These data show the autonomous character of the centriolar region in dividing cells and indicate that two centriolar regions (poles) are not necessary for the formation of the mitotic spindle to begin.

It is evident that centriolar regions may nucleate microtubules where more than two poles are present. In cells with many centriolar complexes, multipolar spindles of a complex form may develop. In other cases, multiple centriolar zones in a cell may produce asters. In cells with only a single pole, a partial spindle may be formed. These data indicate that the type and shape of the spindle produced by a centriolar region in a mitotic cell is often changed by the presence or absence of other centriolar regions but that the presence of a centriolar complex does not necessarily result in spindle formation (i.e., microtubule organization).

X. Conclusions

It has not been the purpose of this chapter to draw specific conclusions. We have attempted to present observations and a discussion that serve to emphasize a rather poor understanding of the centriolar region.

Certainly, with respect to the structure and architecture of the centriolar region, a rather definitive picture can be presented. But even here, one must be careful, because the picture obtained may be quite variable depending upon (1) the species or cell type examined, (2) the fixation and other chemical treatments used, (3) the plane and angle of sectioning, and (4) the phase of the cell cycle.

The biochemical composition of the centriolar zone is far from resolved. The only molecular constituent that is clearly present in the region is tubulin. The immunological and morphological observations are strong with respect to this point. However, other proteins are undoubtedly present, as suggested by the CDR antibody experiments (Marcum et al., 1978). It is also probable that nucleic acids are present in the centriolar region, but whether they are located in the pericentriolar material, the centriole proper, or in both regions, is not resolved. Whether the nucleic acid is RNA, DNA, or both, also has not been determined, though the evidence strongly suggests that at least RNA is present. The function of the nucleic acid is entirely unknown. It may be involved in the actual organization of the spindle, or it may serve a function in centriole replication. The mode of centriole replication is peculiar. The growth of a procentriole from its parent is unique. Just how a nucleic acid would function in this process is unclear though the results of biochemical inhibitor experiments (Went, 1977a,b) have been used to suggest an RNA-directed (reverse-transcriptive) process.

In addition to the still unresolved questions of biochemical organization and replication is the question concerning basic function of the centriolar region. The occurrence of entire groups of organisms that do not contain a cell division structure even resembling a centriole, and the observations on naturally occurring or experimentally produced animal cells without centrioles, seriously raises the question of the need for this organelle in cell division. However, virtually all spindle poles have some structure whether it is an amorphous electron-dense cloud or a more organized entity, such as the microtubular center or polar complex of the diatom.

Even when there are centrioles present, their function in relation to the cloud is obscure. Experimental evidence suggests that the cloud may be the active microtubule organizing component. If this is the case, then what is the role of the centriole? In senescent cells, cilia are often observed growing out of one of the centriolar duplexes. Is it possible that the centriole is in the cell merely to generate cilia (or flagellae) if cellular function requires these structures? It is inherently difficult for an animal cell biologist, who for years has accepted centrioles as a functional part of the division spindle pole, to admit to the possibility that they may not play an active role in the cell division process. This possibility certainly cannot be denied.

How the centriolar zones migrate to establish opposite spindle poles is also unknown. The path that the duplexes follow to attain opposite positioning is quite variable, and the process may involve the migration of both duplexes or only one. The nature of the forces involved in the separation of the duplexes is unknown, though some type of microtubule growth and/or interaction seems likely.

If one conclusion is to be made in this chapter, it is that the organization and function of the centriolar region is far from established. In making this conclusion and discussing many of the areas in question, it has been our desire to focus attention or hopefully stimulate further studies on a truly fascinating cell component.

ACKNOWLEDGMENTS

The authors would like to acknowledge the assistance of the members of Dr. Berns' laboratory during the writing of this chapter. A special thanks to Elaine Kato for her excellent secretarial assistance in the preparation of this chapter. S. P. P. was supported by NIH Training Grant GM 07311-05. Part of the research reported was supported by the following grants: NIH HL 15740, GM 23445, GM 22754, U. S. Air Force AFOSR-77-3136.

REFERENCES

Anderson, R. G. W., and Brenner, R. M. (1971). *J. Cell Biol.* **50**, 10.
Berkowitz, S. A., Katagiri, J., Binder, H. K., and Williams, R. C., Jr. (1977). *Biochemistry* **16**, 5610.
Berns, M. W., and Richardson, S. M. (1977). *J. Cell Biol.* **75**, 977.

Berns, M. W., Olson, R. S., and Rounds, D. E. (1969). *Nature (London)* **221**, 74.
Berns, M. W., Rattner, J. B., Brenner, S., and Meredith, S. (1977). *J. Cell Biol.* **72**, 351.
Bielek, E. (1978). *Cytobiologie* **16**, 480.
Blackburn, G. R., Barrau, M. D., and Dewey, W. C. (1978). *Exp. Cell Res.* **113**, 183.
Brenner, S., Branch, A., Meredith, S., and Berns, M. W. (1977). *J. Cell Biol.* **72**, 368.
Brinkley, B. R., and Stubblefield, R. (1966). *Chromosoma (Berlin)* **19**, 28.
Brinkley, B. R., and Stubblefield, R. (1970). *In* "Advances in Cell Biology" (D. M. Prescott, L. Goldstein, and E. McConkey, eds.), Vol. 1, pp. 119–185. Appleton, New York.
Bryan, J., Nagle, B. W., and Doenges, K. H. (1975). *Proc. Natl. Acad. Sci. U.S.A.* **72**, 3570.
Byers, D., Shriver, K., and Goetsch, L. (1978). *J. Cell Sci.* **30**, 331.
Cande, W. Z., and Wolniak, S. M. (1978). *J. Cell Biol.* **79**, 573.
Cande, W. Z., Snyder, J., Smith, D., Summers, K., and McIntosh, J. R. (1974). *Proc. Natl. Acad. Sci. U.S.A.* **71**, 1559.
Cande, W. Z., Lazarides, E., and McIntosh, J. R. (1977). *J. Cell Biol.* **72**, 522.
Cleveland, D. W., Hwo, S.-Y., and Kirschner, M. W. (1977a). *J. Mol. Biol.* **116**, 207.
Cleveland, D. W., Hwo, S.-Y., and Kirschner, M. W. (1977b). *J. Mol. Biol.* **116**, 227.
Connolly, J. A., and Kalnins, V. I. (1978). *J. Cell Biol.* **79**, 526.
De Brabander, M., De Mey, J., Joniau, M., and Givens, S. (1977). *J. Cell Sci.* **28**, 283.
Deutch, A. H., and Shumway, L. K. (1973). *Protoplasma* **76**, 387.
Dietz, R. (1959). *Z. Naturforsch.* **14b**, 749.
Dippell, R. V. (1976). *J. Cell Biol.* **69**, 622.
Eckert, B. S., and Snyder, J. S. (1978). *Proc. Natl. Acad. Sci. U.S.A.* **75**, 334.
Forer, A. (1965). *J. Cell Biol.* **25**, 95.
Forer, A. (1974). *In* "Cell Cycle Controls" (G. M. Padilla, I. L. Cameron, and A. Zimmerman, eds.), pp. 319–336. Academic Press, New York.
Forer, A., and Jackson, W. T. (1976). *Cytobiologie* **12**, 199.
Franke, W. W., Seib, E., Heith, W., Osborn, M., and Weber, K. (1977). *Cell Biol. Int. Rep.* **6**, 75.
Fujiwara, K., and Pollard, T. D. (1976). *J. Cell Biol.* **71**, 848.
Fulton, C. (1971). *In* "Origin and Continuity of Cell Organelles" (W. Beerman, J. Reinert, and H. Ursprung, eds.), pp. 170–221. Springer-Verlag, New York.
Gaskin, F., and Kress, Y. (1977). *J. Biol. Chem.* **252**, 6918.
Ghosh, S., Paweletz, N., and Ghosh, I. (1978). *Chromosoma (Berlin)* **65**, 293.
Gibbons, I. R., Cosson, M. P., Evans, J. P., Gibbons, B. H., Houck, B., Martinson, K. H., Sale, W. S., and Tang, W. Y. (1978). *Proc. Natl. Acad. Sci. U.S.A.* **75**, 2220.
Gould, R. R. (1975). *J. Cell Biol.* **65**, 65.
Gould, R. R., and Borisy, G. G. (1977). *J. Cell Biol.* **73**, 601.
Heidemann, S. R., and Kirschner, M. W. (1975). *J. Cell Biol.* **67**, 105.
Heidemann, S. R., and Kirschner, M. W. (1978). *J. Exp. Zool.* **204**, 431.
Heidemann, S. R., Sander, G., and Kirschner, M. W. (1977). *Cell* **10**, 337.
Herman, I. M., and Pollard, T. D. (1978). *Exp. Cell Res.* **114**, 15.
Himes, R. H., Burton, P. R., and Gaito, J. M. (1977). *J. Biol. Chem.* **252**, 6222.
Inoué, S. (1961). *In* "The Encyclopedia of Microscopy" (G. L. Clarke, ed.), pp. 480–485. Reinhold, New York.
Inoué, S., and Ritter, H. (1978). *J. Cell Biol.* **77**, 655–684.
Inoué, S., Borisy, G. G., and Kiehart, D. P. (1973). *Biol. Bull.* **145**, 441.
Kane, R. E. (1962). *J. Cell Biol.* **12**, 47.
Kato, K. H., and Sugiyama, M. (1971). *Dev. Growth Diff.* **13**, 359.
Klein, I., Willingham, M., and Pastan, I. (1978). *Exp. Cell Res.* **114**, 229.
Kobayashi, T., Martensen, T., Nath, J., and Flavin, M. (1978). *Biochem. Biophys. Res. Commun.* **81**, 1313.

Kubai, D. F. (1975). *Int. Rev. Cytol.* **43**, 167.

Lazarides, E. (1976). *J. Supramol. Struct.***5**, 531.

Lazarides, E., and Weber, K. (1975). *Proc. Natl. Acad. Sci. U.S.A.* **71**, 2268.

Lee, J. C., Tweedy, N., and Timasheff, S. N. (1978). *Biochemistry* **17**, 2783.

Loeb, J. (1913). "Artificial Parthogenesis and Fertilization." Chicago Univ. Press, Chicago, Illinois.

Maekawa, S., and Sakai, H. (1978). *J. Biochem.* **83**, 1065.

Maller, J., Poccia, D., Nishioka, D., Kidd, P., Gerhardt, J., and Hartman, H. (1976). *Exp. Cell Res.* **99**, 285.

Manton, I., Kowallik, K., and von Stosch, H. A. (1970a). *J. Cell Sci.* **6**, 131.

Manton, I., Kowallik, K., and von Stosch, H. A. (1970b). *J. Cell Sci.* **7**, 407.

Marcum, J. M., Dedman, J. R., Brinkley, B. R., and Means, A. R. (1978). *Proc. Natl. Acad. Sci. U.S.A.* **75**, 3771.

Margolis, R. L., and Wilson, L. (1978). *Cell* **13**, 1.

Margolis, R. L., Wilson, L., and Kiefer, B. I. (1978). *Nature (London)* **272**, 450.

Marx, J. L. (1973). *Science* **181**, 1236.

Mazia, D. (1961). *In* "The Cell" (J. Brachet and A. E. Mirsky, eds.), Vol. 3, p. 134. Academic Press, New York.

McGill, M., Highfield, D. P., Monahan, T. M., and Brinkley, B. R. (1976). *J. Ultrastruct. Res.* **57**, 43.

Miki-Noumura, T. (1977). *J. Cell Sci.* **24**, 203.

Moens, P. B., and Rapport, E. (1971). *J. Cell Biol.* **50**, 344.

Molè-Bajer, J. (1975). *Cytobios* **13**, 117.

Naruse, H., and Sakai, H. (1978). *J. Biochem.* **83**, 1265.

Olmstead, J. B., and Borisy, G. G. (1973). *Biochemistry* **12**, 4782.

Pera, F., and Rainer, B. (1973). *Chromosoma (Berlin)* **42**, 71.

Pera, F., and Schwarzacher, H. G. (1969). *Chromosoma (Berlin)* **26**, 337.

Peterson, J. B., and Ris, H. (1976). *J. Cell Sci.* **22**, 219.

Peterson, J. B., Gray, R. H., and Ris, H. (1972). *J. Cell Biol.* **53**, 837.

Peterson, S. P., and Berns, M. W. (1978). *J. Cell Sci.* **34**, 289.

Peterson, S. P., and Berns, M. W. (1979) *Exp. Cell Res.* **120**, 223.

Phillips, S. G., and Rattner, J. B. (1976). *J. Cell Biol.* **70**, 9.

Pickett-Heaps, J. D. (1969). *Cytobios* **6**, 257.

Pickett-Heaps, J. D., and Bajer, A. S. (1977). *Cytobios* **19**, 171.

Pickett-Heaps, J. D., and Tippit, D. H. (1978). *Cell* **14**, 455.

Randall, J. T., and Disbrey, C. (1965). *Proc. R. Soc. B.* **162**, 473.

Rattner, J. B., and Berns, M. W. (1976a). *Chromosoma (Berlin)* **54**, 387.

Rattner, J. B., and Berns, M. W. (1976b). *Cytobios* **15**, 37.

Rattner, J. B., and Phillips, S. G. (1973). *J. Cell Biol.* **57**, 359.

Rebhun, L. I., Rosenbaum, J., Lefebvre, P., and Smith, G. (1974). *Nature (London)* **249**, 113.

Rieder, C. L. (1979). *J. Cell Biol.* **80**, 1.

Ritter, H., Inoué, S., and Kubai, D. (1978). *J. Cell Biol.* **77**, 638.

Robinow, C. F., and Marak, J. (1966). *J. Cell Biol.* **29**, 129.

Roos, U.-P. (1973). *Chromosoma* **40**, 43.

Sloboda, R. D., and Warbasse, L. H. (1978). *J. Cell Biol.* **79**, 301a.

Sluder, G., and Begg, D. A. (1978). *J. Cell Biol.* **79**, 299a.

Smith-Sonneborn, J., and Plaut, W. (1967). *J. Cell Sci.* **2**, 225.

Snell, W. J., Dentler, W. L., Haimo, L. T., Binder, L. I., and Rosenbaum, J. L. (1974). *Science* **185**, 357.

Snyder, J. A., and McIntosh, J. R. (1975). *J. Cell Biol.* **67**, 744.

Stubblefield, E., and Brinkley, B. R. (1966). *J. Cell Biol.* **30**, 645.
Stubblefield, E., and Brinkley, B. R. (1967). *In* "The Origin and Fate of Organelles" (K. B. Warren, ed.), pp. 175–218. Academic Press, New York.
Telzer, B. R., Moses, M. J., and Rosenbaum, J. L. (1975). *Proc. Natl. Acad. Sci. U.S.A.* **72**, 4023.
Tucker, R. W., Sanford, K., and Frankel, F. R. (1978a). *Cell* **13**, 629.
Tucker, R. W., Fujiwara, K., Stiles, C. D., Scher, C. D., and Pardee, A. B. (1978b). *J. Cell Biol.* **79**, 302a.
Weatherbee, J. A., Luftig, R. B., and Weihing, R. R. (1978). *J. Cell Biol.* **78**, 47.
Weingarten, M. D., Lockwood, A. H., Hwo, S.-Y., and Kirschner, M. W. (1975). *Proc. Natl. Acad. Sci. U.S.A.* **72**, 1858.
Weisenberg, R. C. (1972). *Science* **177**, 1104.
Went, H. A. (1977a). *Exp. Cell Res.* **108**, 63.
Went, H. A. (1977b). *J. Theor. Biol.* **68**, 95.
Wheatley, D. N. (1974). *J. Gen. Virol.* **24**, 395.
Wilson, E. B. (1925). "The Cell in Development and Heredity." Macmillan, New York.
Wissinger, W., and Wang, R. J. (1978). *J. Cell Biol.* **79**, 284a.
Wolfe, J. (1972). *Adv. Cell Mol. Biol.* **2**, 151.
Zackroff, R. V., Rosenfeld, A. C., and Weisenberg, R. C. (1977). *J. Supramol. Struct.* **5**, 577.

INTERNATIONAL REVIEW OF CYTOLOGY, VOL. 64

The Structural Organization of Mammalian Retinal Disc Membrane

J. Olive

Laboratoire de Microscopie Electronique, Institut de Recherche en Biologie Moléculaire du C.N.R.S., Université Paris VII, Paris, France

I.	Introduction	107
II.	General Aspects of the Organization of Photoreceptor Membranes	108
	A. General Morphology	108
	B. Outer Segment Structure	109
	C. Isolation of Outer Segments and Disc Membranes	109
	D. Chemical Composition of the Disc Membrane	111
	E. Rhodopsin Properties	121
III.	Organization of the Rhodopsin Molecules into the Lipid Bilayer	124
	A. Transverse Organization of the Rhodopsin within the Bilayer	124
	B. Tangential Organization	139
	C. Size and Nature of the Intramembranous Particles	143
IV.	Dynamic Aspects of the Disc Membrane	148
	A. Rhodopsin Mobility	148
	B. Lipid Fluidity	149
	C. Conformational Changes during Bleaching	150
	D. Renewal of the Membrane	153
V.	Conclusion	156
	References	162

I. Introduction

Visual function in mammals relies on the presence of two types of photoreceptor cells that, because of their shape, are called cones and rods. It has been known for some time that photoreceptor cells contain colored substances that absorb light selectively and respond to a signal by changing color. Several molecular species of visual pigments with absorption maxima within a fairly narrow range, have been identified in association with these substances. The molecular structures of these pigments, called rhodopsins, are very similar. They contain water, insoluble glycoprotein (opsin), and a chromophore (retinal). The visual pigment molecules in the disc membranes, which form the outer segment of photoreceptor cells are closely associated with lipids.

The rhodopsin molecules are the initial site of interaction with light. The first effect observed is the isomerization of the chromophore from the 11-cis to the all-trans form. This isomerization triggers a series of reactions culminating in the hyperpolarization of the outer segment plasma membrane and the transmission of a signal by the synapses at the base of the photoreceptor cells. The major mechanisms of the transduction and amplification of the light signal into a nervous impulse remain to be elucidated, and several hypotheses have been put forward for this purpose. For instance, Yoshikami and Hagins (1971, 1973) have proposed that the permeability of the outer segment plasma membrane is controlled by the release of CA^{2+} ions from the discs. Yet the possible role of the photopigment in the regulation of CA^{2+} release is not easily apparent. Conversely, new experimental data stress the role of cyclic guanosine monophosphate (cGMP) as the transmitter molecule. In any case, to clarify these mechanisms, whatever the molecular mechanism could be, prior knowledge of the molecular organization of disc membranes, and particularly of the structure of the visual pigment and of the changes induced by light absorption, are required.

Most of the investigations on the molecular organization of disc membranes and the organization of visual pigment have been carried out, either on isolated rods, or on whole retinas, without any distinction being made between cones and rods. For this reason, this article is restricted to the discussion of the latter type of cells, except for certain specific points such as renewal. In addition, when experiments have not yet been carried out with mammalian retinas, relevant results from studies of amphibians have also been included.

II. General Aspects of the Organization of Photoreceptor Membranes

A. GENERAL MORPHOLOGY

Mammalian retinas generally possess two types of photoreceptor cells, rods and cones, which have complementary functions. Rods are very sensitive to light and give scotopic, achromatic vision in low intensity light, whereas cones function under daylight conditions (photopic vision) with a sense of color (cf. Dartnall, 1972). There are three types of cones, containing pigments with spectral maxima at 440, 535, and 575 nm that correspond to the receptor-peak sensitivities in the blue, green, and red regions, respectively. Rods contain only one photopigment with a maximum at 500 nm (Wald and Brown, 1958).

Rods and cones are specialized elongated cellular elements that are closely associated with a network of neuronal cells by means of synaptic terminals (metabolic requirements are provided by the retinal blood vessels) (Fig. 1). The study of the synaptic connections between receptor cells and neurons in the layer of the bipolar cells has revealed rather extensive contact relations (Sjöstrand, 1953, 1958, 1961; De Robertis and Franchi, 1956; Dowling, 1970). Using freeze–fracture technique, Raviola and Gilula (1975) have shown that, in the synapses of cone pedicles and rod spherules, the horizontal cell dendrites are interconnected and connected to the plasma membrane of the photoreceptor endings by "gap junctions." Moreover, the synaptic endings of cone cells make specialized junctions with each other and with the endings of rod cells. These junctions probably mediate electronic coupling between neighboring photoreceptor cells (Raviola and Gilula, 1973).

The rods and cones are axially oriented to the incident light. As shown on the schematic drawing (Fig. 1), they consist of two distinct cellular segments: the inner and outer segments, connected by a cilium. In both rods and cones, the outer segment consists primarily of stacked photoreceptor membranes, but neither mitochondria nor other subcellular organelles dealing with protein synthesis are present. In contrast, the inner segment possesses the complete metabolic machinery of the cell (nucleus, Golgi complex, mitochondria, endoplasmic reticulum), mostly concentrated in the upper part of the inner segment, close to the connecting cilium.

B. OUTER SEGMENT STRUCTURE

The fine structure of the outer segment has been reviewed by Cohen (1972), Knowles and Dartnall (1977), and Rosenkranz (1977).

The outer segment contains a pile of hundreds of stacked, parallel, flat sacs or discs. These sacs are formed by invagination of the plasma membrane near the base of the outer segment. In rods, the discs are membranous structures, enclosed by the plasma membrane, and free within the outer segment except for those close to the connecting cilium, which are still continuous with the invaginating plasma membrane. The disc membrane in the rod, therefore, separates two compartments: the intradisc space inside the sacs, and the interdisc cytoplasmic space lying between them. Conversely, in cones, the plasma membrane is invaginated over the entire length of the outer segment so that the pile of parallel stacked membrane sheets is formed by a single continuous membrane (Fig. 1).

Fluorescent dyes, such as procion yellow, have been shown to penetrate into the parallel invaginations of the cone outer segments, but not into the rod disc membranes as would be expected of such an organization (Laties and Liebman, 1970). In a similar way, lanthanum permeates through the complex invagination of the plasma membrane characteristic of cones, but does not penetrate into the intradisc spaces of rods (Cohen, 1968).

C. ISOLATION OF OUTER SEGMENTS AND DISC MEMBRANES

In order to study the chemical composition of a biological membrane fraction, it is essential to start with a pure membrane preparation. Several techniques for the isolation and purification of the outer segments and the disc membranes have been described.

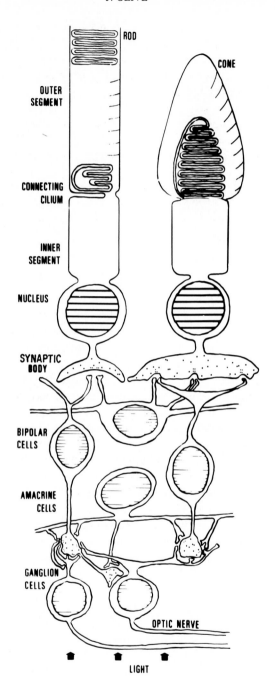

The outer segments are relatively fragile and can be sheared from the retina fairly easily at the ciliar connection. It is only necessary to shake the retinas in an appropriate buffer, by hand or in a mechanical shaker. Then, the larger retinal debris can be removed by filtrating the suspension through a screen (De Grip *et al.*, 1972). The buoyant density of the resulting free outer segments is low (1.08–1.10) compared to that of other membranous organelles of the cell. This density difference favors the purification of the outer segments from other cellular components, either by flotation in discontinuous sucrose gradients (Saito, 1938; Papermaster and Dryer, 1974), in equilibrium density gradients (Collins *et al.*, 1952; McConnell, 1965; DeGrip *et al.*, 1972; Anderson *et al.*, 1975; Knowles, 1976), or in sucrose step-gradients (Papermaster and Dryer, 1974). Ficoll gradients have also been used (Lolley and Hess, 1969; Bownds *et al.*, 1974).

Free photoreceptor disc membranes can be obtained from the isolated outer segments by osmotic shock lysis of the plasma membrane with water or low salt medium. Osmotically intact discs have been purified from protein contaminents, fragmented discs, and other membranous debris by flotation on 5% Ficoll (Smith *et al.*, 1975).

D. CHEMICAL COMPOSITION OF THE DISC MEMBRANE

The disc membranes consist of lipids and proteins in a bilayer configuration, where amphipatic proteins are both exposed at the membrane surfaces and integrated within the hydrophobic lipid core of the leaflets. In Table I, the composition of disc membranes and other membranes are compared. From this comparison, the photoreceptor membrane can be seen to be characterized by a relatively low protein: lipid ratio. However, since different authors obtain large variations in measuring the relative protein and lipid contents of the photoreceptor membrane (Table II), some reservation on the validity of this conclusion seems prudent. Nevertheless, it is clear that during the last fifteen years, progress has been made with respect to better characterized and less contaminated disc membrane preparations.

1. *Lipids*

The most striking aspects of the disc membrane lipid composition are the high content of unsaturated fatty acids and the very low sphingomyelin content. The

---

FIG. 1. Diagrammatic presentation of rod and cone cells connected to the different nerve cells. Light passes from the bottom to the top across the nerve cells and the inner segments of the photoreceptor cells before reaching the outer segments. The opening in the cone outer segment discloses the pile of parallel membrane sheets that are all part of one continuous (plasma) membrane. In the basal part of the rod outer segment is shown the region of the invaginated plasma membrane, whereas in the upper part, the pile of free-floating discs surrounded by the plasma membrane is revealed.

TABLE I

Composition of Various Biological Membranes

	Membrane component (% dry weight)			
	Proteins	Lipids	Phospholipids	Cholesterol
Microsomes[a]	69	31	29	0.6
Golgi[b]	57	43	38	0.9
Mitochondria[c]				
External membrane	63	37	33	0.4
Internal membrane	75	25	—	2.2
Liver plasma membrane[d]				
(canalicular subfraction)	47.4	52.6	36.5	7.3
Disc membranes[e]	44	56	46	<3.0

[a] De Pierre and Ernster (1977).
[b] Zambrano et al. (1975).
[c] Levy and Sauner (1968).
[d] Kremmer et al. (1976).
[e] De Grip et al. (1979).

cholesterol content seems intermediate between that of plasma membranes and intracellular membranes (Table II). On a weight basis, at least 80% of the total lipids are phospholipids; cholesterol accounts for less than 3% (Table I); and several minor components have been detected (Mason et al., 1973). The major phospholipids are phosphatidylethanolamine, phosphatidylcholine, and phosphatidylserine: 44, 36, and 15%, respectively, expressed as percentage of lipid phosphorus (Table III).

Two main characteristics of the fatty acid composition of the disc membrane are that more than 50% are polyunsaturated and that a high proportion of docosahexaenoic fatty acid ($C_{22:6}$) is present (Table IV). Moreover, most of the phospholipids in the disc membrane have been shown to contain one polyunsaturated fatty acid chain that is almost always located at the 2-position of the glycerol moiety, while the other, at the 1-position, is more saturated (Anderson and Sperling, 1971). The occurrence of phospholipids with two polyunsaturated fatty acid chains should also be taken into consideration (F. J. M. Daemen, personal communication).

Another interesting aspect of the molecular properties of phospholipids in the disc membrane is that all do not show similar solubility properties. For instance, only 50% of them are easily extracted by hexane. Phosphatidylethanolamine is extracted preferentially, whereas phosphatidylcholine remains associated with the hexane insoluble fraction (Borggreven et al., 1970).

As in other membrane types, an attempt to demonstrate the asymmetric distribution of phospholipids in the bilayer has been carried out. Raubach et al.

(1974) have labeled the membrane amino groups, both with the membrane permeable reagent ethylacetimidate hydrochloride and the membrane impermeable reagent sodium isethionyl acetimidate hydrochloride. According to these authors, 70% of the lipid amino groups have been found on the outer surface of the membrane. Therefore, most of the phosphatidylethanolamine (major amino group bearing lipid) seems to be on the outside, while the bulk of the remaining phospholipid, phosphatidylcholine, is associated with the interior half of the bilayer. A confirmation of the asymmetric distribution of the phospholipids has recently been obtained by Crain *et al.* (1978), who have labeled the phospholipids with chemical probes. The major part of phosphatidylethanolamine has been found to be located on the outer surface of the disc membrane.

In spite of the indications for an asymmetric distribution of the phospholipids over the disc membrane, caution should be exerted in the interpretation of these studies. All preparations used were disc vesicles rather than stacked discs, with the inherent danger of inside-outside artifacts. In addition, chemical modification of membrane phospholipids may cause unwanted side effects.

2. Proteins

a. *Rhodopsin.* As indicated in Table II, the photoreceptor membrane contains between 40 and 50% of protein (dry weight). The most striking feature of the protein profile of the disc membrane is that, in contrast with most other biological membranes, 80 to 90% of the total protein content is represented by a single major component: rhodopsin. This conclusion has been derived from results of gel filtration of photoreceptor membrane protein solubilized in detergent (Hall *et al.*, 1969), and by the molar ratio between the amino acid content and the retinyl group (Bownds *et al.*, 1971). Furthermore, SDS-polyacrylamide gel electrophoresis of the photoreceptor membrane proteins clearly indicated the presence of a single major protein constituent (Daemen *et al.*, 1972); Robinson *et al.*, 1972; Heitzman, 1972). The molecular weight (MW) of rhodopsin has been calculated from different experimental approaches (Table V). From the

TABLE II

CHEMICAL COMPOSITION OF PHOTORECEPTOR MEMBRANES

Membrane component (% dry weight)		
Protein	Lipid	Reference
40	60	Fleischer and McConnell (1966)
52	48	Nielsen *et al.* (1970)
61	39	Borggreven *et al.* (1970)
44	56	De Grip *et al.* (1979)

TABLE III

PHOSPHOLIPID COMPOSITION OF BOVINE DISC MEMBRANES AS PERCENTAGE OF LIPID PHOSPHORUS

Phospholipid component[a] (% lipid phosphorus)

PC	PE	PS	PI	Sph	Reference
36	45	16	1.5	1	Anderson et al. (1975)
36	44	15	<1	<1	E. H. S. Drenthe (personal communication)

[a] Abbreviations: PC, phosphatidylcholine; PE, phosphatidylethanolamine; PS, phosphatidylserine; PI, phosphatidylinositol; Sph, sphingomyelin.

difference in sedimentation rate of digitonin micelles and digitonin micelles containing bovine rhodopsin, it seems to correspond to 40,000 (Hubbard, 1954). However, using gel filtration chromatography on agarose, Heller (1968) found the MW of bovine rhodopsin to be 28,000–30,000, whereas Robinson et al. (1972) found that of the bullfrog to be 36,000–40,000. More conclusive results are those of Daemen et al. (1972), who have determined the MW of bovine rhodopsin by two independent methods. In the first case, rhodopsin, solubilized in detergent and purified by gel filtration chromatography, was analyzed by quantitative amino acid analysis after hydrolysis. A MW of 39,000 ± 900 was obtained. Second, calibrated SDS-polyacrylamide gel electrophoresis of rod outer segment (ROS) preparations, both before and after enzymatic delipidation gave a value of 39,600 ± 1,000. More recently, the MW of bovine rhodopsin has

TABLE IV

FATTY ACID COMPOSITION OF BOVINE DISC MEMBRANES AS PERCENTAGE OF TOTAL FATTY ACIDS

Phospholipid	Fatty acid (mole percent)							
	$C_{16:0}$	$C_{18:0}$	$C_{18:1}$	$C_{18:2}$	$C_{20:4}$	$C_{22:5}$	$C_{22:6}$	$C_{24:(4/5)}$
Total phospholipids[a]	21	25	4	—	3	5	35	2.2
Phosphatidylcholine	36	22	5	1	4	6	26	—
Phosphatidylethanolamine	16	30	5	1	3	5	40	—
Phosphatidylserine	1	20	1	—	1	8	47	14
Total phospholipids[b]	15	20	3	1	5	7	46	—
Total phospholipids[c]	18	22	4	1	3	3	44	2
Phosphatidylcholine	31	19	5	1	3	2	36	—
Phosphatidylethanolamine	13	25	4	1	2	3	50	—
Phosphatidylserine	4	21	2	—	4	5	48	13

[a] Anderson et al. (1975).
[b] Hendriks et al. (1976).
[c] E. H. S. Drenthe, personal communication.

TABLE V

MOLECULAR WEIGHT DETERMINATION FOR VISUAL PIGMENTS

Method[a]	Molecular weight	Reference
UC	40,000	Hubbard (1954)
GE	35,000	Cavanagh and Wald (1959)
AAA; GFC	27,700	Heller (1968)
AAA	28,000	Shichi et al. (1969)
AAA; GE	38,950 ± 900	Daemen et al. (1972)
GE	38,000 ± 2000	Heitzmann (1972)
UC	35,000 ± 2000	Lewis et al. (1974)

[a] Methods: UC, Ultracentrifugation; AAA, Amino acid analysis; GFC, Gel filtration chromatography; GE, Gel electrophoresis.

been measured as 35,000 by analytical ultracentrifugation (Lewis et al., 1974). This value is in good agreement with those derived from other analytical methods.

The amino acid composition of rhodopsin has been the object of several investigations (Heller, 1968; Shichi et al., 1969; Zorn and Futterman, 1971; Robinson et al., 1972; De Grip, 1974). The results of amino acid analysis show that rhodopsin, in common with other intrinsic membrane proteins (Component III of the glycophorin, for example), contains a high proportion of nonpolar amino acids (cf. Table IV).

Amidination of about 97% of the free membrane protein lysine groups has little effect on the 500 nm absorption or on the regeneration capacity of rhodopsin with 11-cis retinal. The protein amino groups react slightly with uncharged reagents, even with large substituents, as long as the positive charge of the modified lysine residues is retained (De Grip et al., 1973a). This may indicate that the bulk of the amino groups are externally located and make an important contribution to the tertiary structure. Recently, it has been proved that the amino terminus of bovine rhodopsin is blocked and has the sequence of X-Met-Asn(CHO)-Gly-Thr-Glu-Gly-Pro-Asn-Phe-Tyr-Val-Pro-Phe-Ser-Asn(CHO)-Lys-Thr-Gly-Val-Val-Arg. According to Hargrave and Fong (1977), the CHO-groups represent the site of carbohydrate attachment exposed to the hydrophilic environment. The same authors have determined that the C-terminal sequence of rhodopsin is Val-Ser-Lys-Thr-Glu-Thr-Ser-Gln-Val-Ala-Pro-Ala.

In common with other amphipatic membrane proteins engaged in transmembrane operations, rhodopsin is a glycoprotein. A single glycopeptide containing nine amino acid residues has been obtained from chromatography of a peptic digest of retinal pigment (Heller and Lawrence, 1970; Plantner and Kean, 1976). The oligosaccharide moiety of this glycopeptide consists of three residues of

TABLE VI
PROPERTIES OF MEMBRANE TRANSPORT PROTEINS[a]

	Na^+-K^+-ATPase	Ca^{2+}-ATPase	Anion-exchange protein	Acetylcholine receptor	Rhodopsin
Molecular weights of component polypeptides	α—90,000 β—40,000	α—100,000	α—90,000	α—40,000 β—48,000 γ—58,000 δ—64,000 ϵ—105,000	α—38,000
Glycoproteins	β		α	$\alpha(\beta$—$\epsilon)$?	α
Probable structure and molecular weight of the protein part of the enzyme	$\alpha_2\beta_2$ 260,000	$(\alpha_2\,?)$ (200,000?)	α_2 180,000	ϵ_2 or $\alpha_2\beta_2$ 240,000	$(\alpha_2$ to $\alpha_4\,?)$ (76,000–152,000?)
Transmembrane arrangement	α		α		α
Detergent binding (mg/mg of protein)	0.28	0.20	0.77	0.7	1.10
Relative hydrophobic surface area of subunit	0.20–0.24	0.2–0.25	0.5–0.65	0.5–0.6	0.54

[a] From Guidotti (1977).

glucosamine and three residues of neutral sugars [mannose, not galactose as previously proposed by Heller (1968)]. More recently, it has been shown that the N-terminal fragment contains two oligosaccharides (Hargrave and Fong, 1977).

The rhodopsin molecule has receptors that bind Concanavalin A (Con A) and wheat germ agglutinin (WGA). Bound Con A and WGA have been used as specific markers to determine the vectorial and asymmetric organization of the rhodopsin molecules within the intact disc membrane and in detergent-rhodopsin micelles (Renthal et al., 1973; Steinemann and Stryer, 1973). From these observations, however, it is not clearly apparent whether the lectin binds to the cytoplasmic or to the intradisc membrane surface. The cytochemical experiments of Röhlich (1976), using ferritin-labeled Con A (Ft-Con A) on isolated and disrupted retinal rods, provide a strong indication that the lectin receptors are exposed at the intradisc surface of the disc membrane (cf. Fig. 9). However, even these elegant results suffer from uncertainty, since it is difficult to prove that the original sidedness and vectorial organization of the disc membrane have been retained after disruption and incubation of the discs. The only conclusion that can therefore be drawn from this sort of experiment is that the glycopeptide moiety of the rhodopsin probably extends into the hydrophilic environment of the membrane bilayer. The discussion on the vectorial organization and sidedness of the rhodopsin molecules will be continued in a later chapter.

b. *Other Proteins and Enzymatic Activity of Disc Membrane.*   The polyacrylamide gel electrophoresis pattern of isolated disc membranes is characterized by a major component, the rhodopsin, at 39,000 MW (Daemen et al., 1972). Other minor polypeptide bands are found in polyacrylamide gel electrophoretic experiments, particularly in the region of 50,000 MW. These minor polypeptides account for less than 15% of the proteins that remain when rhodopsin has been extracted from the total protein content of disc membrane.

The question remains whether or not some of all of these minor protein components are associated with a specific function of the photoreceptor membrane, in particular, with the phototransduction process. At least some of these minor polypeptides are associated with enzyme activities that can be detected in isolated rod membrane preparations.

1. The rod outer segments contain highly active, light-dependent enzymes that synthesize and degrade 3′,5′-cyclic-guanosine monophosphate (cGMP): respectively, guanylate cyclase and phosphodiesterase (Pannbacker et al, 1972; Miki et al., 1973; Goridis and Virmaux, 1974; Manthorpe and McConnell, 1975; Virmaux et al., 1976; Goridis et al., 1977; Sitaramayya et al., 1977a; Wheeler et al., 1977). High guanylate cyclase activity has been found both in mouse photoreceptor cells isolated by microdissection (Schmidt and Lolley, 1973), and in purified outer segment fractions (Zimmerman et al., 1976).

The phosphodiesterase is activated by light in the presence of guanosine triphosphate (GTP) only when the enzyme is bound to the disc membrane; in fact, the eluted form is inactive. It is estimated that the molar ratio of phosphodiesterase to rhodopsin in the rod outer segment is approximately 1:900 (Miki et al., 1975). Practically all the phosphodiesterase molecules are already activated when only 0.1% of the pigment is bleached, and consequently the cGMP level decreases very rapidly in the receptor (Goridis et al., 1977). A correlation between light-dependent changes in cGMP level and the plasma membrane permeability has been established (Woodruff et al., 1977), and very recent kinetic studies of phosphodiesterase activation and deactivation suggest that this enzyme has a key role in the transduction and amplification mechanism linking the bleaching of the rhodopsin with the hyperpolarization of the photoreceptor plasma membrane. The credibility of this assumption is enhanced by the elegant experiment of Yee and Liebman (1978). These authors, by continuously monitoring the stoichiometric proton release associated with cyclic nucleotide hydrolysis, have found that the activation of phosphodiesterase is amplified by a serial mechanism, which results in a highly specific gain of cGMP hydrolysis. One molecule of bleached rhodopsin may serially trigger the activation of as many as 500 phosphodiesterase molecules, and the highly specific GTP requirement appears to be stoichiometric with the relative amount of disc membrane used in the experiment. These experiments point to the role of the proton release in the coupling mechanism of the light signal with the electrical change of the plasma membrane.

2. The presence of an adenylate cyclase (Miller et al., 1971; Bitensky et al., 1972; Bounds et al., 1974) has been contested by Hendriks et al. (1973). The low activity that has been found seems more likely to be due to a contaminant than to an intrinsic disc membrane-associated activity. In agreement with this latter affirmation, analyses of retinal homogenate fractions have shown that adenylate cyclase and cAMP phosphodiesterase activities are absent in the outer segment bands obtained by isopycnic centrifugation in density gradients of sucrose (Zimmerman et al., 1976).

3. The evidence is stronger for the presence of a NADP-dependent retinoldehydrogenase that converts retinol to retinaldehyde (or vice versa) (Bridges, 1962; Futterman, 1963; De Pont et al., 1970; Kissun et al., 1972; Zimmerman et al., 1975). Retinoldehydrogenase activity has been shown to be the same in native and photolysed rhodopsin, indicating that occupation of the chromophoric site by 11-cis retinaldehyde has no influence on the enzymatic conversion of retinaldehyde. Therefore, this enzymatic reduction must occur at a site distinct from the chromophoric site (Rotmans, 1973). Evidence has been presented that an amino group plays an essential role in the binding of the substrate, and a specific sulfhydryl group also appears to be necessary for the binding of the coenzyme (De Grip et al., 1975). Proof that the enzyme is an entity distinct from

rhodopsin derives from the fact that it can be separated from rhodopsin by preferential extraction with a nonionic detergent (Etingof et al., 1972).

4. It has been reported that rhodopsin is phosphorylated after exposure to light, by means of an ATP-dependent reaction (Bownds et al., 1972; Kühn and Dreyer, 1972), and a water soluble kinase that can phosphorylate rhodopsin has been detected in ROS fragments (Kühn et al., 1973; Chader et al., 1976). Light does not activate protein kinase, and only the bleached form of rhodopsin seems to be a substrate of the kinase (Weller et al., 1975a; Frank and Buzney, 1975). Although there is general agreement that the protein kinase is a membrane bound enzyme, conflicting evidence exists with regard to substrate specificities of the enzyme. Different groups report that the enzyme phosphorylates several kinds of substrate with ATP, such as histones or protamines (Kühn et al., 1973; Chader et al., 1976). But more recently, a purified kinase, isolated from bovine ROS has been shown to be specific for rhodopsin (Shichi and Somers, 1978).

5. The presence, concentration, and localization of Na^+ and K^+ pumps in ROS is a controversial question and contradictory results have been obtained in the identification of NA^+-K^+-ATPase. The presence of this latter activity has been detected in ROS (Bonting et al., 1964; Etingof et al., 1972; Hemminski, 1975); but according to Daemen (1973) and Hendriks et al., (1973), the activity of the Na^+-K^+-ATPase is low, compared with the properties of similar pumps in other membranes and with the amount of rhodopsin protein present in the outer segment; probably too low to allow its effective functioning in such a role. Recently, strong evidence has been presented that the Na^+-K^+-ATPase activity in the retina is not associated with the disc membrane (Zimmerman et al., 1976).

On the other hand, it is obvious that the maintenance of an ionic gradient in the disc–interdisc space is required for the phototransduction sequence involving electrical changes of the ROS membrane (Etingof et al., 1970; Govardovskii, 1971). By lysis of the ROS, Mg^{2+}- and Na^+-K^+-ATPase have been found to be higher than in intact rods. It has therefore been proposed by Hemminski (1975) that the active sites for ATP hydrolysis are associated with the external surface of the disc membrane. This model would then propose that the Na^+ ions accumulate within the discs, whereas the K^+ ions are transported outside them. It remains to be established whether the Na^+-K^+-ATPase associated with the rod plasma membrane would in turn control Na^+ and K^+ transport between the cytoplasmic domain of the rod and the outside. The precise mechanisms and regulation linking the distribution of Na^+ and K^+ ions in the outer segment with amplification of the phototransduction process are also not fully understood.

3. *Ion Content*

We have already stressed the possible role of proton release associated with the activation of phosphodiesterase in relation with the transduction and amplification mechanisms of the light signal. We must recall that before the recent conclusions of Yee and Liebman (1978), several candidates, and in particular calcium ions, were put forward for the role of the messenger in the phototransduction. At this point, we think that it is important to refer briefly to the problem of the Ca^{2+} binding to the disc membrane, since it may still be relevant to the phototransduction mechanism. A number of reports have described the calcium content and translocation of isolated ROS in darkness and the light-induced calcium release upon photoexcitation of rhodopsin (Liebman, 1974; Hendriks et al., 1974; Hagins and Yoshikami, 1975; Yoshikami and Hagins, 1976; Farber and Lolley, 1976; Fishman et al., 1977; Hendriks et al., 1977; Schnetkamp, et al., 1979; Kaupp et al., 1979).

Although widely divergent results have been obtained concerning the Ca^{2+} content of the ROS, which, according to different groups, may vary between 0.1 and 4 mole Ca/mole rhodopsin, most of the experimental evidence points to the fact that Ca^{2+} is localized within the disc membrane. This conclusion has been reached mainly by a direct visualization of a precipitate of calcium antimonate within the disc (Fishman et al., 1977).

It has been reported that calcium may occur in two different forms: one being strongly bound to a sedimentable material containing rhodopsin-rich membranes; the other, soluble (Hendriks et al., 1974). It has been further demonstrated that during illumination of ROS, two mechanisms seem to regulate calcium transport across the membrane: a passive ATP-independent translocation by means of a sodium-calcium (Na–Ca) exchange carrier system; and a net transport of calcium into the rod sacs, which seems to require ATP (Schnetkamp et al., 1977). The major ATP-independent transport mechanisms are Ca–Ca exchange and Na-stimulated Ca efflux, presumably by Na–Ca exchange. Illumination of intact ROS causes a significant shift from the bound fraction to the soluble one, whereas the total calcium content does not vary significantly.

Moreover, the results on calcium release are conflicting. In fact, Szuts and Cone (1977) reported stoichiometries of calcium release: rhodopsins bleached ranging from 10–1000; Hemminski (1975) found 1 calcium released per 6 rhodopsins bleached; and Weller et al. (1975a), about 1 calcium released for every 100 rhodopsins bleached. A recent study using the calcium indicator dye Arsenazo III, together with a sensitive flash kinetic photometric technique to detect fast calcium release, unambiguously established that photoexcitation of rhodopsin results in calcium release from intradiscal binding sites (Kaupp et al., 1979). Since, according to these authors, calcium release does not appear in the cytoplasmic domain unless the disc membrane is made permeable by calcium ionophores, we may then conclude that the role of the calcium ions as a transmitter is rather improbable.

Apart from the calcium, magnesium also seems to be present in (frog) rod outer segments at appreciable levels (Hendriks et al., 1974).

E. Rhodopsin Properties

1. *Absorption Spectrum*

Rhodopsin consists of a protein opsin and a chromophoric group that is the aldehyde of Vitamin A, 11-cis retinal (Wald and Brown, 1950).

Rhodopsin has a typical absorption spectrum characterized, before illumination, by three peaks with maxima at 500 nm (α), 350 nm (β), and 278 nm (γ) (Fig. 2). The γ peak is due to aromatic amino acids of the opsin part of the molecule, whereas the α and β peaks are associated with the chromophoric group. The maximum at 500 nm results from the association of the chromophoric group and opsin and is lost when these components are separated on bleaching. It is replaced by a peak with a maximum at 380 nm, accompanied by a color change from red to yellow.

The binding of opsin to 11-cis retinaldehyde is very specific, and it enhances the stability of opsin to heat and chemical attack. Hubbard (1954) has established that the ratio between the chromophoric group and opsin is one retinaldehyde group per molecule of opsin, and it has been shown that the chromophoric group is attached exclusively to one lysine residue in the opsin molecule (Heller, 1968; Fager *et al.*, 1972; De Grip *et al.*, 1973b) An early suggestion (Collins, 1953; Morton and Pitt, 1957) that retinal is linked by a protonated aldimine bond in rhodopsin has been confirmed by Raman spectroscopy data (Rimai *et al.*, 1970;

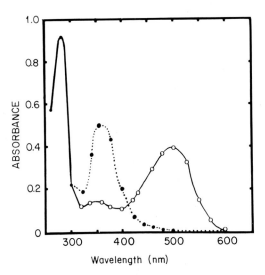

Fig. 2. Absorption spectrum of rod photoreceptor membrane solubilized in 1% digitonin solution, before (solid line) and after (dashed line) illumination. The presence of hydroxylamine during this light exposure led to the formation of retinylidene oxime (Re—$\overset{C}{\underset{H}{}}$=N—OH, λ_{max} = 365 nm). (From Van Breugel, 1977.)

Lewis *et al.*, 1973; Oseroff and Callender, 1974: Mathies *et al.*, 1976; Callender *et al.*, 1976). In rhodopsin, the aldimine bond appears to be inaccessible to various reagents such as hydroxylamine and is presumably shielded in a hydrophobic region of the protein (Wald and Hubbard, 1960; Bownds and Wald, 1965; Poincelot *et al.*, 1970; Daemen *et al.*, 1971).

2. *Photolysis*

Irradiation of rhodopsin isomerizes retinal from the 11-cis to the all-trans configuration (Hubbard and Wald, 1952), yielding a series of transient intermediates, ending with the hydrolysis of retinal from opsin (Fig. 3). The intermediates have been spectrally identified by freezing the solutions by inhibiting the reactions by chemical means, or by picosecond kinetic studies.

The first four steps proceed very rapidly, metarhodopsin II being formed in about 1 millisecond at physiological temperature. The apparent risetime of prelumirhodopsin, the first metastable intermediate produced in the visual pathway, has been observed by picosecond studies of rhodopsin in low-temperature glasses (Peters *et al.*, 1977) or at room temperature (Sundstrom *et al.*, 1977). The data suggest that the initial photochemical steps in the visual process is not cis-trans isomerization but rather proton translocation toward the Schiff base nitrogen of the retinal chromophore. The following reactions are slower, taking minutes at room temperature. It is generally agreed that in the metarhodopsin II stage the all-trans retinal remains linked to the original lysine residue (Rotmans *et al.*, 1974), whereas, upon decay of metarhodopsin II, the chromophore migrates to other sites.

The transition, metarhodopsin I to metarhodopsin II, is accompanied by relatively large conformational changes of the protein during which hidden hydrophilic groups of the protein become exposed to the surrounding medium, and possibly by a change of the electric dipole moment of the pigment protein (Petersen and Cone, 1975). Metarhodopsin III formation occurs parallel to the decay of metarhodopsin II into retinaldehyde or retinal and opsin (Matthews *et al.*, 1963; Baumann, 1972). The specific NADP-dependent retinol oxidoreductase, present in the photoreceptor membrane (Lion *et al.*, 1975; Zimmerman *et al.*, 1975; Zimmerman, 1976), reduces the liberated chromophore to retinal, which appears to diffuse to the pigment epithelium where it is stored as the palmitate or stearate ester (Hubbard and Dowling, 1962; Futterman and Andrews, 1964).

3. *Shape oj the Rhodopsin Molecule*

The shape of the rhodopsin molecule embedded in the lipid bilayer is still a matter of debate. Freeze-fracture experiments and even X-ray diffraction, though giving some indication as to the positioning of rhodopsin within the bilayer, are still insufficient to outline the shape of the photoreceptor unambiguously. Further aspects of this problem will be discussed later.

Attempts to delineate the shape of the rhodopsin molecule in a lipid-like environment have been carried out using rhodopsin-detergent complexes. The shape and size of the detergent micelles and an estimation of the dimensions of the protein and detergent regions of the complex have been established using both energy transfer techniques (Wu and Stryer, 1972; Steineman *et al.*, 1973a; Renthal *et al.*, 1973) and X-ray and neutron scattering studies (Sardet *et al.*, 1976). Energy transfer fluorescence results in bovine rhodopsin-digitonin solution showed an apparent distance between the energy donor site and 11-cis

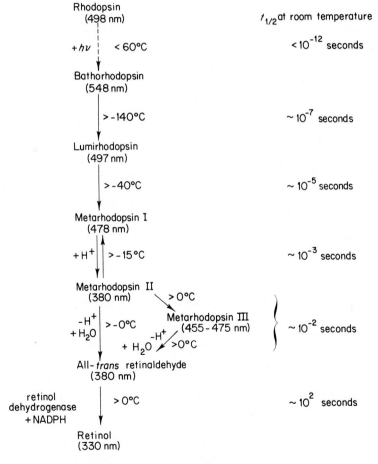

FIG. 3. Sequence of intermediates after illumination of vertebrate (bovine) rhodopsin *in vitro*. The dotted arrow represents the photoreaction. All other arrows indicate thermal dark reactions. The intermediates are characterized by their absorbance maximum (nm). Also indicated are the temperatures above which the reactions can proceed, and on the right, approximate half-life times at room temperature. (From Van Breugel, 1977.)

retinal in the rhodopsin molecules of at least 75Å. This kind of result is consistent with a model where rhodopsin in detergent has an elongated shape.

This model received further support from results of X-ray and neutron scattering techniques applied to rhodopsin complexed with dodecyldimethylamine oxide (DDAO), Triton X-100, or Ammonyx. The value of the radius of gyration of the DDAO micelles (Osborne *et al.*, 1978) and of the Ammonyx micelles (Yeager *et al.*, 1976) has been found to be around 23 Å. Since the radius of gyration, the volumes of the protein and detergent portions, and their relative position within the complex were known, models could be tested. Sardet *et al.* (1976) and Sardet (1977), for DDAO-solubilized rhodopsin, proposed a shape intermediate between a prolate cylinder and a dumbbell. Those authors assumed that in all cases the rhodopsin molecule would be highly anisometric with one of its dimensions necessarily exceeding 80 Å.

From their studies of detergent-rhodopsin micelles, Osborne *et al.* (1978) also proposed a cylindrical model where the protein volume is asymmetrically located on both sides of the membrane. Only a rather small volume of the protein should be on the intradisc side of the membranous zone. The question remains as to whether or not the shape and dimensions of the molecule observed in detergent solution are retained when the photoreceptor protein is embedded within the lipid bilayer. Several pieces of evidence exist that support the idea that the rhodopsin shape is preserved in the membrane, i.e., the kinetics of proton exchange (Osborne, 1976; Osborne and Viala, 1978) and the results of proteolytic cleavage (Pober and Stryer, 1975), which produce the same effect regardless of whether the rhodopsin is complexed with detergent or part of the disc membrane. In addition, the ultraviolet light circular dichroism spectrum is unaltered by rhodopsin solubilization in nonionic detergents (Shichi and Shelton, 1974).

III. Organization of the Rhodopsin Molecules into the Lipid Bilayer

A. Transverse Organization of the Rhodopsin within the Bilayer

1. *Electron Microscopy of Thin Sections*

The interpretation of the ultrastructure of the rod disc membrane was distorted by the dominating postulate of the pure lipid bilayer model for all biological membranes. According to this model, the membrane core consists essentially of lipids, the protein being associated primarily with the hydrophilic surfaces of the leaflets. The unit membrane concept, which originated from studies of the peripheral nerve myelin, assumed that thin sections of membrane appear in electron micrographs as two electron dense layers each about 20 Å thick, separated by a light interzone of about 35 Å, making an unit that is about 75 Å thick.

The repeating lamellar unit seen by electron microscopy of myelin did correspond to the electron density profile derived from X-ray diffraction analysis.

In apparent contradiction with those data and with the unit membrane model, Sjöstrand (1953), Fernandez-Morán (1962), and then Nilsson (1964, 1965) have observed an intrinsic globular fine structure in retinal outer segment disc membrane. Even if their interpretation, that globular structures are lipoproteic micelles, is now considered erroneous, they were the first to establish that not all membranes show similar patterns. Later electron microscopic observations have shown that either globular structures or trilamellar patterns can be obtained according to the fixation and postfixation procedures used. For instance, retinal disc membranes have a uniform triple-layered structure if they are fixed with glutaraldehyde and OsO_4, whereas permanganate fixation and heavy impregnation with uranyl salts result in a globular appearance of the same type of membrane.

In general, the triple-layered pattern of different types of membrane is not symmetric. In some instances, the electron-dense layer close to the cytoplasm is thicker and stains more intensely than the outer half of the bilayer (Sjöstrand and Elfvin, 1962; Sjöstrand, 1963a). In other instances, the asymmetric appearance of the membrane elements consists of a thicker layer at the outer surface of the plasma membrane (Staehelin et al., 1972). The asymmetric appearance of the plasma membrane in thin sections probably reflects the vectorial organization and distribution of the membrane protein and lipid components (Benedetti et al., 1977). In contrast, disc membranes unvariably stain symmetrically (Fig. 4). From this observation, one could assume that there is no preferential distribution nor vectorial organization of the membrane constituents, consistent with the triple-layered aspect. This assumption is relevant to the question of whether the rhodopsin polypeptide chain is almost completely embedded within the hydrophobic core, and hence inaccessible to electron staining, or if it is associated with both leaflets of the bilayer. We must therefore conclude that electron microscopic observations on thin sections do not provide sufficient data to clarify the molecular organization of the photoreceptor membrane. In this respect, X-ray diffraction contributed, to an important extent, to the knowledge on the distribution of the various components in the photoreceptor membrane.

2. X-ray Diffraction

The periodicity of the disc membranes in the native cell makes them very suitable for studies by X-ray diffraction. Such studies have been performed in vivo (Webbs, 1972), on dissected "functioning" retinas (Chabre and Cavaggioni, 1973; Chabre, 1975; Schwartz et al., 1975), on strips of retina (Gras and Worthington, 1969; Blaurock and Wilkins, 1969; Corless, 1972), and on stacked discs (Blasie et al., 1969; Blasie and Worthington, 1969). X-ray diffraction has also been applied to artificially reconstituted rhodopsin lipid water systems (Chabre et al., 1972).

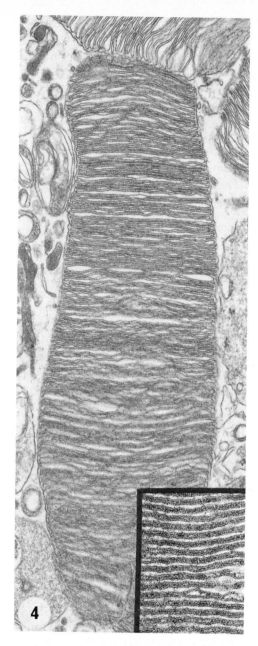

FIG. 4. Thin section of a bovine rod outer segment. This longitudinal section shows the pile of regularly stacked discs, which are not continuous with the plasma membrane. ×25,500. In the inset, higher magnification shows that electron-dense material is symmetrically distributed between the two halves of one membrane element. ×93,500. (The material has been embedded in Vestopal and stained with uranyl acetate and lead citrate.)

From the diffraction along the line normal to the plane of the long axis of the outer segment, an electron density profile of a single layer or unit cell can be derived. The interpretation of the molecular arrangement of the protein and the lipid within the cross section of the disc membrane was not readily apparent at early stages of X-ray diffraction analysis because of the lack of knowledge of other physical and biochemical features of the membrane constituents. Moreover, the interpretation of the X-ray diffraction pattern was difficult because of ambiguous definition of the phase to be attributed to the lamellar organization. Therefore, similar X-ray diffraction profiles were interpreted as indications of symmetric (Blaurock and Wilkins, 1969; Corless, 1972; Chabre and Cavaggioni, 1973) and asymmetric (Gras and Worthington, 1969; Worthington, 1971) models. These discrepancies were also due to the fact that either Fourier analysis or model parameters were used in order to extrapolate the relatative distribution of lipids and proteins in the bilayer. As Blaurock (1977) has recently summarized, most of the profiles obtained in different laboratories are consistent with the conclusion that the membrane profile is close to symmetry. In fact, as Blaurock (1977) stressed, an asymmetric distribution of protein in the bilayer would force an asymmetric distribution of the lipids.

Since both the lipid headgroups and the proteins contribute to the electron density profile that is characterized by two fairly symmetric peaks in each single membrane profile, it remains to be established whether the rhodopsin spans the membrane or is associated with the two halves of the membrane as two distinct molecules. Neutron diffraction analysis of the photoreceptor membrane has been used in an attempt to answer this important question.

3. *Neutron Diffraction*

The low contrast of electron densities among the components of the lipid-protein-water structure and the difficulty of placing electron density profiles on an absolute scale are limitations on the information given by X-ray diffraction studies. Neutron scattering, which has been developed more recently, has the advantage that the range of contrast between the cytoplasm and the membrane can be varied from positive to negative by varying the H_2O/D_2O content of the Ringer solution. Furthermore, contrast variation data, together with the existence of a clearly defined water layer, provide an absolute scale of neutron scattering density (Saibil et al., 1976).

We have already mentioned the experiments of Yeager et al. (1976), Sardet (1977), and Osborne et al. (1978) on detergent-rhodopsin complexes. They assumed that the transmembranous cylindrical shape of the rhodopsin in detergent micelles is comparable with the conformation of the photoreceptor protein in the membrane lipid bilayer. Saibil et al. (1976) have carried out neutron diffraction experiments on isolated rod outer segments. They calculated from the density profiles that at least 50% of the volume of the rhodopsin molecule was located in the hydrophobic region of the membrane. The profiles also indicated a slight asymmetry in the membrane, with protein probably extending into the cytoplasmic space, but not into the intradisc space. These results indicate that the electron density profile obtained from X-ray diffraction should be interpreted as the electron density variation of the lipid bilayer, superimposed on a fairly uniform protein contribution. In the same way, in thin sections, electron-dense layers can be assumed to be protein and phospholipid material. From these results, it appears that rhodopsin molecules penetrate deeply into the lipid bilayer. However, neither X-ray nor neutron diffraction analysis provides sufficient information as to the vectorial polarity of the photoreceptor molecule upon the exposure of the molecules on one or both sides of the disc membrane. These questions have been approached by the application of freeze-fracture techniques and cytochemical methods.

4. *Freeze-Fracture Technique*

Freeze-fractured membranes are split along an interior plane, exposing face views of the hydrophobic interior of the lipid bilayer (Branton, 1966). In most of the biological membranes studied so far, the exposed fracture faces are characterized both by smooth areas and by areas covered with particles (Fig. 5).

FIG. 5. Replica of freeze–fractured photoreceptor outer segment (fixed in glutaraldehyde and impregnated with 20% glycerol). The fracture plane occurs parallel to the axis of the ROS in the plasma membrane, showing its two fracture faces, EF and PF. As in most plasma membranes, the intramembranous particles appear asymmetrically distributed. PF is covered with a large number of IMPs, whereas in EF, fewer particles are present and are separated by smooth areas. ×37,800.

Whereas, when freeze-fracturing is applied to an artificial pure lipid bilayer, the exposed surfaces are smooth and no particles are visible. These results have prompted the conclusion that the particles exposed by the cleavage correspond to the hydrophobic segment of amphipatic membrane proteins.

Disc membranes, which contain a major amphipatic protein rhodopsin, would be expected to yield hydrophobic surfaces covered with particulate entities upon fracture. All the observations carried out by different groups are consistent with this expectation (Clark and Branton, 1968; Leeson, 1970, 1971; Chen and Hubbell, 1973; Olive and Benedetti, 1974; Raubach *et al.*, 1974; Jan and Revel, 1974; Olive and Recouvreur, 1977; Tonosaki *et al.*, 1978). If the fracture plane is parallel to the long axis of the outer segment, the transverse lamellae are seen to be closely packed with narrow interdisc, or interlamellar spaces (Fig. 6). In a more oblique or in a transverse fracture, each disc membrane appears to be split into two asymmetric halves: one rough surface is covered by randomly dispersed particles, whereas the other fracture face is almost completely smooth (Fig. 7).

Experimental evidence has been obtained, using the double replica technique (Olive and Benedetti, 1974; Tonosaki *et al.*, 1978), that indicates that these two surfaces are complementary to one another. In some favorable situations, it appears that the rough surface corresponds to the half of the membrane adjacent to the intradisc space (P face; PF) and the smooth fracture face to the half of the membrane adjacent to the interdisc space (E face; EF). It is generally admitted that these intramembranous particles are associated with the rhodopsin molecules. Freeze-fractured rhodopsin-lipid recombinant membranes (Chabre *et al.*, 1972; Hong and Hubbell, 1972; Chen and Hubbell, 1973), in contrast to pure lipid phases, display bumpy fracture faces similar to those seen in disc membranes, except that in reconstituted membranes both complementary fracture faces are similar.

After fracture of disc membranes, the rhodopsin-associated particles preferentially remain attached to the cytoplasmic half membrane. The problem of how deep the particles penetrate into the bilayer is not easy to resolve. The average diameter of the particles is about 80 Å, and the thickness of the bilayer is of the same order of magnitude. Hence, if the rhodopsin spans the bilayer, one would expect to find pits and depressions corresponding to the protruding entities associated with the P face on the complementary E face. In several types of membranes, even if there exist good reasons to suspect the presence of transmembrane proteins, this result has seldom been observed. The lack of pits on the E face has been interpreted as being due to contamination and an inappropriate shadowing angle. However, this interpretation is not completely satisfactory. A hypothesis of Ververgaert and Verkleij (1978) suggests that the presence of complementary impressions on the opposite face is dependent on the chemical composition of the particles, especially if these are lipid-rich structures like micelles, rather than simply protein.

FIG. 6.   Replica of freeze-fractured mouse ROS (fixed in glutaraldehyde and impregnated with 20% glycerol). The cleavage runs perpendicularly through the outer segment and the stacking of the disc membranes is visible. Note the penetrating cilium (arrows). ×23,100.

FIG. 7   Replica of freeze–fractured mouse ROS (fixed in glutaraldehyde and impregnated with 20% glycerol). Each disc membrane appears to be split into two asymmetric halves: a rough surface (PF) covered by randomly dispersed particles; and another fracture face (EF) that is almost completely smooth. In the disc membrane, the asymmetrical distribution of IMPs is enhanced in comparison with the distribution of IMPs on EF and PF of rod plasma membrane (Fig. 5). ×85,000.

In the disc membrane, the evidence for pits in the EF is questionable, and contradictory results have been produced. Chen and Hubbell (1973) have shown that the pits on the E face are found only if the isolated bovine disc membranes are deeply etched after fracturing. Incidentally, pits and depressions can be identified in the micrographs of Jan and Revel (1974) and of Krebs and Kühn (1977).

Our results clearly indicate that the number of pits, which are occasionally found on the E face, is much smaller than the estimated number of particles found on the corresponding P face. From this observation, one can assume that not all of the particles are located at the same level within the bilayer width.

Another alternative arising from the hypothesis of Ververgaert and Verkleij (1978) could be that not all of the rhodopsin molecules have, at the same time, identical lipid and protein interactions. In other words, it could be that only a fraction of rhodopsin-associated particles will remain associated with boundary lipids, producing pitted images on the E face.

In order to determine the sidedness and vectorial organization of the rhodopsin molecules in the disc membrane, deep etching experiments have been carried out on isolated discs to establish whether or not rhodopsin is exposed on the etched surface (Fig. 8). Unfortunately, the results of this experiment are deceptive and probably misleading, since only very seldom does deep etching demonstrate the presence of protruding particles at one of the true membrane surfaces. Most of the etched membrane fragments are characterized by smooth surfaces. From these results, it could be concluded that the rhodopsin molecules are not visible at the etched true membrane surfaces, whereas part of each molecule is accessible and exposed at the hydrophilic surface, since it can be specifically labeled and cleaved by proteolytic action.

The most probable explanation of this finding is that the physical mass of the rhodopsin polypeptide segment exposed at the hydrophilic surface is insufficient to appear as a particulate entity after freeze-etching. This conclusion is consistent

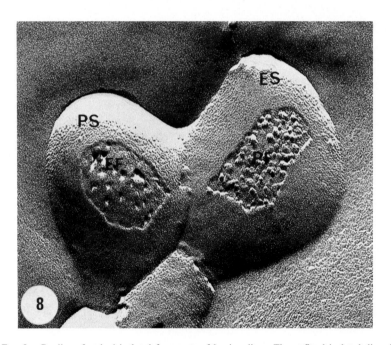

FIG. 8. Replica of etched isolated fragments of bovine discs. The unfixed isolated discs have been rapidly frozen, according to the technique of Gulik-Krzywicki and Costello (1978). They have been fractured at −150°C and etched at −100°C during 1.5 minutes. The true external surface (ES) and protoplasmic surface (PS) of the disc membrane, revealed by the etching, appear completely smooth. ×130,000.

with the general finding that only large oligomeric polypeptides, or a lattice of repeating subunits, are detected on the exposed etched face (Cartaud *et al.*, 1978; Staehelin, 1976), whereas smooth etched surfaces are observed, even when it is known that transmembrane proteins are exposed at both outer and inner membrane surfaces (Pinto da Silva *et al.*, 1973; Pinto da Silva and Nicolson, 1974). Freeze-fracture and freeze-etch analyses only provide evidence that the rhodopsin polypeptide penetrates the hydrophobic core of the bilayer. However, it does not provide sufficient information to allow a conclusion about whether the molecules span the entire width of the disc membrane, or are temporarily associated preferentially with one-half of the photoreceptor membrane domain. These entities are all oriented in the same way, since they are caught on the cytoplasmic side of the membrane, probably by polar interactions between some hydrophilic part of the rhodopsin and some cytoplasmic components such as actin.

5. *Labeling*

A direct proof of the accessibility of rhodopsin from the hydrophilic environment of the membrane has been obtained both by chemical labeling techniques and enzymatic digestion of the rhodopsin.

When labeled antibodies against rhodopsin have been used, the receptor molecules associated with the rhodopsin polypeptide are accessible on the cytoplasmic side of isolated disc membranes uniformly (Blasie *et al.*, 1969; Dewey *et al.*, 1969; Blasie and Worthington, 1969).

More ambiguous are the results concerning the labeling of the intradisc surface of the photoreceptor membrane. The difficulty arises from the fact that closed vesicles derived from intact discs are almost completely impermeable to the labeling ligands and that it is not easy to obtain inside-out vesicles from the discs. Jan and Revel (1974), using Fab antibodies cross-linked to peroxidase, stressed the fact that the intradisc space is stained much more lightly by the ligand than the cytoplasmic side of the disc membrane.

Electron microscopic studies by Godfrey (1973) have been carried out on aldehyde-fixed retina, embedded in a water-containing resin of glutaraldehyde and urea in order to prevent dehydration. The stained, ultrathin sections show that a thin layer containing mucopolysaccharide separates the two disc membranes. However, no evidence is presented that these polysaccharides are part of the rhodopsin molecules. Similarly, Engelhardt and Storteir (1975) have visualized, by ruthenium red methods, a carbohydrate moiety in the intradisc space. We are unable at this point to determine clearly whether or not rhodopsin spans the membrane. But the partial exposure of rhodopsin to the aqueous medium is definitely confirmed.

Fluorescein-labeled Con A binds tightly to disc membranes with a stoichiometry of one Con A monomer per retinal group. The binding involves the sugar recognition sites of Con A, since α-methyl-D-mannoside is a competitive inhibitor of the binding (Renthal *et al.*, 1973; Steinemann and Stryer, 1973; Steinemann *et al.*, 1973a,b). Using Con A-methacrylate microsphere conjugates, labeled Con A binding sites have been visualized by scanning electron microscopy on bovine disc membranes (Molday, 1976). Wheat germ agglutinin, specific for N-acetyl-D-glucosamine, was also found to bind to disc membranes (Renthal *et al.*, 1973). Shaper and Stryer (1977) reported that the carbohydrate moiety is accessible to enzymatic and chemical modifications. These results suggest that the carbohydrate moiety of rhodopsin is located at the cytoplasmic interface, but the possible inversion of the disc membrane during hypotonic disruption has not been taken into consideration. The experiments of Röhlich (1976), described previously, give a more direct proof of the localization of Con A binding sites at the intradisc surface of the disc membrane (Fig. 9).

Wu and Stryer (1972) were able to label one rhodopsin sulfhydryl site in ROS with two fluorescent isomers of N-(iodoacetaminoethyl)-1-aminonaphthalene sulfonic acid, which presumably were membrane impermeable, and Raubach *et al.* (1974) found that half of the rhodopsin amino groups react with the impermeable reagent ethylacetimidate hydrochloride.

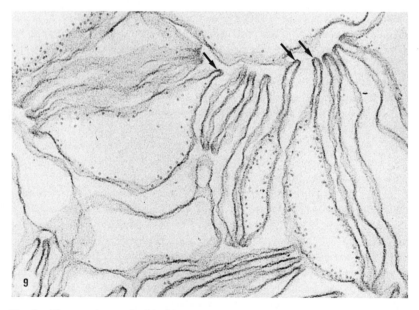

FIG. 9.   Electron micrograph of ferritin (Ft)-labeled concanavalin A disc-derived vacuoles. The Con A-Ft is bound to the internal surface of the vacuole membrane. No conjugate can be found at the external surface. The arrows show persisting membrane loops of the disc edges directly continuous with the interior of the disc-derived vacuoles, indicating the original sideness of the disc membranes. ×84,500. (From Röhlich, 1976.)

In view of the above studies, it is clear that at least some portion of the rhodopsin molecule is localized in the cytoplasmic leaflet of the disc membrane, partially exposed to the cytoplasmic matrix. In contrast, it is not convincingly established whether rhodopsin protrudes or is exposed on the intradisc space.

6. *Proteolytic Digestion*

Another experimental approach used to investigate whether rhodopsin molecules span the lipid bilayer has been to treat the disc membranes with proteolytic enzymes, assuming that the action of these enzymes will be limited to regions of the polypeptide accessible from the aqueous environment. As with other amphipatic intrinsic membrane proteins, rhodopsin is digested by proteases only to a limited extent. After this treatment, there is no loss of 500 nm absorbance or change in the absorption spectrum.

The first step in rhodopsin digestion seems to be similar for all proteolytic enzymes (Table VII); except for trypsin, which does not affect rhodopsin at all. Most attention has been paid to the polypeptide segment, which remains unaffected and membrane bound after proteolysis, rather than to the fragment released. The first distinct digestion product, still membrane bound, appears to be a large fragment of about 28,000 MW. It has been obtained after pronase treatment (Bonting *et al.*, 1974; Van Breugel *et al.*, 1975; Klip *et al.*, 1976), papain digestion (Trayhurn *et al.*, 1974a; Van Breugel *et al.*, 1975; Towner *et al.*, 1977; Fung and Hubbell, 1978; Albert and Litman, 1978), as well as after subtilisin digestion (Saari, 1974; Van Breugel *et al.*, 1975). Similar results have also been obtained with thermolysin (Saari, 1974; Pober and Stryer, 1975) and α-chymotrypsin treatment (Van Breugel *et al.*, 1975; Klip *et al.*, 1976). Further digestion with these proteases gives rise to a second main membrane polypeptide fragment, with a MW around 12,000 in the case of papain treatment, and around 21,000 with other enzymes. Smaller peptides have also been obtained, but only when the disc membranes are submitted to a prolonged digestion with papain or pronase (Klip *et al.*, 1976; Towner *et al.*, 1977; Albert and Litman, 1978).

It is striking that the large membrane bound polypeptide fragments produced by the proteolytic digestion carry the carbohydrate moiety that has been shown to be in the intradisc space (Röhlich, 1978). From these results, we may then conclude that proteolysis of intact disc membranes affects primarily, or exclusively, the photoreceptor protein segment exposed at the cytoplasmic surface. If proteolytic digestion of the disc membranes is preceded by detergent or phospholipase C treatment, then proteolytic degradation is more complete and the polypeptide carrying the carbohydrate residue is digested under these circumstances (Radding and Wald, 1958; Bonting *et al.*, 1974; Van Breugel *et al.*, 1975).

As we have mentioned in the preceding paragraph, the photoreceptor membranes that have been used for most of the protease studies consist primarily of

TABLE VII

POLYPEPTIDE FRAGMENTS OF RHODOPSIN OBSERVED IN ELECTROPHORETIC ANALYSIS OF PHOTORECEPTOR MEMBRANES TREATED WITH PROTEASES

Enzymes	Polypeptide fragments (apparent MW)	Sugar residue[a]	Chromophore binding site[a]	References[b]
Papain	24,800	+	+	(1)
	12,000			
	28,500	+	+	(2)
	23,000		+	(3)
	15,500			
	6,000			
	34,000	+	+	(4)
	27,000	+	+	
	12,000			
	34,000	+	+	(5)
	26,000	+	+	
	19,000	−		
	10,000			
Pronase	28,500	+	+	(2)
	21,500	+		(6)
	19,000	+		
	15,000	+		
Thermolysin	29,000	+		(7)
	23,000	+		
	30,000	+	−	(8)
	18,000		+	
	30,500	+	+	(9)
	25,000	+	+	
	9,500			
Chymotrypsin	24,400	+		(6)
	19,500	+		
	28,500	+		(2)
	21,500	+		
Subtilisin	28,500	+		(2)
	21,500	+		
	23,000		+	(7)
	20,000			

[a] +, Present; −, absent.

[b] References: (1) Trayhurn *et al.* (1974a); (2) Van Breugel *et al.* (1975); (3) Towner *et al.* (1977); (4) Fung and Hubbell (1978); (5) Albert and Litman (1978); (6) Klip *et al.* (1976); (7) Saari (1974); (8) Pober and Stryer (1975); (9) Hargrave and Fong (1977).

closed disc membranes that conserve the vectorial organization existing "in situ". The action of large protease molecules is confined to those regions of the protein oriented towards the cytoplasm, which are accessible from outside the membrane barrier. With native membranes, from this point of view, it is not easy to determine if the rhodopsin is a transmembranous protein with the two termini accessible, respectively, from the outside or from the inside of the membrane. To overcome this problem, Hubbell *et al.*, (1977) have compared native and reconstituted disc membranes. A significant difference exists between these two classes of membrane vesicles regarding the symmetry of the protein orientational distribution with respect to the central plane of the membrane (Fig. 10). In the native structure, the distribution of the rhodopsin molecules is asymmetric, whereas in the reconstituted membrane, the molecules are randomly and symmetrically distributed. Given the symmetric distribution of the protein in the reconstituted membrane vesicles, it is obvious that, if the protein spans the entire width of the membrane, both ends would be simultaneously exposed both at the outer and at the inner membrane surfaces. These two classes of membranes have been treated either with papain or with thermolysin. The results of this experiment clearly indicate that, in native membrane, all the rhodopsin polypeptides are degraded and yield two fragments of 27,000 and 12,000 MW (Fig. 11). Conversely, in the reconstituted membranes, only 60% of the molecules are degraded and produce the two segments, while the remainder of the protein is apparently inaccessible to the proteases.

These results may be interpreted in two different ways. One explanation could be that the rhodopsin does not span the entire width of the bilayer and is therefore accessible only when it is exposed at the outside of the vesicles. The other alternative is that the molecules do span the bilayer, but, the protease action

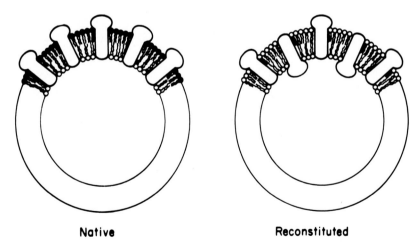

Native   **Reconstituted**

Fig. 10. Schematic representation of vesicles derived from the native and reconstituted disc membrane. The asymmetric rhodopsin molecules have a preferred orientation in the native membrane, whereas in the reconstituted membrane, the orientation is random (or nearly so). (From Hubbell *et al.*, 1977.)

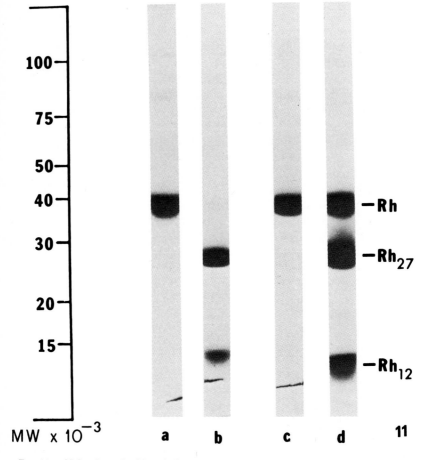

Fig. 11. SDS-polyacrylamide gel electrophoresis patterns of native and reconstituted vesicles before and after papain proteolysis. (a) Native ROS membranes; (b) native ROS membranes after papain proteolysis; (c) reconstituted membranes; (d) reconstituted membranes after papain proteolysis. Rh marks the position of rhodopsin; RH_{27} and Rh_{12} mark the position of rhodopsin proteolytic fragments of molecular weight 27,000 and 12,000, respectively. (From Hubbell *et al.*, 1977.)

being vectorial, no susceptible cleavage points are correctly exposed by the protein in the inverted orientation (relative to the native membrane orientation). While these experiments, using proteolytic cleavage, did not demonstrate the transmembranous nature of rhodopsin conclusively, the later studies of reconstituted membrane vesicles, using iodination with ^{125}I catalyzed by lactoperoxidase, provided evidence that both ends of the rhodopsin molecules are accessible to labeling and that, therefore, the protein is indeed a transmembranous component (Hubbell and Fung, 1978).

B. Tangential Organization

Since the major constituent of disc membranes is a single protein species, we might expect that this class of polypeptides could form a bidimensional lattice of repeating protomers. This lattice is found in other types of membranes characterized by the presence of large amounts of a single protein, or of related polypeptides that can show equivalence in the formation of the lattice repeating units, e.g., in cholinergic synapses (Cartaud *et al.*, 1978), in gap junctions (Benedetti *et al.*, 1976), and in *Halobacterium halobium*, where the photoreceptor protein (bacteriorhodopsin) constitutes a regular lattice of repeating subunits (Blaurock and Stoeckenius, 1971; Henderson and Unwin, 1975).

The problem of the distribution of the photoreceptor protein in the plane of the membrane has been analyzed by various techniques. Small angle X-ray diffraction has been applied by Chabre (1975) to stacked discs from cattle retina. The diffraction pattern observed in the equatorial direction is representative of the molecular organization in the plane of the membrane. Two distinct ranges of reciprocal distances have been obtained: the large angle ranges around 4 Å^{-1}, which corresponds to the correlation distances between the paraffin chains of the lipids; and the small angle ranges between 25 and 100 Å^{-1} which corresponds to electron density fluctuations that are a function of the protein size and of interprotein distances. The latter are not consistent within an array of geometrically packed particles.

If we assume that the intramembranous particles comprise rhodopsin molecules, then the freeze-fracture technique may certainly be considered the most suitable for direct visualization of the photoreceptor on the plane of the disc membrane. This technique, as already mentioned, has been applied with success by different authors to a variety of mammalian photoreceptor membranes, in particular to rat (Leeson, 1970; Jan and Revel, 1974; Olive and Recouvreur, 1977), to rabbit (Leeson, 1971), and to bovine disc membranes (Chen and Hubbell, 1973; Olive and Benedetti, 1974; Raubach *et al.*, 1974; Krebs and Kühn, 1977; Tonosaki *et al.*, 1978). The results of all observations clearly indicate that in most mammalian retinas so far studied, the rhodopsin-associated intramenbranous particles (IMPs) are randomly distributed on the hydrophobic surfaces produced by freeze-cleavage (Fig. 12).

In agreement with the freeze-fracture data, experiments, in which the carbohydrate moiety of rhodopsin is labeled with Con A (Steinemann and Stryer, 1973), or in which the amino groups exposed on the exterior of the membrane are labeled with fluorescent molecules (Raubach *et al.*, 1974), have shown a random distribution of these sites on bovine retina. Even when unfixed or glutaraldehyde-stabilized isolated discs were negatively stained, no geometrically packed array of subunits could be demonstrated. The negatively stained disc membrane surfaces show a granular appearance (Fig. 13).

FIG. 12.   High magnification of the P face of mouse fractured disc membrane showing the randomly dispersed IMPs. The particles show a subunit structure with a central small pit (circles). ×308,000.

Unlike the observations on mammalian photoreceptor membranes, which have unambiguously shown a random distribution of the photoreceptor protein within the plane of the membrane, the investigation of the distribution of rhodopsin in amphibian ROS gave variable and intriguing results. It appears that the rhodopsin-associated globular entities may be distributed in the plane of the membrane in two different ways. The first distribution, which is most frequently observed, is similar to that found in mammalian ROS membranes: rhodopsin is randomly distributed (Corless *et al.*, 1976). In other instances, however, the photoreceptor molecules seem to be distributed in a light-dependent geometrical array of repeating subunits (Fig. 14) (Olive and Recouvreur, 1978). The fracture faces of unbleached frog disc membranes show that the intramembranous particles visualized on the P face form a geometrically packed square array whereas, in the complementary E face, a square array of pits is clearly visible. The center-to-center distance is around 80 Å. Bleached rhodopsin intramembranous particles, however, appear to be randomly distributed on the P face, and the E face does not show pits. This result suggests that not only the organization of the rhodopsin in the plane, but also the penetration of the protein into the width of the bilayer is modified upon bleaching.

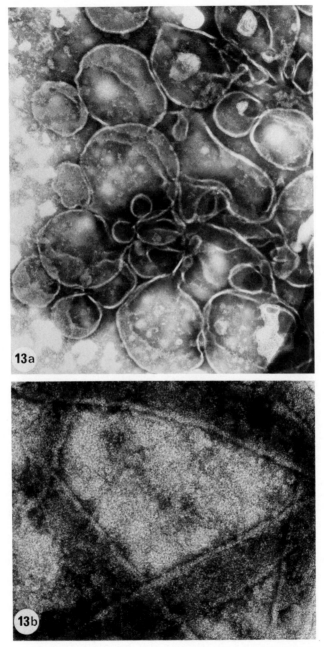

Fɪɢ. 13. (a) Electron micrograph of negatively stained, unfixed, isolated bovine disc membranes, showing that the edge and the surface of the flattened membrane sheets are devoid of budding particulate subunits ×110,000. (b) At higher magnification, the membrane surface displays a granular aspect. ×130,000.

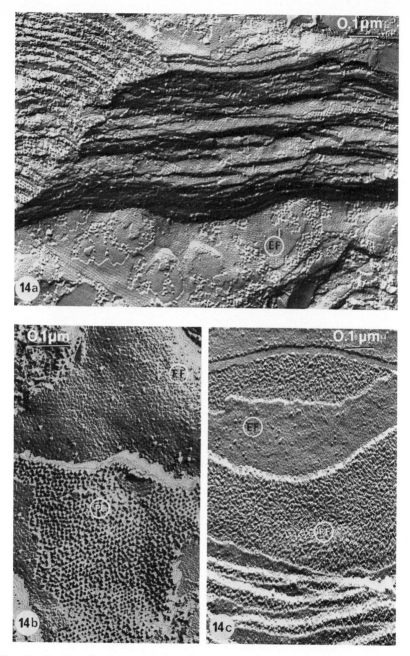

Fɪɢ. 14. Replicas of unfixed and unglycerinated frog retinas, dark-adapted (a and b) or light-adapted (c). In unbleached retinas, a regular array of particles is visualized on the P face, while in the complementary E face, a square array of pits is clearly visible (a and b). In bleached retinas, the IMPs appear to be randomly distributed on the P face, and the E face does not show pits.

C. Size and Nature of the Intramembranous Particles

As already described, the freeze-fracture of disc membranes reveals that the hydrophobic core of the membrane is heterogeneous and that particles are visualized on the fracture faces, particularly on the face closer to the cytoplasmic domain. We have also reported that this asymmetric distribution of the IMPs is one of the main freeze-fracture features of cleaved disc membranes. The nature of the IMPs found in the disc membrane, however, remains to be established.

There is now a growing controversy over the composition of the IMPs found in a variety of biological membranes. These particles were previously thought to consist only of protein. However, the size of IMPs has been observed to vary considerably, even in membranes that contain a single major protein constituent, such as disc membranes. Hence, it appears relevant to discuss here the intriguing problem of the IMPs and whether this parameter reflects a chemical heterogeneity (protein and lipid) of the particles or the multimeric association of identical polypeptides.

Contradictory values of the IMP diameter have been reported as shown in Table VIII. If the particles are proteinaceous, one could derive the number of copies of single polypeptides in the IMPs on the basis of the average size and estimated molecular weight of the protein subunit. The MW of rhodopsin is about 39,000 (Daemen *et al.*, 1972), and therefore, the expected diameter for a spherical IMP formed by a single subunit would be approximately 42 Å. The density of the particles on the fractured disc membranes has been found to be around 5,000 particles/μm^2 (Chen and Hubbel, 1973; Corless *et al.*, 1976; Olive and Recouvreur, 1977), whereas the rhodopsin concentration is about 20,000 molecules/μm^2 (Chen and Hubbell, 1973). Particles 120–140 Å in diameter could contain at least four to five photoreceptor monomers. Such an oligomeric configuration is apparent in high magnification micrographs of freeze-fractured disc membranes (Fig. 12). The variation in size may, therefore, reflect the number of associated subunits. In support of this hypothesis, a number of experiments have

TABLE VIII
Size Values of IMPs in Disc Membranes

Animal	IMP diameter	Reference
Rat	85–150	Leeson (1970)
Bovine	120	Chen and Hubbell (1973)
Mouse	50	Jan and Revel (1974)
Frog	50 (darkness)	Mason *et al.* (1974)
	125–175 (light)	
Frog	120–140	Corless *et al.* (1976)
Mouse	50–115	Olive and Recouvreur (1977)
Frog	125–175	Rosenkranz (1977)

shown that, with rhodopsin, protein-protein interactions and multimeric association of subunits may occur readily. Using the saturation transfer method,
Davoust and Devaux (1978) reported that protein-protein interactions can be
demonstrated in artificial lipid-spin labeled-rhodopsin vesicles if the ratio of lipid
to protein is low, or if the protein is subjected to prolonged illumination.

It is interesting that the conformational change dependent on, or associated
with, photolysis may favor this kind of protein-protein interaction. The observations of Chen and Hubbell (1973) are relevant here. They found that the distribution of IMPs in the plane of freeze-fractured recombinant lipid-protein membranes was a function of the chemical state of the protein and was affected by
bleaching, or by reducing agents such a dithiothreitol (DTT).

These results also indicate that the molecular nature and physiological state of
the lipid hydrocarbon chains probably affect protein interactions and their multimeric association. This is strongly supported by our own observations of IMP
size variation following phospholipase action on isolated disc membranes. If
these photoreceptor membranes are treated with phospholipase C under conditions such that at least 40% of the total phospholipids are removed, an increase in
particle size is detected (Olive et al., 1978).

Modifications that affect the protein content of other types of biological membranes also result in changes in IMP size and density. Mutaftschiev et al. (1977)
have observed such changes in the chl-r mutant of Escherichia coli, which lacks
a multienzyme system associated with the plasma membrane.

A possible relationship between the protein and lipid ratio and the size and
density of IMPs, has also been analyzed using artificial lipid amphipatic protein
vesicles. Segrest et al. (1974) showed that a critical micellar concentration
(CMC) is necessary for the formation of particles, which are then all of the same
size, of about 80 Å. The CMC corresponds approximately to 7,600–16,000 T(is)
hydrophobic segments of red blood cell NM glycophorin. However, below the
CMC, it is not clear whether the particle size diminishes progressively or remains
the same, and only the density of IMPs decreases. Other experiments by Grant
and McConnell (1974), using the complete glycophorin polypeptide (not only its
T(is) segment), provided evidence that, compared with those observed by Segrest, smaller sized particles (40 Å) are produced, even at low protein: lipid ratio.

A correlation between the rhodopsin concentration and the IMP distribution on
the fracture faces has also been found with native membranes. In ROS of vitamin
A-deficient mice, Jan and Revel (1974) have observed the development of
smooth regions, devoid of particles. The decreased rhodopsin concentration is
accompanied by reduction of particle density.

IMP size is probably dependent, not only on the relative amount of proteins
within the lipid bilayer, but also on the nature of the lipids and their metabolic
variation. During postnatal differentiation of the mouse retinal disc membrane, a
progressive decrease of large particles (107 Å) and a correlated increase of

smaller particles (67 Å) have been attributed to the relative proportions of protein and lipid. This latter ratio is dependent on the process of differentiation, which favors the aggregation of a variable number of monomeric rhodopsin molecules into particles of different sizes (Olive and Recouvreur, 1977).

Alternatively, it has been postulated for other types of membranes that the particles revealed by freeze-fracture on the hydrophobic surfaces correlate with the formation or the existence of lipid-protein complexes. If this assumption is true, the particle size would be dependent, not on the number of associated monomers, but on the closely associated lipid boundary around the protein molecules. This hypothesis is based on the experimental data obtained on *Escherichia coli* mutants lacking all glucose- and heptose-bound phosphate in their lipopolysaccharide (LPS) (Verkleij *et al.*, 1977). Reduction in particle density parallels the lack of one major outer membrane protein constituent in the mutant. Loss of LPS, as a result of EDTA treatment, also caused a reduction in particle density. From these results, the authors suggested that the particles consist of LPS aggregates complexed with proteins and stabilized by divalent cations. Unfortunately, in this study, neither the IMP size nor the size variation are mentioned. It is possible that the reduction of particle density is associated with an increase in particle size. In this case, the change of polarity in the membrane, as a result of the loss of polysaccharide, might favor the association of monomeric polypeptides. These results cannot be extrapolated directly to mammalian systems in which no LPS is present. Nevertheless, the hypothesis that the protein-protein association is affected directly or indirectly by the loss of a lipid component is consistent with our results, described previously, on the increase of IMP size after phospholipase C treatment of isolated membranes.

A direct proof that rhodopsin polypeptides alone can form particulate entities is derived from recent experiments on lipid-free, detergent-solubilized rhodopsin. After dialysis to remove the detergent, the protein can be precipitated so as to preserve the particulate organization, as revealed by freeze-facture experiments (Fig. 15) (J. Olive, W. J. De Grip, and F. J. A. Daemen, unpublished). Our experiments (in progress) also show that the rhodopsin particulate organization may vary with different dialysis conditions. We can therefore conclude that rhodopsin molecules, alone or mixed with lipids, may show different degrees of association, and therefore may form oligomeric particles of different sizes.

Another interpretation of the size and variation in size of the IMPs has been developed by Corless *et al.* (1976). They considered the IMPs found on the P face to be artifactual, in the sense that they imply distributions of material within the disc membranes that do not occur at physiological temperatures. They suggest that temperature shift from physiological conditions during the freezing step involves a perturbation and redistribution of material within the disc membrane, resulting in particle formation that is observed exclusively on one fracture face. In addition, Corless *et al.* (1976) have implied that since the E face usually

FIG. 15.   Replica of purified rhodopsin. Bovine disc membranes, solubilized in cationic detergent
dodecyltrimethylammonium, have been delipidated by affinity chromatography over concanavalin
A–Sepharose 4B. Subsequent dialysis of the detergent results in a suspension of lipid- and detergent-
free rhodopsin. In this freeze-fractured material (fixed in glutaraldehyde), a particulate organization is
evident. ×111,650.

fails to show corresponding textual variations, the texture of the P face may be a
postfracture phenomenon.

This kind of interpretation is derived from the knowledge that temperature-
dependent lateral phase separation may occur in different cell membranes if the
freezing rate does not reach the values that prevent lipid-protein segregation. On
unfixed mammalian disc membranes, we have observed fracture features that are
indicative of phase separation. Under these conditions, clustering of IMPs and
smooth plaque formation are observed (Fig. 16). In order to improve the dissipa-
tion of heat and the freezing rate of biological specimens, several techniques
have been developed (Bachmann and Schmitt, 1971; Gulik-Krzywicki and Costello, 1977, 1978; Gross et al., 1978). When a very rapid freezing technique is
applied to unfixed mammalian disc membranes, no perturbation of the fractured
bilayers occurs, and the membrane splits into two asymmetric halves; one of
these is covered with IMPs, as are the fixed membranes. The X-ray diffraction
pattern of a membrane model system clearly indicates that, using this technique,

FIG. 16.   Incidental aspect of the fractured faces when unfixed, unglycerinated rods are incor-
rectly frozen. The cleavage runs at different levels across the disc membrane, and plaques of various
dimensions of smooth appearance are produced. Probably due to phase separation during the freezing, patches of segregated IMPs (arrows in the inset) are visible. ×80,850.

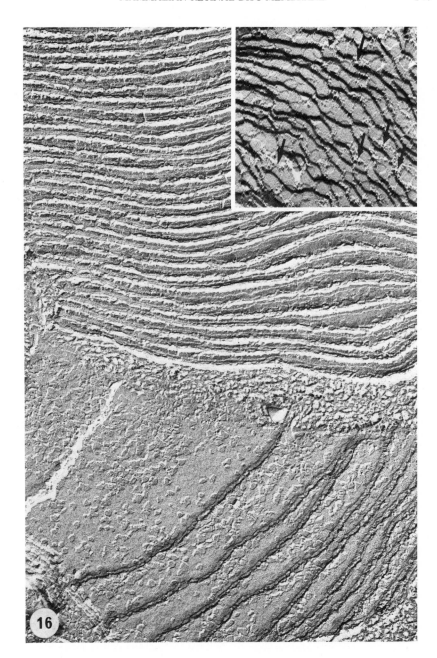

no perturbation of the lamellar phase occurs during the freezing process (Costello and Gulik-Krzywicki, 1976).

IV. Dynamic Aspects of the Disc Membrane

Results of a number of experiments suggest that most components of the photoreceptor membrane, particularly rhodopsin and lipids, have a great deal of mobility. This mobility and its control may be of crucial importance in regulation of physiological membrane phenomena, in particular in the specific interactions between rhodopsin molecules and phosphodiesterase molecules, which have been shown to play a key role in the transduction mechanism. Phosphodiesterase may also control rhodopsin-lipid association and thus the rate of metarhodopsin II production, which has been shown to be dependent on the association of lipid with the pigment (Williams *et al.*, 1974; Van Breugel *et al.*, 1978). Moreover, conformational changes of the rhodopsin molecules have been observed upon bleaching. These are also probably dependent on the mobility of the membrane constituents.

A. RHODOPSIN MOBILITY

1. *Rotational Mobility*

Schmidt (1938) was the first to note that, when viewed from the side, frog rods are highly dichroic, which means that each rhodopsin molecule spends most or all of the time lying parallel to the surface of the lamellar disc membrane. In spite of the ordered layering of the membranes, rods are not dichroic when viewed end-on, implying that there is no preferred orientation of the chromophores about the long axis of the rods.

Early studies by Hagins and Jennings (1959) suggested, as one of several possibilities, that rhodopsin could rotate about an axis perpendicular to the plane of the membrane. This suggestion has been confirmed by Brown (1971, 1972), who reported that photodichroism can be induced in frog retinas fixed with glutaraldehyde, where cross-linking prevents Brownian rotation. This means that the chromophores aligned with the plane of polarization will be bleached, whereas those oriented perpendicular to this plane will not. Such a photoinduced dichroism has also been induced in bovine retinas at −196°C (Strackee, 1972), but the dichroic ratio reported is too small to permit clear conclusions to be drawn.

Cone (1972) and Poo and Cone (1973) have also shown that rhodopsin is free to undergo Brownian rotation in unfixed frog retinas. Following a flash bleach with polarized light, a normal retina should become dichroic for as long as it takes the rhodopsin molecule to undergo rotational diffusion by Brownian motion. Assuming that the protein is spherical, Cone estimated that the viscosity of the membrane in which the rhodopsin was rotating, was about 2 P (poise). This value was confirmed by Liebman and Entine (1974). The same value for the rotational correlation time has been obtained by Baroin et al. (1977) and Kusumi et al. (1978), who have used the technique of saturation transfer electron paramagnetic resonance applied to the study of spin labeled, membrane bound rhodopsin. Furthermore, this protein bidimensional mobility in membranes has been shown to require phospholipids (Baroin et al., 1979). If 30% or more of the phospholipids are removed, the spin label bound to rhodopsin is strongly immobilized.

2. Translational Mobility

The relaxation time measured for diffusion of the rhodopsin molecules indicates that the membrane site occupied by rhodopsin is highly fluid and suggests that rhodopsin is also free to undergo lateral diffusion in the disc membrane.

Using a very elegant method, Poo and Cone (1973) were able to demonstrate the lateral mobility of rhodopsin. By microspectrophotometry, with a very narrow light beam, they bleached a strip parallel to the long axis of single rods and measured the rate at which the bleached rhodopsin diffused away and was replaced by unbleached rhodopsin. They obtained a value for the membrane viscosity (2 P) similar to that estimated from the relaxation time for rotational diffusion of rhodopsin in frog rods (Cone, 1972).

B. Lipid Fluidity

Both the rotational and lateral diffusion of rhodopsin in the disc membrane imply that the membrane is highly fluid. Such a lipid fluidity might be expected, since the phospholipids of the photoreceptor membrane are polyunsaturated.

The spin label technique has been used to study lipid fluidity in the disc membrane. It has been shown that a high degree of mobility was encountered in both native disc membranes and reconstituted vesicles, but apparently contradictory results have been obtained concerning the fluidity of the lipid boundary layer of rhodopsin due to different experimental conditions used by the authors.

In reconstituted vesicles, with a low lipid:protein ratio, Pontus and Delmelle (1975) detected a protein-associated, immobilized lipid layer forming a "lipid-annulus" in a similar way to that described for Ca^{2+}-ATPase in the sarcoplasmic reticulum membrane (Warren et al., 1975). At a temperature below 5°C, Watts et al. (1978) found, similarly, that a lipid boundary layer is immobilized. Conversely, under physiological conditions in regard to both temperature and lipid:protein ratio, the boundary lipid layer has been found to be highly fluid in native bovine disc membranes and in reconstituted rhodopsin-lipid vesicles

(Davoust and Devaux 1978; Favre *et al.*, 1979). Moreover, these authors have shown that the fluidity in reconstituted membranes depends both on the temperature and on the relative proportion of lipids, so that when the ratio of lipid to protein decreases, protein aggregation is induced and the lipids become immobilized. This latter conclusion can be derived from the electron spin resonance (ESR) spectra.

Nuclear magnetic resonance (NMR) comparative data on ROS disc membranes and liposomes of purified ROS phospholipids again indicated that rhodopsin does not markedly alter the lower frequency segmental motions of phospholipids in in the ROS membrane (Brown *et al.*, 1977 b). These NMR data suggest that, in both the ROS disc membranes and liposomes, domains of phospholipids in in the ROS membrane (Brown *et al.*, 1977b). These NMR data rhodopsin may be preferentially interacting with the more fluid phospholipids. A similar conclusion has been derived from freeze-fracture experiments. The random dispersion of rhodopsin within the lipid bilayer and its regenerability yield depend a great deal on the fluid state of the lipid hydrocarbon chains. If quenching occurs below the lipid transition temperature, a separation of a pure solid phase and segregated, rhodopsin-rich domains is obtained (Hong and Hubbell 1972, 1973; Chen and Hubbell, 1973).

C. CONFORMATIONAL CHANGES DURING BLEACHING

It is well established that photoisomerization of the chromophore from the 11-cis form to the all-trans configuration and the release of the chromophore from the opsin are followed by a number of changes that affect the opsin conformation and, in turn, its interaction with the neighboring lipid molecules. Since these changes occur within a short range domain and are restricted to the protein molecule itself, most of the studies of the conformational changes during the bleaching process have been carried out on detergent-solubilized rhodopsin. In the following section, we summarize the most significant results dealing with conformational changes of the photoreceptor protein in detergent, and we try to compare this information with results obtained directly with intact membranes, to gain a better understanding of the conformational changes that may occur in the intact disc photoreceptor membrane.

1. *Conformational Changes of Detergent-Solubilized Rhodopsin*

The occurrence of conformational changes in the photoreceptor protein on bleaching can be deduced by comparing the solubility properties of rhodopsin and opsin in detergent. Rhodopsin appears to be less soluble in SDS, since it requires a detergent concentration greater than the CMC, whereas opsin is readily extracted from the ROS membrane with the CMC (Virmaux, 1977). Moreover, these results are consistent with earlier reports that opsin is more susceptible than rhodopsin to chemical modification in detergent solution (Ostroy *et al.*, 1966; Zorn and Futterman, 1971). The higher susceptibility of opsin to detergent action, compared with rhodopsin, is probably due to a more open structure in the absence of chromophore, rather than to a light-induced conformational change itself (De Grip *et al.*, 1973a). Opsin behaves differently from rhodopsin during amidination (De Grip *et al.*, 1973a) and covalent bonding of the exposed sulfhydryl groups to *N*-ethylmaleimide (Zorn, 1974). In both cases, opsin loses its recombination capacity.

A conformational change of opsin in delipidated detergent micelles has been inferred from studies of the kinetics of exchange of labile protons. On bleaching, a very large change occurs; all the exchangeable protons become accessible to the solution (Osborne, 1976). This is characteristic of an extensive unfolding of the protein resulting from denaturation. Neutron small angle scattering also confirms that a large perturbation occurs on bleaching such preparations (Osborne *et al.*, 1978).

Spectral studies, such as circular dichroism (CD) and optical rotatory dispersion (ORD), have provided information about the conformation of rhodopsin, both in the membrane and in detergent. Solubilization of rhodopsin with detergent results in a loss of helical structure and the perturbation of aromatic residues. During bleaching, the conformational stability of solubilized rhodopsin is significantly decreased compared with that observed *in situ* (Rafferty *et al.*, 1977; Rafferty, 1977).

2. *Conformational Changes of the Protein in the Intact Disc Membrane*

Different physicochemical techniques have been used to study photoinduced changes in the photoreceptor membrane. The CD measurements have indicated that rhodopsin bound in ROS membranes undergoes little or no conformational change on bleaching by light. However, opsin conformation becomes sensitive to light if ROS are treated with phospholipase A or delipidated with hexane (Shichi, 1971). The thermal stability of bleached ROS is also decreased after partial delipidation; this is due not to denaturation of opsin, but to the loss of phospholipids (Shichi, 1973). Opsin possesses dynamic conformational reversibility, therefore, only when it is in the native state, i.e., associated with phospholipids. Additional evidence that some conformational changes in the membrane, affecting the protein-lipid interaction, occur on bleaching has been provided by the observation that a greater proportion of the phospholipids can be extracted with hexane after photolysis (Poincelot and Abrahamson, 1970). However, this could not be confirmed by Borggreven *et al.* (1970).

We have already described experiments using spin labels, which indicate that bleached rhodopsin must have a somewhat different configuration from dark-adapted rhodopsin. This is consistent with the change of reactivity of the sulfhydryl groups and with the modifications in the ESR spectra of spin labeled rhodopsin detected under the influence of light (Rousselet and Devaux, 1978).

Recent measurements of the diamagnetic asymmetry of rods, which allows their orientation in magnetic fields, have shown that a 6% change in the diamagnetic asymmetry is observed on bleaching. This change is directly related to the conformation and orientation of the protein (Chabre, 1978).

3. Long-Range Reorganization

Conformational changes in the distribution of the components within the membrane have been described in a number of experiments using X-ray diffraction, neutron diffraction, and electron microscopy, both on intact outer segments and isolated disc membranes. Controversial results have been obtained from X-ray diffraction studies. The inward shift of bleached rhodopsin into the lipid hydrocarbon core of the disc membrane, reported by Blasie (1972), has not been confirmed by other authors. According to Chabre (1975), changes in the lamellar diffraction indicate that there is, on bleaching, a small but definite increase of electron-dense material at the cytoplasmic edge of the membrane, in agreement with the previous observations of Corless (1972). The same, light-dependent outward shift of protein towards the cytoplasmic side of the membrane has also been observed by neutron diffraction (Saibil et al., 1976). In addition to the protein shift in the thickness of the bilayer, an increased disorder and slow swelling in isotonic Ringer solution has been described in the disc spacing after bleaching, both in isolated rods and intact cells (Chabre and Cavaggioni, 1975; Schwartz et al., 1975). In calcium-free Ringer, the light-induced shrinkage is considerably enhanced. Results of the freeze-fracture experiments of Mason et al. (1974) have been tentatively interpreted in terms of a change in the penetration of rhodopsin-associated IMPs toward the hydrophobic core of the membrane after bleaching. However, the poor quality of the replicas renders the interpretation rather difficult. As mentioned previously (Section IV,B), we have obtained some results indicating that a square lattice of particles is frequently observed in dark-adapted frog outer segments, whereas this pattern is never found in light-exposed membranes.

In summary, small but definite changes have been observed on bleaching. In all cases, they have been described as an increased disorder of the rhodopsin molecules; usually this appears to be manifest as an outward shift toward the cytoplasmic side of the disc membrane. These changes are consistent with an unfolding of the polypeptide chain and an increased polarity of the protein (accessibility of sulfhydryl groups and other protons). In association with these structural changes, the cell membrane conductance is modified. Unfortunately, no proof has yet been obtained that rhodopsin is itself a light-triggered ion translocator.

D. Renewal of the Membrane

In other membrane systems, it has been postulated that existence of a fluid lipid bilayer is advantageous for membrane renewal and assembly (Palade, 1978). This is probably also true of the renewal process in disc membranes, where the existence of a highly fluid lipid core not only serves the purpose of phototransduction and ionic translocation, but also provides an opportunity for rhodopsin to be renewed rapidly.

It was demonstrated ten years ago that vertebrate ROS continuously produce new membranes (Young, 1967) and intermittently shed groups of old ones from the top of the outer segment (Young and Bok, 1969). In contrast, cone cells restrict their renewal activities to the replacement of molecules in existing membranes (Young, 1971). Recently, however, it was reported that cones, in various species, do have the capacity to shed membranes, indicating that both molecular replacement and membrane replacement occur (Hogan, 1972; Hogan *et al.*, 1974; Anderson and Fisher, 1976; Young, 1977; O'Day and Young, 1978; Young, 1978). In rod cells, it has been clearly demonstrated, using autoradiography techniques, that membrane components are synthesized in the inner segment at one or more sites, then transported to the outer segment, where final membrane assembly takes place. New membranes are assembled by addition of membrane constituents at growth sites on existing outer segments membrane. This takes the form of an invagination of the outer cell membrane at the base of the outer segment.

Protein synthesis occurs predominantly in the myoid portion of the inner segment (Young, 1967; Young and Droz, 1968; Hall *et al.*, 1969; Young, 1973). Shortly after injection of radioactive precursors, large amounts of radioactive protein began to accumulate in rod discs at the base of the outer segment (Fig. 17). Heavily labeled discs are displaced along the ROS toward the top of the cell, by the continuous formation of the newer discs; the degree of labeling of these discs decreases progressively (Young, 1967). Evidence has been provided that glucosamine is first bound within the rod photoreceptor in the ribosome-rich myoid region. Then the labeled product appears in the Golgi zone, where additional glycosylation presumably occurs (Bok *et al.*, 1974, 1977). Finally, glycosylated opsin rapidly traverse the region of massed mitochondria and passes through the connecting cilium to reach the base of the outer segment, where a heavy band of radioactivity can be observed on electron microscope autoradiograms (Fig. 18) (Bok *et al.*, 1974).

Elucidation of the intracellular region for chromophore addition was difficult to establish, because the retinal linkage is labile to conventional histological procedure. Recently, Bok *et al.* (1977) provided some evidence that the rhodopsin glycoprotein reaches the outer segment before the chromophore is added.

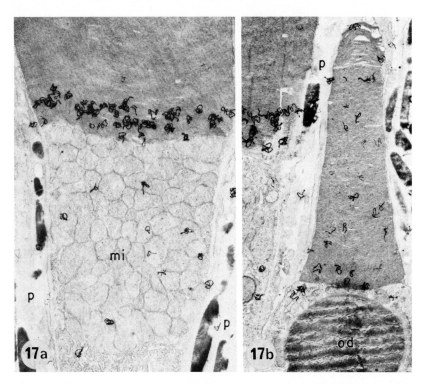

FIG. 17. Electron microscope autoradiography obtained one day after injection of a mixture of radioactive amino acids. (a) In rods, the discs in which the radioactive protein has been incorporated are further displaced from the base of the outer segment. This is believed to result from the continued production of new discs in this region. (b) In cones, protein delivered to the cone outer segment from its site of synthesis in the myoid has become diffusely distributed in that structure. There is no evidence of the neoformation of membranous discs. ×8085. (From Young and Droz, 1968.)

Lipids appear to be renewed in photoreceptor cells by molecular replacement, by fatty acid exchange in phospholipids already *in situ,* and by insertion of new phospholipids into the membranes, balanced by the removal or destruction of comparable molecules (Young, 1973; Bibb and Young, 1974a,b). When glycerol or inositol are supplied to visual cells, they are immediately concentrated in the myoid region of the inner segment. Most of the newly formed phospholipids migrate to the outer segment and participate in the renewal of disc membranes (Bibb and Young, 1974a).

The formation of new membranes at the base of the rod outer segment is coupled with shedding of old membranes from the tip of the cell. The detached discs are then phagocytized and degraded by the adjacent pigment epithelium (Young and Bok, 1969). It has been shown that the rhodopsin polypeptide was

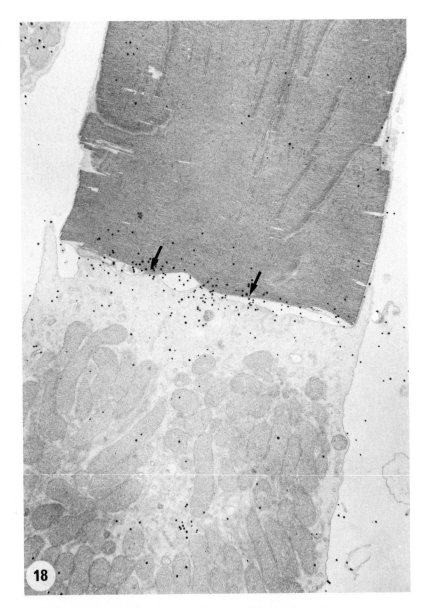

Fɪɢ. 18.   Electron microscope autoradiogram of rod outer segment base from a retina that was pulsed from 1 hour with [³H]glycosamine and chased for 5 hours. Basal outer segment discs (arrows) now show the greatest concentration of radioactivity. ×8010. (From Bok *et al.*, 1974.)

renewed at a rate that corresponds to that of disc membrane renewal (Hall *et al.*, 1969; Bargoot *et al.*, 1969).

Independently of the membrane replacement, protein is also renewed in rod cells by molecular replacement. Diffusely distributed radioactive protein has been detected in ROS shortly after injection of labeled precursors (Bok and Young, 1972). In cone cells, the accumulation of new labeled protein at the base of the outer segment has never been observed. It has been concluded that membrane replacement does not take place in cone visual cells (Young, 1969). A reevaluation of the cone disc renewal process seems to be necessary because of recent reports, previously mentioned, that cones may shed discs from the ends of their outer segment. It has not yet been demonstrated if disc shedding by cones is part of the renewal process of this type of photoreceptor cell.

One important feature of both these renewal processes, membrane replacement and molecular replacement, is the possibility that protein inserted at any point during membrane formation may undergo diffusion within the membrane. This assumes, as discussed earlier, both lipid fluidity and protein mobility phenomena.

It is not readily apparent how the new protein penetrates the existing disc structure in the rods, since no continuity exists between the discs. Some evidence has been provided by autoradiography of protein diffusion within the ciliary matrix and probably along the outer segment periphery, beneath the outer membrane (Young, 1968). Within the membrane itself, other evidence derived from freeze-fracture experiments indicates that, during the differentiation process of mouse rod disc membrane, the distribution of rhodopsin-associated particles is heterogeneous (smooth areas are interspersed with particulate regions), and that the particle size corresponds to larger oligomeric organization of the protein than in mature cells (Olive and Recouvreur, 1977).

V. Conclusion

The main aim of the present article was to depict a molecular model of the mammalian retinal disc membrane on the basis of results derived from the application of methods that allowed the visualization of structure within the plane and the thickness of the membrane element (Fig. 19).

A current hypothesis of the main functional aspect of the photoreceptor membrane claims that the physical signal (photons) generates a transduction mechanism. This step involves the release of a "messenger" that reaches the plasma membrane of the outer segment. In turn, the electrical and permeability properties of the disc membrane are modified, thus transmitting secondary signal to the synapse. This transmembrane control process, which amplifies the light signal at the level of the disc membrane, is associated with an amphipatic glyco-

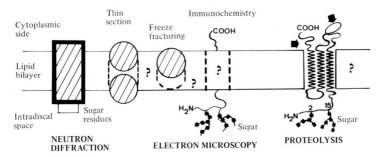

FIG. 19. Gallery of various models for transverse (vectorial) organization of the rhodopsin molecules within the lipid bilayer, as inferred from the results of a panoply of technical approaches.

protein rhodopsin, which contains amino acid sequences that favor the interaction both with hydrophilic and hydrophobic domains of the bilayer.

The problems of how the light signal is actually received by the receptor protein, and how it is transferred either to the cytoplasmic side of the disc membrane, or to the interior of the disc, have not yet been completely resolved. An important step in this transduction mechanism is the dissociation of a covalent bond between the opsin and the chromophore, which seems to involve short range intramolecular conformational change. The second important event is a change in the permeability of the plasma membrane, which may be associated with Ca^{2+} release from the discs following the transformation of rhodopsin into opsin.

We have briefly summarized the hypothesis that stresses the importance of Ca^{2+} ions as a potential transmitter substance. However, it remains to be established (1) whether Ca^{2+} is pumped from the interior of the disc compartment or simply released from the trapping cytoplasmic surface of the disc membrane into the cytosol of the outer segment, and (2) what the role of rhodopsin, as integral membrane protein, is in the calcium translocation. Until now, no clear experimental evidence has been provided for this kind of ionic translocation from the interior of the disc to the cytosol. Furthermore, the most recent data demonstrated that Ca^{2+}, released from the binding sites inside the disc compartment, does not cross the disc membrane unless this latter is made permeable by appropriate ionophore. It could also be that other factors, which are necessary for the ionic translocation, have not yet been identified, simply because they are lost during the isolation of the ROS.

It is well established that interactions between neurotransmitters, peptide hormones, and other regulatory ligands, and specific receptors at the plasma membrane level, involve the stimulation of a cyclase, which amplifies the signal and triggers the intracytoplasmic effect (Kahn, 1976; Shechter et al., 1978). Likewise, it has recently been proposed that these transduction and amplification mechanisms also play an important role in the function of the photoreceptor membrane. In keeping with this hypothesis, it has been demonstrated that the interaction of light signal and photoreceptor protein is associated with the activation of guanylate cyclase and phosphodiesterase. It is also noteworthy that the similarity between the two systems (as stressed by Wheeler and Bitensky, 1977), not only bears on the presence of a cyclase, but also on the dual switch activation mechanism. Light and GTP activate the photoreceptor phosphodiesterase system, as the peptide hormone and GTP activate the adenylate cyclase system, and the required order of the activation steps is light (or hormone) before GTP.

Moreover, a correlation between light-activated phosphorylation of the rhodopsin and activation or inactivation of the GTPase has been established, since GTP seems to be the preferred phosphate donor for phosphorylation (Chader et al., 1976), and inhibitors of phosphorylation affect the light-induced permeability change of the outer segment plasma membrane (Miller et al., 1977). The observed decay of the light effect on rhodopsin phosphorylation has been interpreted by these authors as a consequence of the rearrangement, or conformational changes, in the structure of opsin. Cooperative interactions between bleached and unbleached rhodopsin molecules could explain how a photoactivated rhodopsin molecule makes unbleached rhodopsin molecules susceptible to phosphorylation. In addition, even if Ca^{2+} is not the actual substance that transmits the effect of light from the ROS discs to the outer membrane, it could play an important role; and previous experiments have shown that phosphorylation of bleached rhodopsin affects the Ca^{2+} permeability of disc membranes (Weller et al., 1975b).

We will now draw on ultrastructural data concerning the architecture of the photoreceptor disc membrane, which might shed some light on both the existence of a transmembrane channel and membrane pattern relevant to the existence of cooperativity of the rhodopsin molecules. We have tried to present the various results in favor of or against the transmembrane character for rhodopsin molecules. Most of the experimental data are consistent with a transmembrane location of the rhodopsin, but the data do not necessarily correlate with a role of permeaphore for the polypeptide.

In general, in various types of active pumps, or ionophores (Na^+-K^+-ATPase, Ca^{2+}-ATPase, Na^+ channels) and more complex permeaphores (junctional channels), the transport function is based on oligomeric association of an unique class of proteins of equivalently related polypeptides (Warren et al., 1974; Van Winckle et al., 1976; Benedetti et al., 1976; Vogel et al., 1977; Rojas and Bergman, 1977; Martonosi, 1978; Schoot et al., 1978). In the case of the rhodopsin molecules, it is not yet clearly understood whether channelling across the disc membrane depends on the proper folding of individual rhodopsin polypeptides or on oligomeric association of rhodopsin molecules.

A model has recently been proposed by Hubbell and Fung (1978): it is comparable to that of Henderson and Unwin (1975) for bacteriorhodopsin. The latter model postulates that several α-helical segments of a single molecule could, by themselves, represent a transmembrane gateway. By a proper inwardly directed segregation of polar residues of the α-helices, the core of the gateway could be a hydrophilic channel connecting the two aqueous compartments of the disc membrane directly.

Apart from the experiments of Davoust and Devaux (1978) on rhodopsin-lipid recombinants, the experimental evidence for an oligomeric association of the photoreceptor pigment molecules derives primarily from the study of detergent-solubilized rhodopsin (Makino *et al.*, 1977). In contrast, neutron diffraction results are more consistent with the presence of single molecules embedded in the fluid lipid phase than with the existence of close protein-protein interactions.

Freeze-fracture and freeze-etch experiments have been used to investigate the localization of rhodopsin molecules and whether rhodopsin occurs as randomly dispersed, multifolded polypeptides or as oligomers. Three parameters have been studied: (1) the degree of IMP penetration into both fracture faces (EF and PF), (2) the presence of budding entities on the etched true surfaces of the membrane, and (3) the particle size. We have described the two fracture faces obtained by cleavage of the disc membrane characterized by a marked asymmetry in the distribution of the IMPs. The inner layer of the leaflet, adjacent to the protoplasmic face (P face) is covered with a large number of IMPs, whereas the inwardly directed, outer fracture face of the leaflet close to the disc compartment (E face) is studded with fewer IMPs, and complementary pits are only detected occasionally.

These results pose two different problems. The first concerns the almost completely asymmetric association of the IMPs with the cytoplasmic half of the disc membrane, whereas rhodopsin molecules have been shown by other physical and cytochemical techniques, to occupy a transmembrane position. This casts doubt on the protein nature of the IMPs visualized in the fractured disc membrane. In all experiments using lipid-protein recombinant membranes (including those containing rhodopsin), IMPs occur on the fracture faces only if amphipatic proteins of high degree of hydrophobicity are embedded in the lipid phase. Another contribution to an understanding of the true nature of IMPs has been provided by the results of investigations on chloroplasts, where specific functional entities can be deleted by genetic selection. In this kind of experiment, a direct correlation between the existence, size, distribution, and density of the IMP classes, with the functional properties of the membrane has been established (Olive *et al.*, 1979).

With reasonable optimism, we may therefore conclude that IMPs contain the photoreceptor protein, which may be associated with other constituents such as lipids. But, if so, why this transmembrane protein is found associated with only one half of the membrane after freeze cleavage, remains to be established. It is not impossible that the mobile protein might be able to jump from one half of the membrane to the other, so dipping into the bilayer only temporarily, as well as being able to rotate freely on its axis and to diffuse rapidly in the plane of the membrane. Such a model has been proposed by Martonosi (1978 and personal communication), for the Ca^{2+}-ATPase of sarcoplasmic reticulum.

No complementarity between the fracture faces is found even in the purple membrane, for which the elegant low-dose electron microscopy experiments of Henderson and Unwin (1975) clearly demonstrate that the α-helical segments of the bacteriorhodopsin are embedded in both leaflets of the bilayer. Postcleavage, plastic deformation, and subsequent precipitation of membrane components on one or both fracture faces have been proposed, in order to explain the lack of complementary pits in the face opposite to that covered with the IMPs.

Another intriguing result that emerges from the freeze-etching experiments is the absence of clear particulate entities exposed at the etched surface of isolated disc membrane. The results showing the accessibility of the rhodopsin polypeptide to protease action indicate strongly that specific segments of the molecule emerge from the lipid bilayer and are exposed mostly at the cytoplasmic surface of the disc membrane. This model would then imply that the exposed segments of the rhodopsin molecules could be visualized on deep etched true surfaces of the disc membrane. This seems not to be the case, since the etched surfaces are almost completely devoid of budding entities. It is worth mentioning again that, in other types of biological membranes, where the polypeptides are exposed to the outer surface and can easily be labeled by specific electron-dense probes (antibodies, peroxidase, lectin, hemocyanin, ferritin), the etched surfaces also have a smooth appearance (Shotton et al., 1978). This negative result is probably due to the fact that, during etching, collapse and sublimation of water-labile structures occur readily, and that the resolution of the replica is insufficient for the visualization of the exposed polypeptides and glycoprotein segments.

Nevertheless, it is conceivable that the exposed polypeptides are of insufficient volume to be visualized as particles either by freeze-etching, or by negative staining. When this volume is sufficiently large, particles may be seen as in negatively stained preparations of isolated inner mitochondrial membranes, isolated rat liver plasma membranes, or intestinal brush border (Kagawa and Racker, 1966; Benedetti and Emmelot, 1968; Louvard et al., 1973), or in etched and negatively stained receptors of acetylcholine (Cartaud et al., 1978). Analysis of particle size may indicate whether rhodopsin molecules are monomeric or oligomeric. The results obtained so far have shown a high degree of heterogeneity. In most cases, the particle size is larger than that expected for a single monomeric polypeptide and generally implies that four or five monomers are associated. However, the ideas that the particles are formed of protein molecules closely associated with lipids and that the particle size variation is correlated to lipid boundary cannot be completely excluded. It has to be admitted that freeze-fracture and freeze-etch have so far provided only rather rough and incomplete information with regard to the molecular organization and localization of the photoreceptor protein.

Protease experiments, in which the rhodopsin is artificially cleaved, provide evidence that the C-terminus of rhodopsin is on the cytoplasmic surface, whereas the N-terminus must be located on the opposite side of the membrane. Further, protease action produced consistent, strongly membrane bound fragments. The native polypeptide must therefore cross the membrane more than once. Hubbell and Fung (1978) arrived at the conclusion that approximately 120 amino acid residues would be involved in at least five α-helical segments of seven turns, each spanning the bilayer thickness and giving rise to a natural channel between the helices.

Another interesting approach to the problem of how the intramembranous folding of the rhodopsin polypeptide represents a coupling device between the outer and inner surfaces is provided by the recently postulated signal hypothesis and by the post translational process that occurs in several newly synthetized polypeptide chains (Blobel and Dobberstein, 1975; Furthmayr, 1977; Kreibich et al., 1978; Sturgess et al., 1978). According to this hypothesis, the synthesis of membrane protein, which begins on free polyribosomes, requires that the newly synthesized peptide carries a hydrophobic signal sequence at the N-terminus, which facilitates its crossing through the hydrophobic membrane barrier. This process, elegantly demonstrated by Sabatini and co-workers (Kreibich et al., 1978), involves the recognition of active ribosomes by membrane receptors (or ribophorin I and II) and gives topographic specificity to the protein biosynthetic process. If those nascent polypeptides are programmed to become intrinsic membrane proteins, a second hydrophobic signal segment will be included close to the C-terminus. Final synthesis and processing of newly synthesized protein may involve cleavage of the N-terminal hydrophobic signal sequence by a peptidase or ''signalase''. Moreover, glycosylation may also intervene by the action of specific glycosyltransferase. But for membrane proteins, glycosylation need not necessarily be associated with cleavage of the first hydrophobic signal sequence, and the second C-sequence might be retained as an intrinsic part of the polypeptide. In this case, the amphipatic character of the nascent protein is retained, and the protein remains strongly associated with the hydrophobic membrane core.

It has not yet been established whether this model is also operative for rhodopsin biosynthesis, but if rhodopsin is assembled within the membrane by such a mechanism, it would give support to the idea that at least three helical segments span the bilayer, probably with a channel running between them.

ACKNOWLEDGMENTS

I wish to thank Dr. E. L. Benedetti for helpful discussions and advice during the preparation of the manuscript. I am especially grateful to Drs. F. J. M. Daemen, W. J. De Grip, P. F. Devaux, and J. Davoust, for their comments and criticism on earlier drafts of this manuscript. Many thanks also to Drs. D. Bok, G. Guidotti, W. L. Hubbell, P. Röhlich, P. J. G. M. Van Breugel, and R. W. Young for providing me with micrographs and data, to Drs. F. J. M. Daemen, W. J. De Grip, E. H. S. Drenthe, and Th. Hendriks for allowing me to utilize some of their unpublished data, and to Dr. M. Chabre for helpful advice. I gratefully acknowledge the International Union of Biochemistry, Macmillan Journal Ltd., Academic Press, Inc. Ltd., and Rockfeller University Press for generously allowing me to reproduce Table V and Figs. 9, 10, 11, 17, and 18.

REFERENCES

Albert, A. D., and Litman, B. J. (1978). *Biochemistry* **17**, 3893.
Anderson, D. H., and Fisher, S. K. (1976). *J. Ultrastruct. Res.* **55**, 119.
Anderson, R. E., and Sperling, L. (1971). *Arch. Biochem. Biophys.* **144**, 673.
Anderson, R. E., Maude, M. B., and Zimmerman, W. (1975). *Vision Res.* **15**, 1087.
Bachmann, L., and Schmitt, W. W. (1971). *Proc. Natl. Acad. Sci. U.S.A.* **68**, 2149.
Bargoot, F. G., Williams, T. P., and Beidler, L. M. (1969). *Vision Res.* **9**, 385.
Baroin, A., Thomas, D. D., Osborne, B., and Devaux, P. F. (1977). *Biochem. Biophys. Res. Commun.* **78**, 442.
Baroin, A., Bienvenue, A., and Devaux, P. F. (1979). *Biochemistry* **18**, 1151.
Baumann, C. (1972). *J. Physiol. London* **222**, 643.
Benedetti, E. L., and Emmelot, P. (1968). *In* "The Membranes" (A. J. Dalton and F. Haguenau, eds.), pp. 33–120. Academic Press, New York.
Benedetti, E. L., Dunia, I., Bentzel, C. J., Vermorken, A. J. M., Kibbelaar, M., and Bloemendal, H. (1976). *Biochem. Biophys. Acta* **457**, 353.
Benedetti, E. L., Dunia, I., Olive, J., and Cartaud, J. (1977). *In* "Structural and Kinetic Approach to Plasma Membrane Functions" (C Nicolau and A. Paraf, eds.), pp. 60–76. Springer-Verlag Berlin and New York.
Bibb, C., and Young, R. W. (1974a). *J. Cell Biol.* **61**, 327.
Bibb, C., and Young, R. W. (1974b). *J. Cell Biol.* **62**, 378.
Bitensky, M. W., Gorman, R. E., and Miller, W. M. (1972). *Science* **175**, 1363.
Blasie, J. K. (1972). *Biophys. J.* **12**, 191.
Blasie, J. K., and Worthington, C. R. (1969). *J. Mol. Biol.* **39**, 417.
Blasie, J. K., Worthington, C. R., and Dewey, M. M. (1969). *J. Mol. Biol.* **39**, 407.
Blaurock, A. E. (1977). *In* "Vertebrate Photoreception" (H. B. Barlow and P. Fatt, eds.), pp. 61–76. Academic, New York.
Blaurock, A. E., and Stoeckenius, W. (1971). *Nature (London)* **233**, 152.
Blaurock, A. E., and Wilkins, M. H. F. (1969). *Nature (London)* **223**, 906.
Blobel, G., and Dobberstein, B. (1975). *J. Cell Biol.* **67**, 835.
Bok, D., and Young, R. W. (1972). *Vision Res.* **12**, 161.
Bok, D., Basinger, S. F., and Hall, M. O. (1974). *Exp. Eye Res.* **18**, 225.
Bok, D., Hall, M. O., and O'Brien, P. J. (1977). *Int. Congr. Cell Biol.,* 1st pp. 608–671.
Bonting, S. L., Caravaggio, L. L., and Canady M. R. (1964). *Exp. Eye Res.* **3**, 47.
Bonting, S. L., De Grip, W. J., Rotmans, J. P., and Daemen, F. J. M. (1974). *Exp. Eye Res.* **18**, 77.
Borggreven, J. M. P. M., Daemen, F. J. M., and Bonting, S. L. (1970). *Biochim. Biophys. Acta* **202**, 374.

Bownds, D., and Wald, G. (1965). *Nature (London)* **205**, 254.

Bownds, D., Gordon-Walker, A., Gaide-Huguenin, A. C., and Robinson, W. (1971). *J. Gen. Physiol.* **58**, 225.

Bownds, D., Dawes, J., Miller, J., and Stahlman, M. (1972). *Nature (London)* **237**, 125.

Bownds, D., Brodie, A., Robinson, W. E., Palmer, D., Miller, J., and Shedlovsky, A. (1974). *Exp. Eye Res.* **18**, 253.

Branton, D. (1966). *Proc. Natl. Acad. Sci. U.S.A.*, **55**, 1048.

Bridges, C. D. B. (1962). *Vision Res.* **2**, 210.

Brown, P. K. (1971). *Biophys. J.* **11**, 248a.

Brown, P. K. (1972). *Nature (London)* **236**, 35.

Brown, M. F., Miljanich, G. P., and Dratz, E. A. (1977a). *Proc. Natl. Acad. Sci. U.S.A.* **74**, 1978.

Brown, M. F., Miljanich, G. P., and Dratz, E. A. (1977b). *Biochemistry* **16**, 2640.

Callender, R. H., Doukas, A., Crouch, R., and Nakaniski, K. (1976). *Biochemistry* **15**, 1621.

Cartaud, J., Benedetti, E. L., Sobel, A., and Changeux, J. P. (1978). *J. Cell Sci.* **29**, 313.

Cavanagh, H. E., and Wald, G. (1969). *Fed. Proc.* **28**, 344.

Chabre, M. (1975). *Biochim. Biophys. Acta* **382**, 322.

Chabre, M. (1978). *Proc. Natl. Acad. Sci. U.S.A.* **75**, 5471.

Chabre, M., and Breton, J. (1979). *Vision Res.* **19**, 1005.

Chabre, M., and Cavaggioni, A. (1973). *Nature (London)* **244**, 118.

Chabre, M., and Cavaggioni, A. (1975). *Biochim. Biophys. Acta* **382**, 336.

Chabre, M., Cavaggioni, A., Osborne, H. B., Gulik-Krzywicki, T., and Olive, J. (1972). *FEBS Lett.* **26**, 197.

Chader, G. J., Fletcher, R. T., O'Brien, P. J., and Krishna, G. (1976). *Biochemistry* **15**, 1615.

Chen, Y. S., and Hubbell, W. L. (1973). *Exp. Eye Res.* **17**, 517.

Clark, A. W., and Branton, D. (1968). *Z. Zellforsch. Mikrosk. Anat.* **91**, 586.

Cohen, A. I. (1968). *J. Cell Biol.* **37**, 424.

Cohen, A. I. (1972). *In* "Handbook of Sensory Physiology" (M. G. F. Fuortes, ed.), Vol. VII/2, pp. 63–110. Springer-Verlag Berlin and New York.

Collins, F. D. (1953). *Nature (London)* **171**, 469.

Collins, F. D., Love, R. M., and Morton, R. A. (1952). *Biochem. J.* **51**, 292.

Cone, R. A. (1972). *Nature (London)* **236**, 39.

Corless, J. M. (1972). *Nature (London)* **237**, 229.

Corless, J. M., Cobbs, W. H., Costello, M. J., and Robertson, J. D. (1976). *Exp. Eye Res.* **23**, 295.

Costello, M. J., and Gulik-Krzywicki, T. (1976). *Biochim. Biophys. Acta* **455**, 412.

Crain, R. C., Marinetti, G. V., and O'Brien, D. F. (1978). *Biochemistry* **17**, 4186.

Daemen, F. J. M. (1973). *Biochim. Biophys. Acta* **300**, 255.

Daemen, F. J. M., Jansen, P. A. A., and Bonting, S. L. (1971). *Arch. Biochem. Biophys.* **145**, 300.

Daemen, F. J. M., De Grip, W. J., and Jansen, P. A. A. (1972). *Biochim. Biophys. Acta* **271**, 419.

Daemen, F. J. M., Schnetkamp, P. P. M., Hendriks, Th., and Bonting, S. L. (1977). *In* "Vertebrate Photoreception" (M. B. Barlow and P. Fatt, eds.), pp. 29–40. Academic Press, New York.

Dartnall, H. J. A. (1972). *In* "Handbook of Sensory Physiology" (H. J. A. Dartnall, ed.) Vol. II/I pp. 122–145. Springer Verlag, Berlin and New York.

Davoust, J., and Devaux, P. F. (1978). *Int. Biophysic Congr., 6th, Kyoto* p. 174.

De Grip, W. J. (1974). Ph.D. Thesis, Catholic University of Nijmegen, Nijmegen, The Netherlands.

De Grip, W. J., Daemen, F. J. M., and Bonting, S. L. (1972). *Vision Res.* **12**, 1697.

De Grip, W. J., Van De Laar, G. L. M., Daemen, F. J. M., and Bonting, S. L. (1973a). *Biochim. Biophys. Acta* **325**, 315.

De Grip, W. J., Bonting, S. L., and Daemen, F. J. M. (1973b). *Biochim. Biophys. Acta* **303**, 189.

De Grip, W. J., Bonting, S. L., and Daemen, F. J. M. (1975). *Exp. Eye Res.* **21**, 549.

De Grip, W. J., Daemen, F. J. M., and Bonting, S. L. (1979). *Methods Enzymol.* (in press)

De Pierre, J. W., and Ernster, L. (1977). *Annu. Rev. Biochem.* **46**, 201.
De Pont, J. J. H. M., Daemen, F. J. M., and Bonting, S. L. (1970). *Arch. Biochem. Biophys.* **140**, 275.
De Robertis, E., and Franchi, C. M. (1956). *J. Biophys. Biochem. Cytol.* **2**, 307.
Dewey, M. M., Davis, P. K., Blasie, J. K., and Barr, L. (1969). *J. Mol. Biol.* **39**, 395.
Dowling, J. E. (1970). *Invest. Ophtalmol.* **9**, 655.
Engelhardt, P., and Storteir, S. (1975). *J. Ultrastruct. Res.* **50**, 369.
Etingof, R. N., Berman, A. L., Govardovskii, V. I., and Leont'ev, V. G. (1970). *Biochim. Biophys. Acta* **205**, 459.
Etingof, R. N., Sobota, A., and Ostapenko, I. A. (1972). *Biokhimiya* **37**, 1172.
Fager, R. S., Sejnowski, P., and Abrahamson, E. W. (1972). *Biochem. Biophys. Res. Commun.* **47**, 1244.
Farber, D. B., and Lolley, R. N. (1976). *Exp. Eye Res.* **22**, 219.
Favre, E., Baroin, A., Bienvenue, A., and Devaux, P. F. (1979). *Biochemistry* **18**, 1156.
Fernandex-Morán, H. (1962). *Circulation* **26**, 1039.
Fishman, M. L., Oberc, M. A., Hess, H. H., and Engel, W. K. (1977). *Exp. Eye Res.* **24**, 341.
Fleischer, S., and McConnell, D. G. (1966). *Nature (London)* **212**, 1366.
Frank, R. N., and Buzney, S. M. (1975). *Biochemistry* **14**, 5110.
Fung, B. K. K., and Hubbell, W. L. (1978). *Biochemistry* **17**, 4396.
Furthmayr, H. (1977). *J. Supramol. Struct.* **7**, 121.
Futterman, S. (1963). *J. Biol. Chem.* **238**, 1145.
Futterman, S., and Andrews, J. S. (1964). *J. Biol. Chem.* **239**, 81.
Godfrey, A. J. (1973). *J. Ultrastruct. Res.* **43**, 228.
Goridis, C., and Virmaux, N. (1974). *Nature (London)* **248**, 57.
Goridis, C., Urban, P. F., and Mandel, P. (1977). *Exp. Eye Res.* **24**, 171.
Govardovskii, V. I. (1971). *Nature (London)* **234**, 53.
Grant, C. W. M., and McConnell, H. M. (1974). *Proc. Natl. Acad. Sci. U.S.A.* **71**, 4653.
Gras, W. J., and Worthington, C. R. (1969). *Proc. Natl. Acad. Sci. U.S.A.* **63**, 233.
Gross, H., Bas, E., and Moor, H. (1978). *J. Cell Biol.* **76**, 712.
Guidotti, G. (1977). *J. Supramol. Struct.* **7**, 489.
Gulik-Krzywicki, T., and Costello, M. J. (1977). *Proc. Annu. EMSA Meet., 35th* pp. 330–333.
Gulik-Krzywicki, T., and Costello, M. J. (1978). *J. Microsc.* **112**, 103.
Hagins, W. A., and Jennings, W. H. (1959). *Trans. Faraday Soc.* **27**, 180.
Hagins, W. A., and Yoshikami, S. (1975). *Ann. N.Y. Acad. Sci.* **264**, 314.
Hall, M. O., Bok, D., and Bacharach, A. D. E. (1969). *J. Mol. Biol.* **45**, 397.
Hargrave, P. A., and Fong, S. L. (1977). *J. Supramol. Struct.* **6**, 559.
Heitzman, H. (1972). *Nature (London)* **235**, 114.
Heller, J. (1968). *Biochemistry* **7**, 2906.
Heller, J., and Lawrence, M. A. (1970). *Biochemistry* **9**, 864.
Hemminski, K. (1975). *Exp. Eye Res.* **20**, 79.
Henderson, R., and Unwin, P. N. T. (1975). *Nature (London)* **257**, 28.
Hendricks, Th., De Pont, J. J. H. M., Daemen, F. J. M., and Bonting, S. L. (1973). *Biochim. Biophys. Acta* **330**, 156.
Hendriks, Th., Daemen, F. J. M., and Bonting, S. L. (1974). *Biochim. Biophys. Acta* **345**, 468.
Hendriks, Th., Klompmakers, A. R., Daemen, F. J. M., and Bonting, S. L. (1976). *Biochim. Biphys. Acta* **433**, 271.
Hendricks, Th., Van Haard, P. M. M., Daemen, F. J. M., and Bonting, S. L. (1977). *Biochim. Biophys. Acta* **467**, 175.
Hogan, M. J. (1972). *Trans. Am. Acad. Ophthalmol. Oto-Lar.* **76**, 64.
Hogan, M. J., Wood, I., and Steinberg, R. H. (1974). *Nature (London)* **252**, 305.

Hong, K., and Hubbell, W. L. (1972). *Proc. Natl. Acad. Sci. U.S.A.* **69**, 2617.

Hong, K., and Hubbell, W. L. (1973). *Biochemistry* **12**, 4517.

Hubbard, R. (1954). *J. Gen. Physiol.* **37**, 381.

Hubbard, R., and Dowling, J. E. (1962). *Nature (London)* **193**, 341.

Hubbard, R., and Wald, G. (1952). *Science* **115**, 60.

Hubbell, W. L., and Fung, B. K. K. (1978). *In* "Membrane Transduction Mechanism" (R. A. Cone and J. Dowling, eds.). Raven, New York.

Hubbell, W., Fung, K. K., Hong, K., and Chen, Y. S. (1977). *In* "Vertebrate Photoreception," (H. B. Barlow and P. Fatt, eds.), pp. 41–59. Academic Press, New York.

Jan, L., and Revel, J. P. (1974). *J. Cell Biol.* **62**, 257.

Kagawa, Y., and Racker, E. (1966). *J. Biol. Chem.* **241**, 2475.

Kahn, C. R. (1976). *J. Cell Biol.* **70**, 261.

Kaupp, U. B., Schnetkamp, P. P. M., and Junge, W. (1979). *Biochim. Biophys. Acta* **552**, 390.

Kissun, R. D., Graymore, C. N., and Newhouse, P. J. (1972). *Exp. Eye Res.* **14**, 150.

Klip, A., Darszon, A., and Montal, M. (1976). *Biochem. Biophys. Res. Commun.* **72**, 1350.

Knowles, A. (1976). *Biochem. Biophys. Res. Commun.* **73**, 56.

Knowles, A., and Dartnall, H. J. A. (1977). *In* "The Eye" (H. Davson, ed.), Vol. 2B. Academic Press, New York.

Krebs, W., and Kühn, H. (1977). *Exp. Eye Res.* **25**, 511.

Kreibich, G., Czakó-Graham, M., Grebenau, R., Mok, W., Rodriguez-Boulan, E., and Sabatini, D. S. (1978). *J. Supramol. Struct.* **8**, 279.

Kremmer, T., Wisher, M. H., and Evans, W. H. (1976). *Biochim. Biophys. Acta* **455**, 655.

Kühn, H., and Dreyer, W. J. (1972). *FEBS Lett.* **20**, 1.

Kühn, H., Cook, J. H., and Dreyer, W. J. (1973). *Biochemistry* **12**, 2495.

Kusumi, A., Ohnishi, S., Ito, T., and Yoshizawa, T. (1978). *Biochim. Biphys. Acta* **507**, 539.

Laties, A. M., and Liebman, P. A. (1970). *Science* **168**, 1475.

Leeson, T. S. (1970). *Can. J. Ophthalmol.* **5**, 91.

Leeson, T. S. (1971). *J. Anat.* **108**, 147.

Levy, M., and Sauner, M. T. (1968). *Chem. Phys. Lipids* **2**, 291.

Lewis, A., Fager, R. S., and Abrahamson, E. W. (1973). *J. Raman Spectrosc.* **1**, 465.

Lewis, M. S., Kreig, L. C., and Kirk, W. D. (1974). *Exp. Eye Res.* **18**, 29.

Liebman, P. A. (1974). *Invest. Ophtalmol.* **13**, 700.

Liebman, P. A., and Entine, G. (1974). *Science* **185**, 457.

Lion, F., Rotmans, J. P., Daemen, F. J. M., and Bonting, S. L. (1975). *Biochim. Biophys. Acta* **384**, 283.

Lolley, R. N., and Hess, H. N. (1969). *J. Cell. Physiol.* **73**, 9.

Louvard, D., Maroux, S., Baratti, J., Desnuelle, P., and Mutaftschiev, S. (1973). *Biochim. Biophys. Acta* **291**, 747.

Makino, M., Hamanaka, T., Orii, Y., and Kito, Y. (1977). *Biochim. Biophys. Acta* **495**, 299.

Manthorpe, M., and McConnell, D. G. (1975). *Biochim. Biophys. Acta* **403**, 438.

Martonosi, A. (1978). *FEBS Fed. Eur. Biochem. Meet., 11th Copenhagen 1977* **45**, 135.

Mason, W. T., Fager, R. S., and Abrahamson, E. W. (1973). *Fed. Proc.* **32**, 327.

Mason, W. T., Fager, R. S., and Abrahamson, E. W. (1974). *Nature (London)* **247**, 188.

Mathies, R., Oseroff, A. R., and Stryer, L. (1976). *Proc. Natl. Acad. Sci. U.S.A.* **73**, 1.

Matthews, R. G., Hubbard, R., Brown, P. K., and Wald, G. (1963). J. Gen. Physiol. **47**, 215.

Miki, N., Keirns, J. J., Marcus, F. R., Freeman, J., and Bitensky, M. W. (1973). *Proc. Natl. Acad. Sci. U.S.A.* **70**, 3820.

Miki, N., Baraban, J. M., Keirns, J. J., Boyce, J. J., and Bitensky, M. W. (1975). *J. Biol. Chem.* **250**, 6320.

Miller, W. H., Gorman, R. E., and Bitensky, M. W. (1971). *Science* **174**, 295.

Miller, J. A., Brodie, A. E., and Bownds, M. D. (1975). *FEBS Lett.* **59,** 20.
Miller, J. A., Paulsen, R., and Bownds, M. D. (1977). *Biochemistry* **16,** 2633.
Molday, R. S. (1976). *J. Supramol. Struct.* **4,** 549.
Morton, R. A., and Pitt, G. A. J. (1957). *In* "Fortschr. Chem. Organ. Naturstoffe" (W. Herz, H. Grisebach, and G. W. Kirby, eds.), Vol. XIV, pp. 244–316. Springer, Wien.
Mutaftschiev, S., Olive, J., Bertrand, J. C., and Azoulay, E. (1977). *Biol. Cell.* **29,** 17.
Nielsen, N. C., Fleischer, S., and McConnell, D. G. (1970). *Biochim. Biophys. Acta* **211,** 10.
Nilsson, S. E. G. (1964). *J. Ultrastruct. Res.* **11,** 581.
Nilsson, S. E. G. (1965). *J. Ultrastruct. Res.* **12,** 207.
O'Day, W. T., and Young, R. W. (1978). *J. Cell Biol.* **76,** 593.
Olive, J., and Benedetti, E. L. (1974). *Mol Biol. Rep.***1,** 245.
Olive, J., and Recouvreur, M. (1977). *Exp. Eye Res.* **25,** 63.
Olive, J., and Recouvreur, M. (1978). *Int. Congr. Eye Res. 3rd* Osaka **VII-10,** 131.
Olive, J., Benedetti, E. L., Van Breugel, P. J. G. M., Daemen, F. J. M., and Bonting, S. L. (1978). *Biochim. Biophys. Acta* **509,** 129.
Olive, J., Wollman, F. A., Bennoun, P., and Recouvreur, M. (1979). *Mol. Biol. Rep.* **5,** 139.
Osborne, H. B. (1976). *FEBS Lett.* **67,** 23.
Osborne, H. B., and Nabedryk-Viala, E. (1978). *Eur. J. Biochem.* **44,** 383.
Osborne, H. B., Sardet, C., Villaz, M. M., and Chabre, M. (1978). *J. Mol. Biol.* **123,** 177.
Oseroff, A. R., and Callender, R. H. (1974). *Biochemistry* **13,** 4243.
Ostroy, E. O., Rudney, H., and Abrahamson, E. W. (1966). *Biochim. Biophys. Acta* **126,** 409.
Palade, G. E. (1978). *In* "Molecular Specialization and Symmetry in Membrane Function" (A. K. Solomon and M. Karnovsky, eds.), pp. 1–30. Harvard Univ. Press, Cambridge, Massachusetts.
Pannbacker, R. G., Fleischman, D. E., and Reed, D. E. (1972). *Science* **175,** 757.
Papermaster, D. S., and Dreyer, W. J. (1974). *Biochemistry* **13,** 2438.
Peters, K., Applebury, M. L., and Rentzepis, P. M. (1977). *Proc. Natl. Acad. Sci. U.S.A.* **74,** 3119.
Petersen, D. C., and Cone, R. A. (1975). *Biophys. J.* **15,** 1181.
Pinto da Silva, P., and Nicolson, G. L. (1974). *Biochim. Biophys. Acta* **363,** 311.
Pinto da Silva, P., Moss, P. S., and Fudenberg, H. H. (1973). *Exp. Cell Res.* **81,** 127.
Plantner, J. J., and Kean, E. L. (1976). *J. Biol. Chem.* **251,** 1548.
Pober, J., and Stryer, W. (1975). *J. Mol. Biol.* **95,** 477.
Poincelot, R. P., and Abrahamson, E. W. (1970). *Biochemistry* **9,** 1820.
Poincelot, R. P., Millar, P. G., Kimbel, R. L., and Abrahamson, E. W. (1970). *Biochemistry* **9,** 1809.
Pontus, M., and Delmelle, M. (1975). *Biochim. Biophys. Acta* **401,** 221.
Poo, M. M., and Cone, R. A. (1973). *Exp. Eye Res.* **17,** 503.
Radding, C. M., and Wald, G. (1958). *J. Gen. Physiol.* **42,** 371.
Rafferty, C. N. (1977). *Biophys. Struct. Mech.* **3,** 123.
Rafferty, C. N., Cassim, J. Y., and McConnell, D. G. (1977). *Biophys. Struct. Mech.* **2,** 277.
Raubach, R. A., Nemes, P. P., and Dratz, E. A. (1974). *Exp. Eye Res.* **18,** 1.
Raviola, E., and Gilula, N. B. (1973). *Proc. Natl. Acad. Sci. U.S.A.,* **70,** 1677.
Raviola, E., and Gilula, N. B. (1975). *J. Cell Biol.* **65,** 192.
Renthal, R., Steinemann, A., and Stryer, L. (1973). *Exp. Eye Res.***17,** 511.
Rimai, L., Kilponen, R. G., and Gill, D. (1970). *Biochem. Biophys. Res. Commun.* **41,** 492.
Robinson, W. E., Gordon-Walker, A., and Bownds, D. (1972). *Nature (London)* **235,** 112.
Röhlich, P. (1976). *Nature (London)* **263,** 789.
Rojas, E., and Bergman, C. (1977). *Trends Biochem. Sci.* **2,** 6.
Rosenkranz, J. (1977). *Int. Rev. Cytol.* **50,** 26.
Rotmans, J. P. (1973). Ph.D. Thesis, Catholic University of Nijmegen, Nijmegen, The Netherlands.
Rotmans, J. P., Daemen, F. J. M., and Bonting, S. L. (1974). *Biochim. Biophys. Acta* **357,** 151.

Rousselet, A., and Devaux, P. F. (1978). *FEBS Lett.* **93,** 161.

Saari, J. C. (1974). *J. Cell Biol.* **63,** 480.

Saibil, H., Chabre, M., and Worcester, D. (1976). *Nature (London)* **262,** 266.

Saito, Z. (1938). *Tohoku J. Exp. Med.* **32,** 432.

Sardet, C. (1977). *In* "Structure and Kinetic Approach to Plasma Membrane Functions" (C. Nicolau and A. Paraf, eds.), pp. 91–103. Springer-Verlag Berlin and New York.

Sardet, C., Tardieu, A., and Luzzati, V. (1976). *J. Mol. Biol.* **105,** 383.

Schmidt, W. J. (1938). *Kolloidzeitschrift* **85,** 137.

Schmidt, S. Y., and Lolley, R. N. (1973). *J. Cell Biol.* **57,** 117.

Schnetkamp, P. P. M., Daemen, F. J. M., and Bonting, S. L. (1977). *Biochim. Biophys. Acta* **468,** 259.

Schnetkamp, P. P. M., Klompmakers, A. A., and Daemen, F. J. M. (1979). *Biochim. Biophys. Acta* **552,** 379.

Schoot, B. M., De Pont, J. J. H. M., and Bonting, S. L. (1978). *Biochim. Biophys. Acta* **522,** 602.

Schwartz, S., Cain, J. E., Dratz, E. A., and Blasie, J. K. (1975). *Biophys. J.* **15,** 1201.

Segrest, J. P., Gulik-Krzywicki, T., and Sardet, C. (1974). *Proc. Natl. Acad. Sci. U.S.A.* **71,** 3294.

Shaper, J. H., and Stryer, L. (1977). *J. Supramol. Struct.* **6,** 291.

Shechter, Y., Schlessinger, J., Jacobs, S., Chang, K. J., and Cuatrecasas, P. (1978). *Proc. Natl. Acad. Sci. U.S.A.* **75,** 2135.

Shichi, H. (1971). *J. Biol. Chem.* **246,** 6178.

Shichi, H. (1973). *Exp. Eye Res.* **17,** 533.

Shichi, H., and Shelton, E. (1974). *J. Supramol. Struct.* **2,** 7.

Shichi, H., and Somers, R. L. (1978). *J. Biol. Chem.* **253,** 7040.

Shichi, H., Lewis, M. S., Irreverre, F., and Stone, A. L. (1969). *J. Biol. Chem.* **244,** 529.

Shotton, D., Thompson, K., Wofsy, L., and Branton, D. (1978). *J. Cell Biol.* **76,** 512.

Sitaramayya, A., Virmaux, N., and Mandel, P. (1977a). *Exp. Eye Res.* **25,** 163.

Sitaramayya, A., Virmaux, N., and Mandel, P. (1977b). *Neurochem. Res.* **2,** 1.

Sjöstrand, F. S. (1953). *J. Cell. Comp. Physiol.* **42,** 15.

Sjöstrand, F. S. (1958). *J. Ultrastruct. Res.* **2,** 122.

Sjöstrand, F. S. (1961). *In* "The Structure of the Eye" (G. K. Smelser, ed.), pp. 1–28. Academic Press, New York.

Sjöstrand, F. S. (1963a). *J. Ultrastruct. Res.* **9,** 561.

Sjöstrand, F. S. (1963b). *J. Ultrastruct. Res.* **8,** 517.

Sjöstrand, F. S., and Elfvin, L. G. (1962). *J. Ultrastruct. Res.* **7,** 504.

Smith, H. G., Jr., Stubbs, G. W., and Litman, B. J. (1975). *Exp. Eye Res.* **20,** 211.

Staehelin, L. A. (1976). *J. Cell Biol.* **71,** 136.

Staehelin, L. A., Chlapowski, F. J., and Bonneville, M. A. (1972). *J. Cell Biol.* **53,** 73.

Steinemann, A., and Stryer, L. (1973). *Biochemistry* **12,** 1499.

Steineman, A., Wu, C. W., and Stryer, L. (1973a). *J. Supramol. Struct.* **1,** 348.

Steinemann, A., Renthal, R., and Stryer, L. (1973b). Abstracts *Ann. Meet. Biophys. Soc., 17th* p. 231a.

Strackee, L. (1972). *Photochem. Photobiol.* **15,** 253.

Sturgess, J., Moscarello, M., and Schachter, H. (1978). *Curr. Top. Membr. Transp.* **11,** 15.

Sundstrom, V., Remtzepis, P. M., Peters, K., and Applebury, M. L. (1977). *Nature (London)* **267,** 645.

Szuts, E. Z., and Cone, R. A. (1977). *Biochim. Biophys. Acta* **468,** 194.

Tonosaki, A., Yamasaki, M., Washioka, H., and Mizoguchi, J. (1978). *Int. Congr. Electron Microsc. 9th, Toronto* **II,** 156.

Towner, P., Sale, G. J., and Akhtar, M (1977). *FEBS Lett.* **76,** 51.

Trayhurn, P., Mandel, P., and Virmaux, N. (1974a). *FEBS Lett.* **38**, 351.

Trayhurn, P., Mandel, P., and Virmaux, N. (1974b). *Exp. Eye Res.* **19**, 259.

Trayhurn, P., Habgood, J. O., and Virmaux, N. (1975). *Exp. Eye Res.* **220**, 479.

Van Breugel, P. J. G. M. Ph.D. Thesis, Catholic University of Nijmegen, Nijmegen, The Netherlands.

Van Breugel, P. J. G. M., Daemen, F. J. M., and Bonting, S. L. (1975). *Exp. Eye Res.* **21**, 315.

Van Breugel, P. J. G. M., Geurtz, P. H. M., Daemen, F. J. M., and Bonting, S. L. (1978). *Biochim. Biophys. Acta* **509**, 136.

Vandenberg, G. A., Gaw, J. E., Dratz, E. A., and Swartz, S. (1976). *Abstr. Annu. Meet. Biophys. Soc., 20th* p. 37a.

Van Winkle, W. B., Lane, L. K., and Schwartz, A. (1976). *Exp. Cell Res.* **100**, 291.

Verkleij, A., Van Alphen, L., Bijvelt, J., and Lugtenberg, B. (1977). *Biochim. Biophys. Acta* **466**, 269.

Ververgaert, P. H. J. Th., and Verkleij, A. J. (1978). *Int. Congr. Electron Microsc., 9th, Toronto* **II**, 154.

Virmaux, N. (1977). *Biophys. Struct. Mech.* **3**, 128.

Virmaux, N., Nullans, G., and Goridis, C. (1976). *J. Neurochem.* **26**, 233.

Vogel, F., Meyer, H. W., Grosse, R., and Repke, K. R. H. (1977). *Biochim. Biophys. Acta* **470**, 497.

Wald, G., and Brown, P. K. (1950). *Proc. Natl. Acad. Sci. U.S.A.* **36**, 84.

Wald, G., and Brown, P. K. (1958). *Science* **127**, 222.

Wald, G., and Hubbard, R. (1960). *In* "The Enzymes" (P. D. Boyer, H. Lardy, and K. Myrbäck, eds.), Vol. 3, pp. 369–386. Academic Press, New York.

Warren, G. B., Toon, P. A., Birdsall, N. J. M., Lee, A. G., and Metcalfe, J. C. (1974). *Proc. Natl. Acad. Sci. U.S.A.* **71**, 622.

Warren, G. B., Houslay, M. D., Metcalfe, J. C., and Birdsall, N. J. M. (1975). *Nature (London)* **255**, 684.

Watts, A., Wolatovski, I., and Marsh, D. (1978). *Int. Conf. Magn. Resonance Biol. Syst., 7th Nara, Japan* p. 80.

Webbs, N. G. (1972). *Nature (London)* **235**, 44.

Weller, M., Goridis, C., Virmaux, N., and Mandel, P. (1975a). *Exp. Eye Res.* **21**, 405.

Weller, M., Virmaux, N., and Mandel P. (1975b). *Nature (London)* **256**, 68.

Wheeler, G. L., and Bitensky, M. W. (1977). *Proc. Natl. Acad. Sci. U.S.A.* **74**, 4238.

Wheeler, G. L., Matuo, Y., and Bitensky, M. W. (1977). *Nature (London)* **269**, 822.

Williams, T. P., Baker, B. N., and McDowell, J. H. (1974). *Exp. Eye Res.* **18**, 69.

Woodruff, M. L., Bownds, D., Green, S. H., Morrisey, J. L., and Shedlovsky, A. (1977). *J. Gen. Phys.* **69**, 667.

Worthington, C. R. (1971). *Fed. Proc.* **30**, 57.

Wu, C. W., and Stryer, L. (1972). *Proc. Natl. Acad. Sci. U.S.A.* **69**, 1104.

Yaeger, M., Schoenborn, B. P., Engelman, D. M., Moore, P. B., and Stryer, L. (1976). *Biophys. J.* **16**, 36a.

Yee, R., and Liebman, P. A. (1978). *J. Biol. Chem.* **253**, 8902.

Yoshikami, S., and Hagins, W. A. (1971). *Biophys. J.* **11**, 47a.

Yoshikami, S., and Hagins, W. A. (1973). *In* "Biochemistry and Physiology of Visual Pigments" (H. Langer, ed.), pp. 245–255. Springer-Verlag Berlin and New York.

Yoshikami, S., and Hagins, W. A. (1976). *Biophys. J.* **16**, 35a.

Young, R. W. (1967). *J. Cell Biol.* **33**, 61.

Young, R. W. (1968). *J. Ultrastruct. Res.* **23**, 462.

Young, R. W. (1969). *Inv. Ophthalmol* **8**, 222.

Young, R. W. (1971). *Vision Res.* **11**, 1.

Young, R. W. (1973). *Ann. Ophthalmol.* **5,** 843.

Young, R. W. (1977). *J. Ultrastruct. Res.* **501,** 172.

Young, R. W. (1978). *Inv. Ophthalmol* **17,** 105.

Young, R. W., and Bok, D. (1969). *J. Cell Biol.* **42,** 392.

Young, R. W., and Droz, B. (1968). *J. Cell Biol.* **39,** 169.

Zambrano, F., Fleischer, S., and Fleischer, B. (1975). *Biochim. Biophys. Acta* **380,** 357.

Zimmerman, W. F. (1976). *Exp. Eye Res.* **23,** 159.

Zimmerman, W. F., Lion, F., Daemen, F. J. M., and Bonting, S. L. (1975). *Exp. Eye Res.* **21,** 325.

Zimmerman, W. F., Daemen, F. J. M., and Bonting, S. L. (1976). *J. Biol. Chem.* **251,** 4700.

Zorn, M. (1974). *Exp. Eye Res.* **19,** 215.

Zorn, M., and Futterman, S. (1971). *J. Biol. Chem.* **246,** 881.

INTERNATIONAL REVIEW OF CYTOLOGY, VOL. 64

The Roles of Transport and Phosphorylation in Nutrient Uptake in Cultured Animal Cells

ROBERT M. WOHLHUETER AND PETER G.W. PLAGEMANN

Department of Microbiology, University of Minnesota, Minneapolis, Minnesota

I. Introduction 171
II. Theoretical and Methodological Considerations 173
   A. Nomenclature 173
   B. Transport Models 174
   C. Phosphorylation Models 176
   D. Kinetic Behavior of Transport and Phosphorylation Operating in
      Tandem 177
   E. Dissociating Transport from Phosphorylation 190
III. Uptake of Nucleosides 192
   A. The Nucleoside Transport System 192
   B. Nucleoside Kinases 195
   C. Nucleoside Transport and Phosphorylation in Tandem . . . 200
   D. Does Nonmediated Permeation Contribute to the Uptake of
      Nucleosides? 203
   E. Physiological Regulation of Nucleoside Uptake 204
   F. Effect of Transport Inhibitors on Uptake 207
IV. Uptake of Nucleobases 207
   A. Transport of Purines 208
   B. Phosphoribosylation of Purines 210
   C. Transport and Phosphoribosylation in Tandem 211
   D. Physiological Regulation of Purine Uptake 212
V. Uptake of Hexoses 213
   A. Transport of Hexoses 214
   B. Phosphorylation of Hexoses 217
   C. Hexose Transport and Phosphorylation in Tandem 219
   D. Nutritional and Hormonal Regulation of Hexose Uptake . . 223
   E. Malignant Transformation and Hexose Uptake 227
VI. Uptake of Vitamins and Choline 230
VII. Concluding Discussion 232
   References 234

I. Introduction

Among the many functions of the cell membrane, maintenance of selective permeability is primary in both the temporal and physiological senses. The lipid bilayer surrounding the cell is essentially impermeable to hydrophilic substances

171

(Lieb and Stein, 1974), and thus retains within the confines of the cell those products of biosynthesis essential to cell life. But the permeability barrier is not absolute: waste products of metabolism leave the cell, cells communicate with one another via chemical messengers, and cells exploit a number of essential and/or useful nutrients present in the extracellular space.

To facilitate this exploitation of nutrients, cells have evolved mechanisms by which molecules of nutrient, intrinsically unable to permeate a lipid bilayer, gain entry into the cell. Mammalian cells possess two general types of mechanism for bringing small nutrient molecules into the cell. One of these, exemplified in amino acid transport (Christensen, 1975), is an energy-dependent transport system able to establish a concentration gradient of the nutrient across the cell membrane. The other, best exemplified by monosaccharide, nucleoside, and nucleobase transport, consists of a nonconcentrative transport system coupled with phosphorylation or phosphoribosylation of the substrate, whereby the metabolic alteration prevents the nutrient from exiting the cell (as well as "activates" it for subsequent metabolic purposes). As a result, the anionic derivative of the nutrient molecule is trapped within and is at the disposal of the cell. In this article, we consider this second mode of nutrient uptake as it occurs in cultured mammalian cells.

The uptake of hexoses, nucleosides, and nucleobases, or operationally stated, the incorporation of radiolabeled, exogenous substance into acid-soluble and/or acid-insoluble cellular material, has been studied in a variety of cultured animal cells (see reviews by Plagemann and Richey, 1974; Berlin and Oliver, 1975; Perdue, 1979). In many instances, the rates of uptake have been related to exogenous substrate concentration, and the effects of inhibitors and temperature on nutrient uptake have been described.

Of particular interest to the cell physiologist are the many observations that the rate of uptake of a particular nutrient responds to the physiological demand. Thus, for example, the uptake of uridine in quiescent cells is increased soon after stimulation with serum or hormones (Cunningham and Pardee, 1969; Jimenez de Asua et al., 1974); the uptakes of nucleosides and hexoses are inversely related to the cellular cAMP level, or decrease upon prolonged treatment with dibutyryl-cAMP (Hatanaka, 1974; Sheppard, 1972; Hauschka et al., 1972; Kram et al., 1973); lectin-stimulated lymphocytes increase their uptake of thymidine in preparation of new DNA synthesis (Barlow and Ord, 1975); and viral transformation elicits increases in the uptake of hexoses (Wallach, 1975; Perdue, 1979).

The correlation between enhanced nutrient uptake and stimulation of cell growth or proliferation had led many investigators to postulate that modulation of nutrient uptake is centrally involved in the regulation of cell proliferation (Pardee, 1971; Holley, 1972; Bhargava, 1977), and, more specifically, that modulation of uptake is essentially a membrane phenomenon, i.e., that messages impinging on the cell membrane are more or less directly transduced into altered permeability of the membrane toward specific nutrients. On the other hand, there is considerable evidence that intracellular phosphorylation (or phosphoribosylation) determines the rate of uptake of nucleosides (Wohlhueter *et al.*, 1976; Rozengurt *et al.*, 1977; Heichal *et al.*, 1979), nucleobases (Marz *et al.*, 1979), and hexoses (Colby and Romano, 1975; Graff *et al.*, 1978) in several cell lines and under a variety of conditions. The implication of this evidence is that modulation of uptake is essentially the result of regulation of the activities of those enzymes responsible for trapping the nutrient within the cell.

We endeavor in this article to resolve these conflicting interpretations. In Section II, we approach the question theoretically: What kinetic behavior might be expected of a simple uptake system comprising membrane permeation of substrate followed by intracellular conversion to a nonpermeable derivative? How might the kinetic characteristics of the component steps, permeation and metabolic conversion, influence the overall process of uptake of exogenous substrate into the cell? By what criteria can one distinguish whether permeation or phosphorylation is the rate-determining step?

In Sections III through VI, we examine critically the experimental evidence. Especially relevant to that discussion are numerous successful attempts to dissect uptake, kinetically and/or chemically, so as to reveal the operation of its component parts. This new knowledge is then applied to a reinterpretation of some older uptake data.

Finally, we draw some general conclusions about the relative roles of transport and phosphorylation in the uptake of phosphorylatable nutrients in cultured cells and offer some specific recommendations on the methodological requisites for distinguishing these roles.

II. Theoretical and Methodological Considerations

A. Nomenclature

Our theoretical treatment proceeds from the premise that, in essence, the uptake of sugars, nucleosides, and nucleobases into mammalian cells may be described as a two-step process: the permeation of exogenous substrate across the cell membrane, followed by phosphorylation or phosphoribosylation of intracellular substrate to yield metabolites to which the cell membrane is virtually impermeable. In our attempt to elucidate the kinetic behavior of the overall process, we rely, so far as possible, on established nomenclature of its component parts.

The general term *permeation* means the transfer of substrate from one side of the membrane to the other in chemically unaltered form. Permeation may be *nonmediated,* in the operational sense that specificity or saturability cannot be demonstrated, or *mediated* by some specific, saturable mechanism. The latter case we designate as *transport*. The general term *translocation* implies that the transfer of substrate across the membrane necessarily results in its appearance in chemically altered form. *Uptake,* which is to be fastidiously distinguished from transport, is defined operationally as the appearance within the cell of radioactivity derived from an exogenous substrate. *Incorporation* is the net transfer of radioisotope from one designated compound (or class of compounds) to another.

Transport may be *passive* (also known as "facilitated diffusion" or "nonconcentrative transport") or *active,* by which is implied the capacity to establish an electrochemical gradient across the membrane. Specifically excluded from designation as active transport is the unequal distribution of radioisotope across the membrane resulting from the binding of substrate to intracellular material, the predominance of different ionic forms within and without the cell, and the metabolic conversion of isotopic substrate within the cell.

B. TRANSPORT MODELS

In describing the transport component of nutrient uptake, we use the nomenclature and rate equations developed by Stein and his collaborators (Stein, 1967; Lieb and Stein, 1974; Eilam and Stein, 1974) to describe a simple, carrier-mediated mechanism. The original publications should be consulted for details of the derivations; a brief description of terms and quantities follows. The simple carrier model is described by five fundamental constants: a substrate carrier affinity constant K, and four resistivities, R_{12}, R_{21}, R_{ee}, and R_{oo}, which are proportional to the duration of a round trip of the carrier in four possible modes (respectively, loaded inbound–empty outbound, empty inbound–loaded outbound, loaded in both directions, empty in both directions). The resistivity constants are combinations of the rate constants associated with the various elements of the model. Three can be measured directly as the reciprocal of maximum flux attainable in three different experimental configurations, viz. $1/R_{12}$ is the maximal initial velocity of entry, $1/R_{21}$ is the maximal initial velocity of exit, $1/R_{ee}$ is the maximal velocity of isotope exchange, whereby the chemical concentration of substrate is the same at both the outer (face 1) and inner (face 2) faces of the membrane. R_{oo} is not independent of these, since necessarily $R_{oo} + R_{ee} = R_{12} + R_{21}$. Shorthand notation for four common experimental protocols, and the relation of the Michaelis–Menten parameters obtained with each protocol to each of these fundamental constants are shown in Table I.

Functional symmetry of the carrier system corresponds to redundancy among the R constants. If the operation of the carrier is indifferent with respect to inner and outer face of the membrane, $R_{12} = R_{21}$. If the velocity of carrier "movement" is not dependent on whether the carrier is loaded or empty, $R_{ee} = R_{oo}$. Thus, for the fully symmetrical system, $R_{12} = R_{21} = R_{ee} = R_{oo}$, and the

TABLE I

SUMMARY OF TRANSPORT TERMINOLOGY[a]

Designation	Experimental protocol		Apparent Michaelis–Menten parameters[b]	
	cis-side	trans-side	K_m	V
Zero-trans	Labeled substrate at various concentrations	No substrate	$K_{12}^{zt} = \dfrac{R_{oo}K}{R_{12}}$	$V_{12}^{zt} = \dfrac{1}{R_{12}}$
Infinite-cis (net flux)	Labeled substrate at "infinite" concentration	Labeled substrate at various concentrations	$K_{12}^{ic} = \dfrac{R_{12}K}{R_{ee}}$	$V_{12}^{ic} = \dfrac{1}{R_{12}}$
Infinite-trans	Labeled substrate at various concentrations	Unlabeled substrate at "infinite" concentration	$K_{12}^{it} = \dfrac{R_{21}K}{R_{ee}}$	$V_{12}^{it} = \dfrac{1}{R_{ee}}$
Equilibrium exchange	Labeled substrate at various concentrations	Unlabeled substrate at the same concentration as on cis-side	$K^{ee} = \dfrac{R_{oo}K}{R_{ee}}$	$V^{ee} = \dfrac{1}{R_{ee}}$

[a] Four of the many conceivable designs for transport experiments are designated according to Eilam and Stein (1974). The Michaelis–Menten parameters apparent in a given type of experiment are related to the resistivities (R-constants) and substrate:carrier affinity constant (K) as defined by a simple, carrier model (see text and Eilam and Stein, 1974). For a completely symmetrical system, by definition, all R constants are equal. Consequently, all Michaelis–Menten constants are equal to each other and to K.

[b] The expressions denote the cis-side of the membrane as 1 and the trans-side as 2. We arbitrarily assign 1 to the outside, so the subscript 12 refers to influx. For efflux experiments, subscript 12 is replaced by 21.

Michaelis–Menten constant apparent in any experimental configuration equals K, so that the entire system may be described by two parameters, K and R. In fact, the symmetrical model accounts well for nucleoside transport in five mammalian cell lines (Wohlhueter *et al.*, 1979). Therefore, we feel justified in taking the symmetrical transport model as a point of departure from which to develop equations describing the kinetic behavior of transport and phosphorylation acting in tandem.

C. PHOSPHORYLATION MODELS

As a first approximation to the phosphorylation component of a tandem pathway, we take a simple, unimolecular, Michaelian enzyme, the velocity of which relates to the *intracellular* concentration of its substrate according to the Michaelis–Menten equation, and the product of which is trapped within the cellular compartment but has no influence on the activity of our simple enzyme.

The potential shortcomings of such a model fall into three categories. First, kinases and phosphoribosyltransferases (the physiological counterparts of our model) catalyze bimolecular reactions. Consequently, the apparent K_m and V_{max} with which a given nutrient is phosphorylated (or phosphoribosylated) depends on the local concentration of ATP (or PRPP[1]) in the enzyme compartment (Cleland, 1963a). The Michaelis constants of adenine and hypoxanthine phosphoribosyltransferases are such [see Table V] that cellular PRPP pools must be seriously depleted before the transferases become unsaturated with respect to PRPP (Henderson and Khoo, 1965); that is, the unimolecular approximation is appropriate at the concentrations of PRPP usually encountered. Hexose and nucleoside kinases utilize MgATP as cosubstrate; here, too, the K_m with respect to MgATP is such that typical cellular ATP concentrations (1–3 mM) are sufficient to saturate the enzymes. These generalizations, of course, are not inviolable. Indeed, there is one report that indicates that it is a shift in the kinetics of uridine kinase with respect to ATP that modulates the rate of uridine phosphorylation in serum-stimulated, quiescent cells (Goldenberg and Stein, 1978).

A second potential shortcoming of the simple model is that the products or end products of phosphorylation may not be inert. For example, hexokinase is inhibited by glucose-6-phosphate (Krebs, 1972) and uridine kinase is inhibited by CTP and UTP (Orengo, 1969). Thus, if the UTP pool is enlarged as a result of continued uridine phosphorylation, the effective kinetic constants of uridine kinase will change with time.

Third, it does not necessarily follow that kinetic parameters ascribed to purified enzymes on the basis of experiments *in vitro* pertain to the situation *in*

---

[1]Unusual abbreviation: PRPP, 5-phosphoribosyl-1-pyrophosphate. The system of subscripts and superscripts to kinetic and thermodynamic quantities, used throughout this article, are elucidated in Table I and in the legend to Fig. 1.

situ (see discussion by Sols and Marco, 1970). This uncertainty may actually be turned to advantage, in that, with proper precautions, the kinetic behavior of nutrient uptake by cells may yield information on the kinetic properties of the enzymic component *in situ*.

These problems are considered on a case by case basis in later sections where the uptake of particular substrates is discussed.

D. Kinetic Behavior of Transport and Phosphorylation Operating in Tandem

1. *The Coupled Model*

We explore, in this section, the kinetic behavior of a simple, two-component system comprising symmetrical, carrier-mediated, nonconcentrative transport across a cell membrane coupled with simple, irreversible Michaelian phosphorylation within the cell (see Fig. 1). Although this system may somewhat over-

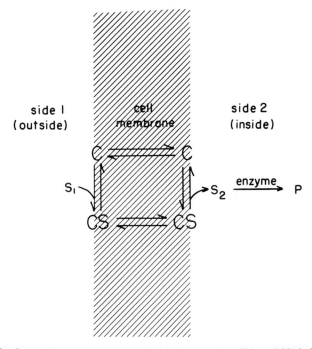

Fig. 1.   Simple model for transport and phosphorylation in tandem. This model is the basis for the steady state and rate equations developed here. It comprises a symmetrical, carrier-mediated (C), nonconcentrative transport system facilitating substrate (S) movement across the cell membrane (side 1 = outside, side 2 = inside) and an irreversible, unimolecular, Michaelian enzyme catalyzing the formation from S_2 of product P, which does not permeate the membrane. The kinetic parameters of the transport system are designated K^t and V^t, those for the enzyme by K^e and V^e.

simplify the physiological realities, nonetheless, we find it instructive to simulate its behavior by means of computer. Specifically, we wish to gain insight into several questions of physiological import: (1) What is the relationship between the rate of intracellular phosphorylation and extracellular substrate concentration? (2) How does this relationship depend on the relative kinetic properties of transport and phosphorylation? (3) How is this relationship affected by inhibitors directed at different targets, for example, by an inhibitor influencing transport velocity only, or transport K_m only? (4) What is the contribution of each step toward determining the overall rate of nutrient uptake into the cell?

Heichal *et al.* (1979) have also treated, at the theoretical level, the question of the kinetic behavior of tandem systems of transport and phosphorylation and applied their analysis to uridine uptake. Their approach is somewhat different than ours, although both treatments lead to the same conclusions.

2. *The Steady State*

If we consider the extracellular compartment infinitely large in comparison to the intracellular (so that extracellular substrate concentration S_1 is constant), our model must eventually attain a steady state in which the rate of enzymatic conversion plus efflux just balances influx. In that steady state, the *concentration* of intracellular free substrate S_2 and the *rate* of substrate phosphorylation are constant, and the latter is the rate of "uptake" of substrate into the cell. Designating the Michaelis–Menten parameters of the transport system as K^t and V^t, and those of the phosphorylating enzyme as K^e and V^e (see Fig. 1), we have:

$$\text{influx} = v^t_{12} = \frac{V^t S_1}{K^t + S_1} \tag{1}$$

$$\text{efflux} = v^t_{21} = \frac{V^t S_2}{K^t + S_2} \tag{2}$$

$$\text{phosphorylation rate} = v^e = \frac{V^e S_2}{K^e + S_2} \tag{3}$$

and

$$\frac{dS_2}{dt} = v^t_{12} - v^t_{21} - v^e = 0 \tag{4}$$

Substituting Eqs. (1–3) into Eq. (4), we may write a quadratic equation in S_2:

$$-(V^t K^t + V^e K^t + V^t S_1)S_2^2 + (K^t V^t S_1 - K^e K^t V^e - K^t V^e S_1 - K^{t2} V^e)S_2 + K^t K^e V^t S_1 = 0 \tag{5}$$

For positive values of the kinetic parameters and S_1, Eq. (5) has a unique, positive root. Thus, for any given values of the kinetic parameters and S_1, we can solve for the intracellular concentration of free substrate S_2 and, in turn, for v^t_{21}

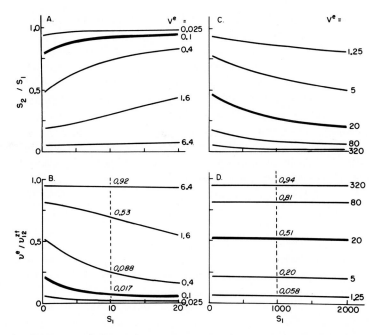

FIG. 2. Steady-state relationship between intra- (S_2) and extracellular (S_1) concentration of substrate, and between the velocities of uptake and entry. For any given set of kinetic parameters and S_1, a value for S_2 was obtained using Eq. (5). Velocities were calculated by substituting S_1 or S_2, as appropriate, into the Michaelis–Menten equation for transport or enzyme (cf. Fig. 1). The bold lines of Panels A and B are generated to approximate the uptake system for thymidine in Novikoff cells: $K^t = 225\ \mu M$, $V^t = 27$ pmole/μl cell water·second, $K^e = 3\ \mu M$, $V^e = 0.1$ pmole/μl cell water·second. The other curves are for the same parameters except for V^e, which is varied from 0.025 to 6.4. Panels C and D are analogous plots, but are modeled on deoxyglucose uptake: $K^t = 650\ \mu M$, $V^t = 20$ pmole/μl cell water·second, $K^e = 600\ \mu M$, and $V^e = 20$ pmole/μl cell water·second (bold line). For the other curves, V^e takes on values between 1.25 and 320. The numbers along the broken line at $S_1 = 10$ (Panel C) and $S_1 = 1000$ (Panel D) are the coefficients of rate determination by transport, as defined in Eq. (6).

and v^e. Figure 2 shows the dependence of the steady state concentration of S_2 on S_1 for several sets of kinetic parameters. The left-hand panels encompass the actual situation with thymidine (see Section III), in that $K^e \ll K^t$; V^t is held constant, while V^e takes on values between 0.09 and 24% of V^t. The right-hand panels encompass the situation with deoxyglucose (see Section V), in that $K^t \cong K^e$; V^e takes on values between 6 and 1600% of V^t. The lower coordinates of Fig. 2 plot the ratio of phosphorylation rate to influx versus exogenous substrate concentration for the same sets of kinetic constants.

These simulations bear out the common sense notions (1) that when V^e increases relative to V^t, the concentration of intracellular free substrate approaches zero (Fig. 2A and C), while the velocity of uptake reflects that of permeation

(Fig. 2B and D); and (2) that when $V^t >> V^e$, intracellular substrate concentrations approach those in the extracellular compartment (Fig. 2A and C), while the velocity of uptake reflects that of phosphorylation (i.e. $v^e/v_{12}^{zt} \rightarrow 0$, see Fig. 2B and D). The figure also demonstrates that there is a middle ground in which the concentration and velocity ratios are dependent not only on the kinetic properties of both the transport and phosphorylation apparatus, but also on the exogenous substrate concentration employed. Within this middle ground it is simply impossible to point to "*the* rate-limiting step." In order to express quantitatively the contribution of the transport component to the determination of the overall flux in our two-component system, we use a "coefficient of rate determination" with respect to transport, d_t:

$$d_t = \frac{\delta V^t}{\delta v^e} \qquad (6)$$

That is, d_t is that portion of an infinitesimal increase in maximal rate of entry that is reflected as an increase in the steady-state rate of uptake. The numbers along the vertical lines (arbitrarily taken at $S_1 = 10$ or 1000) in Fig. 2B and D, give d_t for each set of kinetic parameters employed to generate these curves.

Equation 5 (and the simplistic model is represents) has proven its utility in relating the measured intracellular substrate concentration to exogenous substrate concentration in Novikoff hepatoma cells taking up deoxyglucose (Graff *et al.*, 1978; see also Section V,C). In this case, the *in situ* kinetic parameters V^e and K^e were fitted by nonlinear regression, whereby S_1 was the independent variable, S_2 the measured, dependent variable, and K^t and V^t were fixed at known values.

3. *The Time Course of Uptake*

The foregoing discussion commenced with the observation that "eventually" our model system must come into steady state. We now confront the questions of when such a steady state might be achieved, and whether it is of any operational or physiological significance.

There is now a mass of evidence demonstrating the rapidity of transport of nucleosides (Wohlhueter *et al.*, 1976; Rozengurt *et al.*, 1977; Heichal *et al.*, 1979; Lum *et al.*, 1979), nucleobases (Marz *et al.*, 1979), and hexoses (Graff *et al.*, 1978). In the absence of metabolism, and at concentrations of exogenous substrate less than K_{12}^{zt} these transport systems operate sufficiently rapidly to nearly completely equilibrate substrate across the membrane within 10 (nucleosides and nucleobases) to 60 seconds (hexoses) at 25° (and in about one-third of these times at 37°C). Thus the steady state described in the previous section may be expected within 60 seconds, at the latest, and conceivably within a few seconds at 37°C. Operationally, these data imply that in order to measure directly initial rates of zero-trans entry (where $S_1 < K_{12}^{zt}$), one must employ sampling intervals that are small relative to these times. Conversely, the rates of uptake measured at time intervals large relative to the time of attainment of steady state reflect the accumulation of nonpermeable (phosphorylated) metabolites within the cell, corresponding to v^e of our steady-state model.

To simulate the time course of accumulation of substrate and phosphorylated product intracellularly, we begin with the rate equations for influx, efflux, and enzymatic conversion [Eq. (1), (2), (3), or similar expressions], write the differential with respect to time for the component of interest, and integrate numerically over a specified time interval. This approach, though somewhat demanding of computer time, is mathematically straightforward and easily generalized to more highly ramified metabolic networks, including, for example, alternate metabolic routes, and reactions whose products are themselves transported out of the cell. Asymmetry of a transport system is readily accomodated by enlarging the expressions for influx [Eq. (1)] and efflux [Eq. (2)] (Eilam and Stein, 1974):

$$\text{influx} = v_{12}^t = \frac{KS_1 + S_1 S_2}{K^2 R_{00} + KR_{12} S_1 + KR_{21} S_2 + R_{ee} S_1 S_2} \tag{7}$$

$$\text{efflux} = v_{21}^t = \frac{KS_2 + S_1 S_2}{K^2 R_{00} + KR_{12} S_1 + KR_{21} S_2 + R_{ee} S_1 S_2} \tag{8}$$

And a finite ratio of medium volume to intracellular volume is accommodated by including the relationship

$$S_1 = -(S_2 + P) \left(\frac{\text{intracellular volume}}{\text{medium volume}} \right) \tag{9}$$

Figure 3 presents a series of simulated time courses of uptake at various substrate concentrations. To generate these curves, kinetic constants were chosen to resemble those found experimentally in Novikoff rat hepatoma cells for thymidine uptake (Panels A–C) (Wohlhueter et al., 1979), for hypoxanthine uptake (Panels D–F) (Marz et al., 1979), and for deoxyglucose uptake (Panels G–I) (Graff et al., 1978).

These three situations differ in the following ways. For thymidine, $V^t \gg V^e$, $K^t \gg K^e$, and $V^t/K^t > V^e/K^e$. Thus, at any concentration of exogenous substrate, including those within the first-order range of both transport and phosphorylation processes, entry exceeds phosphorylation, S_2 is substantial relative to S_1, and there is considerable efflux of intracellular thymidine. At high exogenous concentrations of substrate, accumulation of phosphorylated products makes only an insignificant contribution to total cellular radioactivity (Panel C).

For hypoxanthine, although $V^t > V^e$ and $K^t > K^e$, V^t/K^t is less than V^e/K^e, so that when S_1 is small compared to K^t and K^e, the rate of phosphoribosylation is nearly equal to that of entry, and S_2 is very low (Panel D). As the phos-

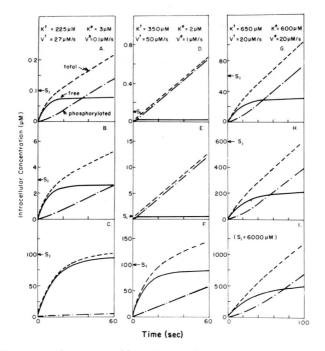

FIG. 3.   Time course of appearance of free substrate and phosphorylated products within the cell. The accumulation with time of each specific entity of the model of Fig. 1 was simulated by numerical integration of Eq. (1) through (3), as appropriate (for example, $dS_2/dt = $ influx − efflux − phosphorylation rate). In Panels A, B, and C, kinetic parameters were assigned (as indicated at the head of the column) to approximate those of thymidine uptake in Novikoff cells, and exogenous thymidine was taken at 0.1, 3, and 100 μM, respectively, as indicated by "S_1" on the ordinates. In Panels D, E, and F, the kinetic parameters were assigned to approximate those for hypoxanthine uptake in Novikoff cells. And in Panels G, H, and I, the kinetic parameters were assigned to approximate those for deoxyglucose uptake in Novikoff cells.

phoribosyltransferase becomes saturated (Panel F), the uptake pattern reverts to something like that for thymidine. For deoxyglucose, the kinetic parameters for transport and phosphorylation are comparable, and the uptake pattern is nearly independent of exogenous concentration.

It is evident from these simulations that, in general, the incorporation of radiolabel derived from an exogenous substrate into total cellular material follows biphasic kinetics. The latter phase corresponds to the steady state (in which $dS_2/dt = 0$); it measures the rate of incorporation of radiolabel into nonpermeable products; and its slope is less than the slope of the initial phase. The initial phase corresponds to the zero-trans influx of substrate. The extent and duration of this initial phase are dependent on the relative values of the kinetic parameters for transport and phosphorylation and on the concentration of exogenous substrate. If, at a given concentration of exogenous substrate, influx $>>$ phosphorylation rate, the extent of this initial phase will be maximal, in the sense that the steady-state concentration of free substrate in the cell approaches that in the medium (see Panels A–C); the duration of the initial phase is determined by the absolute rate of entry. If, as in Panels D and E, the intracellular steady-state concentration of free substrate approaches zero; the extent and duration of the initial phase become diminishingly small.

One criterion to demonstrate the existence of an initial phase is failure of the long-lived, second-phase slope to extrapolate through the origin (exemplified in Fig. 3). Operationally, this criterion is not always easy to apply, for it suffers from three types of experimental error: (1) The precise location of the plot origin depends, in practice, on the precision with which radioisotope associated with the cell, but external to it, can be estimated. (2) Improper handling of cell samples allows intracellular, free substrate to be leached out, thus obliterating the initial phase. (3) Measurements may be made at a time when the concentration of radioactivity within the cells greatly exceeds that in the medium; extrapolation of such measurements to a positive y-intercept, the value of which cannot exceed the concentration of radioactivity in the medium, may not be distinguishable from extrapolation to zero.

A second criterion for the existence of an initial phase is the demonstration of a significant (with respect to S_1) intracellular pool of free substrate. The accuracy with which this pool is estimated is influenced also by the three errors enumerated above, as well as a fourth: namely, the extent to which the experimental design permits intracellular metabolism to proceed after influx of exogenous substrate has been terminated.

A kinetic differentiation of two phases of uptake presupposes an experimental design with which one or preferably all of these criteria can be rigorously examined. It is obvious that early sampling times, large intracellular volumes, and a wide concentration range favor such examination, as do accurate estimations of extra and intracellular volumes, and methods to effectively halt influx, efflux, and intracellular metabolism at precise times. Kinetic differentiation of these two phases, or rigorous proof that there are not two phases, is essential to an assessment of the roles of transport and phosphorylation in the uptake of nutrients into cultured cells.

Failure to detect two kinetic phases at a given exogenous concentration of substrate does not necessarily mean that the *maximum* velocity of transport is less than that of phosphorylation, nor that biphasic kinetics might not obtain at another concentration (compare Panels D and F).

This virtual disappearance of biphasic kinetics at low substrate concentration may be of special physiological relevance. In this case, it is the relationship of V^t/K^t to V^e/K^e that determines the relative rates of transport and phosphorylation, since, for a Michaelis–Menten process, V/K_m is equal to the rate constant apparent in the first-order range. Regardless of V^t, when $V^t/K^t < V^e/K^e$, the rate of uptake will be determined predominantly by the rate of transport. Even so, the "uptake kinetics" will largely reflect those of the kinase, because, to obtain a K_m for uptake, one must raise the exogenous substrate concentration above K_m of the kinase, in which range the kinase again predominates in the determination of uptake rate (see Panel F).

4. On the Relationship between K_m^{uptake} and the Kinetic Constants of Transport and Phosphorylation

There have been numerous investigations of the kinetics of nutrient uptake in cultured cells in which the data have been analyzed in terms of the relationship between long-term, linear uptake rates (doubtless corresponding to the second, steady-state phase discussed in the previous section) and exogenous substrate concentration. Frequently this relationship has been perceived as hyperbolic (or the sum of a hyperbolic and a linear term), and, accordingly, has been expressed as a Michaelis–Menten constant for uptake (see review by Plagemann and Richey, 1974).

We want to enquire, at the theoretical level, what information the kinetic constants of uptake can yield about the kinetic properties of its component parts, viz. of transport and phosphorylation. In the steady-state model under discussion, the rate of phosphorylation (and thus of substrate entrapment within the cell) was considered to obey Michaelis–Menten kinetics with respect to the *intracellular* concentration of substrate S_2. The relationship of S_2 to *extracellular* concentration of substrate S_1 is dependent, in a rather complicated way, on the kinetic constants of transport and of the phosphorylating enzyme. The relationship of phosphorylation rate to S_1 is, in the general case, correspondingly complicated, but calculable with the aid of Eq. (5).

The two limiting situations are intuitively clear: (1) If transport is negligibly slow in comparison to phosphorylation capacity, the rate of phosphorylation will be paced by the rate of entry, so the Michaelis–Menten constant and maximal velocity apparent in (long-term) uptake studies will be those of the transport step. (2) If transport is very rapid with respect to phosphorylation, S_2 will always equal S_1, so that the kinetic parameters apparent in uptake studies will be those of the phosphorylating enzyme *in situ*.

Figure 4 illustrates (for the case where $K^t \gg K^e$) how the double reciprocal plots [$1/$(steady-state uptake velocity) versus $1/S_1$] change as one begins at the latter extreme and progresses toward the former. The plots, in fact, are somewhat curved, but the experimentally relevant question is to what extent real data falling on these theoretical curves might reasonably be perceived as falling on straight lines and used, thus, to estimate an apparent V and K_m. Moreover, we can draw certain qualitative generalizations as to errors involved in approximating these

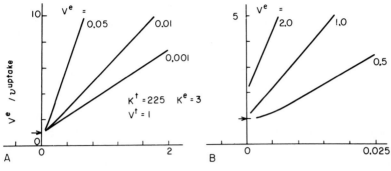

I / (exogenous concentration of substrate)

FIG. 4.   The relationship of steady-state uptake velocity to exogenous substrate concentration, as seen in double reciprocal plots. The Michaelis–Menten constants ($K^t = 225$, $K^e = 3$) were taken to approximate those for thymidine uptake in Novikoff cells (in dimensions of micromolar). The maximum velocity of the enzymatic reaction (V^e) relative to that for transport was taken at 0.001 to 0.05 (Panel A) and 0.5 to 2 (Panel B). The steady-state uptake velocity (v^{uptake}) is normalized to V^e to illustrate the extent of saturation of the enzyme. The ordinate is the reciprocal of the normalized uptake velocity, V^e/v^{uptake}, and the arrow on the ordinate indicates saturation of enzyme, where $v^{uptake} = V^e$. The abscissa corresponds to exogenous concentrations down to 0.5 μM (Panel A) or 40 μM (Panel B).

curves by straight lines. For example, when $V^t/V^e = 0.001$, K_m^{uptake} is approximately 3, but as the maximal velocity of phosphorylation approaches or exceeds that of transport, the apparent K_m^{uptake} would progressively overestimate the true K^e and approach that for transport. The lower limit of K_m^{uptake} is K^e, approached as $V^e \ll V^t$. But in the absence of rigorous proof that one or the other of these limiting situations exists (for practical purposes), kinetic constants based on long-term uptake studies cannot be construed as measures of either transport or phosphorylation kinetics. A quantitative reinterpretation of uptake data to yield information on either of the component parts of uptake, thus, is problematic, and requires, at the least, independent information on the kinetic characteristics of that step that is less well defined by uptake kinetics.

5. *Kinetic Patterns of Inhibition of Uptake*

An important corollary to the observation that K_m^{uptake} and V^{uptake} do not necessarily reflect the analogous parameters for transport or phosphorylation, is that changes of K_m^{uptake} and/or V^{uptake} cannot be judged according to the usual paradigms of enzyme inhibition patterns such as those developed by Cleland (Cleland, 1963b).

A conceptually simple example is afforded by an uptake system whose transport component operates with K^t and V^t large relative to K^e and V^e of the kinase component. Since the uptake behavior of this system will reflect predominantly the kinase component, a range of S_1 appropriate for the study of uptake kinetics encompasses K^e, but is wholly within that range that is first-order with respect to the transport component. Now, imagine an inhibitor I, whose sole effect is to decrease the maximum velocity of transport; a decrease in velocity of transport caused by the inhibitor can be overcome by an increase in S_1 (since transport rate increases linearly with S_1). This is a phenomenon that would be described as competitive inhibition according to the enzyme paradigm.

The hypothetical situation is fairly closely approached in reality by the thymidine uptake system (see Section III,C), the kinetic parameters of which serve as the basis for the simulations in Figs. 5 and 6. In Fig. 5, Panels A through D demonstrate the Lineweaver–Burk patterns expected for long-term uptake velocities as a function of exogenous substrate concentration with each of four

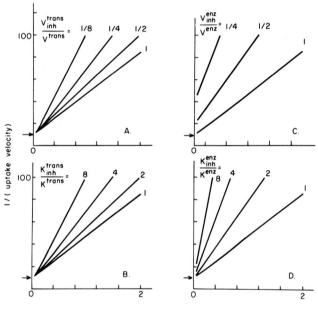

FIG. 5.   Patterns of inhibition of uptake for various types of inhibitors. Double reciprocal plots were generated for uptake in a system (see model of Fig. 1) approximating that for thymidine uptake in Novikoff cells: $K^t = 225\ \mu M$, $V^t = 27$ pmole/μl cell water·second, $K^e = 3\ \mu M$, $V^e = 0.1$ pmole/μl cell water·second. The abscissas correspond to exogenous concentration $\geqq 0.5\ \mu M$; the ordinates are 1/(steady-state rate of uptake), as calculated from Eqs. (3) and (5). The arrows on the ordinates correspond to the reciprocal of the maximum velocity of enzyme in the absence of inhibitor. In each panel, the postulated inhibitor influences only one of the four kinetic parameters of uptake, as indicated by changing ratios of a given parameter in the presence of inhibitor to that parameter in the absence of inhibitor.

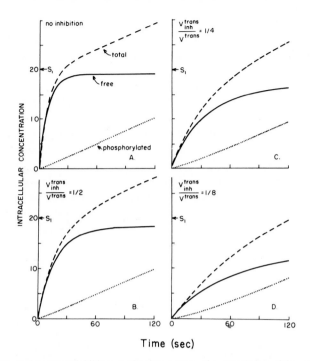

Time (sec)

FIG. 6.   Effect of transport inhibitor on the time course of accumulation of free substrate and phosphorylated product within a cell. An inhibitor is assumed to affect only the maximal velocity of transport of a thymidine-like uptake system (kinetic parameters of uptake are those given for Fig. 5A). The concentration of exogenous thymidine was taken at 20 μM. Panel A represents the uninhibited situation; Panel B, 50% inhibition; Panel C, 75% inhibition; Panel D, 87.5% inhibition. The time axis is in seconds; the ordinate gives the intracellular concentration of isotopic substrate and nonpermeable derivatives of it. The arrows on the ordinate indicate the concentration of substrate in the extracellular space.

hypothetical inhibitors, whose mode of action affects only one of the four kinetic parameters determining the steady-state uptake rate according to Eq. (5).

As noted above, the relationship of 1/(uptake velocity) versus 1/(exogenous substrate concentration) is complex, so that the curves are not linear. But, again, straight lines could be imposed on real data points described by such curves in the belief that we are working with a simple, single-step Michaelian process. We might then draw conclusions corresponding to those straight lines. To do so would lead to false conclusions regarding the mechanism of inhibition. For example, Pattern A represents an inhibitor that lowers the maximum velocity of transport (say, by increasing the microviscosity of the lipid bilayer). We might be tempted to regard this inhibitor as a competitive inhibitor of uptake of a substrate whose apparent $K_m^{\text{uptake}} = 4$. Whether the pattern is strictly "simple, linear competition" (i.e., whether replots of K_m versus [I] are linear), would depend on the relationship between [I] and the change in microviscosity. In fact, steady-state uptake inhibition patterns yield little information on the mechanism of inhibition at the transport level (compare Fig. 5A and B). If an inhibitor is known to act at the enzyme level, then the uptake inhibition pattern is apt to correspond qualitatively to that for the enzyme (see Fig. 5, C and D), but even here quantitation of inhibition kinetics (K_i and a determination of linearity of inhibition with I) is not, in general, valid.

All this does not imply that kinetic studies of uptake are useless to decipher mechanisms of uptake inhibition; it implies only that more complicated models need be considered, and that, usually, more information needs to be gathered. Figure 6 traces the time courses (obtained be numerical integration as described in Section II,D,3) of intracellular accumulation of free substrate and phosphorylated products for a single concentration of extracellular substrate and increasing concentrations of an inhibitor acting strictly noncompetitively at the transport level (analogous to the situation in Fig. 6A). Additional information on the

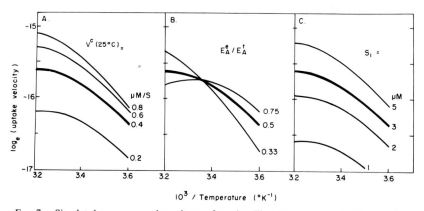

10^3 / Temperature ($°K^{-1}$)

FIG. 7. Simulated temperature dependence of uptake. The simple uptake model was assigned kinetic and thermodynamic characteristics approximating those for thymidine uptake in Novikoff cells, as far as these were available. For a reference temperature of 25°C, $K^t = 225 \mu M$, $V^t = 27$ pmole/μl cell water·second; $K^e = 3 \mu M$, $V^e = 0.4$ pmole/μl cell water·second. The Arrhenius energy for V^e was taken at 18.3 kcal/mole, and enthalpy for substrate: carrier dissociation was taken at 9.3 kcal/mole [values reported by Wohlhueter *et al.* (1979)]. The corresponding thermodynamic parameters for thymidine kinase were arbitrarily set at 9.15 kcal/mole for V^e (equal to one half that for V^t), and 13.95 kcal/mole for K^e (equal to 1.5 times that for K^t). In all cases, these thermodynamic characteristics were considered constant over the whole temperature range. Using these parameter values and taking exogenous substrate at 3 μM (all values expressed in molarity or moles/ liter·second), we generated the Arrhenius plots indicated by the boldface curve in each panel. The additional curves in each panel were generated by varying only one of the eight parameters necessary to describe such curves. In Panel A, the maximal enzymatic velocity at 25°C (V^e) was assigned various values as indicated; in Panel B, the Arrhenius activation energy of the enzymatic reaction (E_A^e) was assigned values equal to 0.75, 0.5, and 0.33 times that of transport (E_A^t); in Panel C, exogenous concentration of substrate (S_1 was taken at 1, 2, 3, and 5 μM).

steady-state intracellular level of free substrate, and on the time course of attainment of that steady state are required to diagnose the mechanism of inhibition. Given the temporal resolution of the two kinetic phases of uptake available here, we immediately discover that it is the initial phase (that corresponding to transport) that is subject to inhibition. The latter phase (corresponding to enzymatic conversion) is apparently unaffected (Panel B) until the inhibition of transport has become severe (Panel D). If such data were available for a series of exogenous substrate and inhibitor concentrations, a rather complete conclusion regarding the site and strength of inhibition could be drawn.

Figure 6 is based upon the kinetic characteristics of nucleoside uptake, for which K^t and K^e are quite disparate. If $K^t \cong K^e$ (as, for example, with hexoses), uptake velocity shows greater sensitivity to inhibition at the transport level. But the principles involved and the information required to elucidate the site and strength of inhibition are the same.

6. Temperature Dependence of Uptake

Similar considerations apply to an evaluation of the effect of temperature on uptake. In general, the temperature dependence of the four kinetic parameters appearing in Eq. (5) will be different. In traversing a wide range of temperature, qualitatively different patterns of the steady state may emerge, giving rise to apparent curvature in an Arrhenius plot of the uptake system. Figure 7 provides some examples of nonlinear Arrhenius plots for uptake for a hypothetical situation in which the four kinetic parameters determining uptake rates are all thermodynamically well behaved, i.e., for each parameter, the logarithm of the parameter value changes linearly with respect to 1/(absolute temperature).

7. Physiological Modulation of Uptake

The four panels of Fig. 6 depict the progression from a nearly fully phosphorylation-limited uptake system (Panel A, $d_t = 0.006$) to one in which transport is a significant codeterminant of the uptake rate (Panel D, $d_t = 0.074$). The coefficient of rate determination [the computation of which presupposes a knowledge of the kinetic parameters of transport and phosphorylation: see Eq. (5) and (6)] bears crucially on the feasibility of physiological regulation of uptake, and on the likelihood of genetic alteration of a given component of the uptake system in response to treatment with a toxic drug taken up (and activated) by that system.

The rationale is precisely that used to discern "key regulatory enzymes" in a metabolic pathway (Newsholme and Smart, 1973). To the extent that carrier-mediated transport establishes a "near-equilibrium reaction" (i.e., to the extent that the steady-state S_2 approaches S_1), modulation of carrier activity is an ineffective means of controlling uptake velocity. By the same token, a partial, mutational incapacitation of the carrier is unlikely to be of selective advantage in reducing the rate at which an activated drug accumulates within the cell. In fact, it may well be that, even in the complete absence of a transport system, non-mediated permeation of the drug suffices to establish toxic levels within the cell.

There are innumerable reports of the regulation of nutrient uptake in cultured cells in response to treatment with hormones, serum factors, and lectins. At least for nutrients transported into the cell by facilitated diffusion and entrapped there by phosphorylation, an elucidation of the mechanism of such regulation requires, as a first step, an assessment of the relative rates of transport and phosphorylation *in situ* with and without the modulating signal. Such an assessment requires, in turn, methodology permitting concentration measurements of intracellular substrate and phosphorylated products and resolution of the (potentially) biphasic kinetics of uptake.

E. Dissociating Transport from Phosphorylation

We have been concerned up to this point with an understanding of the kinetic behavior of transport and phosphorylation (or phosphoribosylation) acting in tandem as a system to take up nutrients into the animal cell. Recent work has provided experimental demonstration of biphasic kinetics of uptake of hexoses (Graff *et al.*, 1978), nucleosides (Wohlhueter *et al.*, 1976; Rozengurt *et al.*, 1977; Marz *et al.*, 1977; Plagemann *et al.*, 1978b; Heichal *et al.*, 1979), and nucleobases (Marz *et al.*, 1979) consistent with the ideas developed here. These experiments, which are discussed in later sections, illustrate the practicability of kinetic dissection of uptake processes. Still, it is evident from the foregoing simulations, and more so from the experiments themselves, that, for even the simplest, two-step, Michaelis–Menten systems, the relationship between substrate availability and rate of entrapment within the cell is complex.

It would be of obvious advantage to separate the transport and metabolic components of an uptake system by physical or chemical means. With respect to the metabolic component, this is accomplished by characterizing enzymatic behavior in cell-free preparations. A discussion of the technology employed is, of course, beyond the scope of this article. But we call to the reader's attention the problems and assumptions involved in extrapolating the results of *in vitro* enzymology to the enzyme *in situ* (Sols and Marco, 1970).

With respect to the transport component, experimental manipulation is limited by the requirement that the compartments separated by the cell membrane must be preserved. A variety of techniques have been devised that meet this requirement (see Wohlhueter *et al.*, 1978a), and we present here only a brief summary of them. One approach has been to construct mutant cell sublines devoid of the enzymatic activity(ies) responsible for metabolism of intracellular substrate. Thymidine transport, for example, has been characterized in Novikoff rat hepatoma cells lacking thymidine kinase (Wohlhueter *et al.,* 1979). This approach is limited to nonessential nutrients such as nucleosides and nucleobases, the salvage enzymes for which are dispensible. It is more easily applied when a single enzyme is involved in metabolism but has been employed in combination with other techniques to block metabolism in more ramified pathways. Adenosine kinase-negative mutants of Chinese hamster ovary cells, treated with an adenosine deaminase inhibitor (deoxycoformycin) are rendered incapable of metabolizing adenosine and are thus good subjects for studying adenosine transport (Lum *et al.,* 1979).

Another tactic is to choose substrate analogs that are transported, but not metabolized. This approach has been particularly useful in the study of hexose transport, since hexokinase-negative cell lines are not available. 3-O-Methyl-D-glucose has been shown to be transported by the same carrier as glucose, but is not phosphorylated by hexokinase (Crane *et al.,* 1957; Regen and Morgan, 1964). 5'-Deoxyadenosine, which is neither a phospho-group acceptor nor a substrate for adenosine deaminase, has been employed in an analogous capacity to study nucleoside transport (Kessel, 1978).

Kinase activity is halted in cells depleted of ATP by treatment with KCN and iodoacetate (Plagemann *et al.,* 1976). The same treatment apparently depletes also the pool of PRPP, since the phosphoribosyltransferases are also inoperative. ATP-depleted cells survive and exclude trypan blue for at least two hours (Wohlhueter *et al.,* 1978a), with little or no apparent alteration of the transport systems for hexoses, nucleosides, and nucleobases. ATP-depletion is, thus, a useful experimental device for functionally segregating transport from phosphorylation.

A fourth methodological approach is to prepare vesicles from cells that are largely devoid of cytoplasmic enzyme activities (Quinlan and Hochstadt, 1974, 1976). The approach should be generally useful, limited only by the experimenter's ability to prepare, from a given cell type, sealed vesicles uncontaminated by cytoplasmic enzymes.

None of these experimental approaches obviates the need for assay methodology whose speed is consistent with that of the transport process. And by all accounts (reviewed by Plagemann and Wohlhueter, 1980, transport velocity is considerable for all of the substrates treated in this article. Functional isolation of the transport apparatus does not, by itself, address the question of the relative roles of transport and phosphorylation in nutrient uptake. It does, however, allow a much more rigorous characterization of the transport process, for, in the absence of metabolism, a greater variety of experimental designs become feasible, including efflux and isotope exchange studies (Eilam and Stein, 1974). It is the behavior of the transport systems per se that has prompted our theoretical considerations of how transport and phosphorylation might be expected to operate in tandem in the uptake of nutrients into cells.

III. Uptake of Nucleosides

Mammals, as intact organisms, do not have a nutritional requirement for purines or pyrimidines (or nucleosides thereof) by virtue of the capacity of most cells and tissues to synthesize purine and pyrimidine nucleotides *de novo*. There are, however, certain tissues that lack the synthetic pathways for purines and are, apparently, dependent on an external source i.e., either dietary or the endogenous production of other tissues (Pritchard *et al.*, 1970; Murray *et al.*, 1970; Murray, 1971; Roux, 1973). Whether capable of *de novo* synthesis or not, most tissues and cells appear to be able to utilize purine and pyrimidine nucleosides via salvage pathways.

Cultured cell lines, as a rule, possess the *de novo* synthetic pathways. They show no nutritional requirement for preformed purines or pyrimidines or their nucleosides, nor does an exogenous source augment the growth rates of cultured cells (see, for example, Marz *et al.*, 1977b). But if exogenous sources are available, they are utilized to apparent advantage, in that they spare, presumably, a comparable amount of costly *de novo* synthesis. The pathways of utilization may be considered to comprise mediated transport plus phosphorylation to the nucleoside monophosphate, whereby, "phosphorylation" may be direct (as for example with thymidine), or more circuitous (for example, inosine → hypoxanthine → IMP), or even branched (for example, adenosine → AMP and adenosine → inosine → hypoxanthine → IMP).

Although these salvage pathways appear to be simple economy measures, severe dysfunction is associated with their congenital absence in humans. Thus, the absence of hypoxanthine phosphoribosyltransferase (hypoxanthine + PRPP → IMP) is associated with the Lesch–Nyhan syndrome of mental retardation (Rosenbloom *et al.*, 1967), and the absence of adenosine deaminase (adenosine → inosine) or purine phosphorylase (inosine → hypoxanthine + ribose-1-phosphate) is associated with a deficient immune system (Meuwissen *et al.*, 1975).

A. The Nucleoside Transport System

It has been established that the transport of nucleosides into several mammalian cell lines (cultured in suspension or in monolayer) involves a saturable, nonconcentrative mechanism consistent with carrier-mediated permeation. [Nucleoside and nucleobase transport is the subject of a recent review by the authors (Plagemann and Wohlhueter, 1980, which should be consulted for a more detailed account.] Such transport has been reported in Novikoff rat hepatoma cells for thymidine (Wohlhueter *et al.*, 1976, 1979), uridine (Plagemann *et al.*, 1978b), deoxycytidine, cytosine arabinoside (Plagemann *et al.*, 1978a); in untransformed and SV40-transformed NIL-8 hamster fibroblasts for uridine and cytosine arabinoside (Koren *et al.*, 1978, 1979); in L1210 Leukemia cells for deoxycytidine (Kessel and Shurin, 1968) and 5'-deoxyadenosine (Kessel, 1978); in P388 leukemia cells for uridine (Plagemann *et al.*, 1978b) and adenosine (Lum *et al.*, 1979); in mouse L-cells for thymidine (Wohlhueter *et al.*, 1979), and uridine (Plagemann *et al.*, 1978b); in untransformed and SV40-transformed 3T3 cells for uridine (Koren *et al.*, 1978); in HeLa cells for thymidine (Wohlhueter *et al.*, 1979) and uridine (Plagemann *et al.*, 1978b); in Chinese hamster ovary cells for thymidine (Wohlhueter *et al.*, 1979) and uridine (Plagemann *et al.*, 1978b); and in chemically transformed hamster MCT fibroblasts for cytosine arabinoside (Koren *et al.*, 1978). Common to all of these studies is the considerable technical effort made by the investigators to ensure that the initial flux of substrate into the cells was measured, rather than some longer lived uptake process dependent on cellular metabolism. In most cases this was accomplished by excluding metabolism by one or more of the techniques outlined in Section II,E (viz. by the use of kinase-deficient sublines or ATP-depleted cells) and necessitated the use of short sampling intervals, usually < 10 seconds.

These studies are distinguished from nucleoside uptake studies (summarized in Plagemann and Richey, 1974; Wohlhueter and Plagemann, 1980) not only by their technical approach, but, more significantly, by their results. Wherever experimental design has paid rigorous attention to very early substrate influx (or, in the terminology developed in Section II, wherever nucleoside transport, as opposed to uptake, has been estimated) the kinetic parameters relating influx to exogenous substrate concentration are vastly different than those relating uptake velocity to exogenous substrate concentration. Table II provides some examples of experiments where the kinetics of transport and uptake of a given substrate has been measured in the same cell line. The greater-than-100-fold discrepancy between the kinetic parameters apparent in transport and in uptake experiments are analyzed in Section III,C in terms of the kinetic behavior of tandem processes developed previously. We will first consider more closely the kinetic characteristics of nucleoside transport per se. Our own experience suggests that some of the variation in maximal velocities and Michaelis–Menten constants of transport, evident in Table II may be attributed to the various procedures employed to estimate initial velocities at a given substrate concentration. Graphical estimation of initial rates has often encompassed only a very few early time points and may underestimate initial velocity. If, instead, we calculate initial slopes by fitting integrated rate equations to data describing the whole time course of approach to transmembrane equilibrium (Wohlhueter *et al.*, 1979), we find that significant deviations from linearity accrue even within a few seconds of exposure to exogenous substrate.

Even so, the data are sufficiently accurate to support several generalizations regarding nucleoside transport in cultured mammalian cells. All of the transport studies cited at the beginning of this section are consistent with the view that

TABLE II
KINETIC CHARACTERISTICS OF NUCLEOSIDE TRANSPORT AND UPTAKE IN VARIOUS CELL LINES[a]

Substrate	Cell line	Transport[b]			Uptake[c]		
		K_{12}^{z1} (μM)	V_{12}^{z1} (pmole/μl cell water · second)	References[d]	K_m^{uptake} (μM)	V^{uptake} (pmole/μl cell water · second)	References[d]
Adenosine	N1S1-67, Novikoff hepatoma	—	—		7.7	1.9	1,2
	P388, murine leukemia	123	29	3	—	—	5
2'-Deoxycytidine	N1S1-67, Novikoff hepatoma	672	39	4	1.2	0.04	1,2
Thymidine	N1S1-67, Novikoff hepatoma	225	27	6	0.5	0.12	1,2
	Chinese hamster ovary	103	6.8	6	2.0	0.06	7
Uridine	N1S1-67, Novikoff hepatoma	250	25	8	14	1.6	1,2,9
	Chinese hamster ovary	169	5.6	10	—	—	—
	3T3, mouse fibroblasts	220	21	10,12	6	0.5	11,13

[a] Transport was measured by rapid sampling techniques with cells incapable of metabolizing the transport substrate either by virtue of enzyme deficiency, ATP depletion, specific inhibition (of adenosine deaminase), or a combination of these, except in the case of 3T3 cells, where first-phase velocities (see text, Section II) were measured in metabolizing cells. Uptake was measured as the second-phase, steady-state velocity of accumulation of total radioactivity within the cell. Where a range of values was given in the original literature, the median of that range is tabulated.

[b] Transport studies were conducted at 20 to 25°C.

[c] Uptake studies were conducted at 37°C. Original data were generally expressed per 10⁶ cells or per mg cell protein, and have been converted to cell water basis by assuming 1.3 μl water/10⁶ cells.

[d] References are: (1) Plagemann (1971a); (2) Plagemann (1971b); (3) Lum et al. (1979); (4) Plagemann et al. (1978a); (5) Plagemann and Erbe (1974b); (6) Wohlhueter et al. (1979); (7) Plagemann et al. (1975); (8) Plagemann et al. (1978b); (9) Plagemann and Roth (1969); (10) Koren et al. (1978); (11) Jiminez de Asua et al. (1974); (12) Stein and Rozengurt (1976); (13) Weber and Rubin (1971).

transport of nucleosides is saturable, nonconcentrative (and energy-independent), and sufficiently rapid to establish transmembrane equilibrium within a minute. Recently we have examined in detail the kinetics of thymidine transport in Novikoff cells (Wohlhueter *et al.*, 1979), and reached the further conclusions that: (1) Transport data can be adequately described by a simple, carrier-mediated transport model (Eilam and Stein, 1974) that behaves symmetrically with respect to the two sides of the cell membrane. (2) There is probably a single carrier of broad specificity, capable of transporting all of the natural purine and pyrimidine ribonucleosides and deoxyribonucleosides, as well as a number of artificial nucleoside analogs.

The adequacy of the simple carrier model has been observed in five cell lines (Novikoff hepatoma, Chinese hamster ovary, mouse L, human HeLa, and P388 murine leukemia) and is illustrated in Fig. 8, which also emphasizes the rapidity of transmembrane equilibrium and the utility of integrated rate equations. That the transporter has broad specificity is based on inhibition of thymidine transport by other nucleosides and is confirmed by an emerging pattern of mutual inhibition among nucleosides (Plagemann *et al.*, 1978a,b; Koren *et al.*, 1978).

In most of these characteristics, the nucleoside transporter of cultured cells resembles that observed in human erythrocytes (Oliver and Paterson, 1971; Cass and Paterson, 1972, 1973; Cabantchik and Ginsburg, 1977). Moreover, it is, in reality, quite similar to the simple carrier model we have used as the transport component of uptake for purposes of simulating the kinetic behavior of uptake systems (see Section II) and thereby justifies adoption of that model.

B. Nucleoside Kinases

Having permeated the cell membrane, nucleosides may be phophorylated by any of a number of kinases. In contrast to the nucleoside transporter of mammalian cells, which appears to have broad specificity, there are several nucleoside kinases, each with different substrate specificities. Table III summarizes the specificities and Michaelis–Menten constants of the nucleoside kinases commonly found in mammalian cells.

An elaborate review of the enzymology of nucleoside kinases would exceed the scope of the present article; the reader is referred to other reviews on the subject (Henderson and Paterson, 1973; Anderson, 1973; Cihak and Rada, 1976; Littlefield, 1977; Fox and Kelley, 1978). Some aspects of their enzymology are pertinent to an understanding of the relationship between transport and phosphorylation of nucleosides and warrant comment.

1. Specificities of the kinases are rather narrow as compared to that of the nucleoside transporter. Thus, frequently, two nucleosides may compete at the level of transport, but not at the level of phosphorylation.

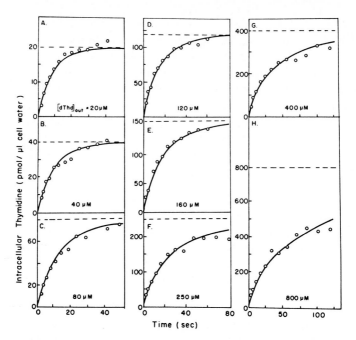

FIG. 8.   Zero-trans influx of thymidine into thymidine kinase deficient Novikoff rat hepatoma cells. The time course of accumulation of free thymidine within cells incubated at 23°C with various concentrations of exogenous [*methyl*-³H]thymidine (indicated in each panel) was followed at short intervals by rapid mixing and centrifugation through silicone oil, as described by Wohlhueter *et al.* (1979), from which the figure is reproduced with permission. The integrated rate equation for a fully symmetrical, carrier-mediated transport model (i.e., the integral of Eq. (7) and (8), where all R-constants are held equal) was fitted by the method of least squares to the data, whereby time and exogenous thymidine concentration (S_1) were treated as independent variables, and intracellular concentration of thymidine (S_2) as the dependent variable. The best fitting kinetic constants were $K = 168 \pm 7.8$ μM and $R = 0.0458 \pm 0.0007$ $\mu l \cdot$ second/pmole ($V_{12}^{zt} = 22$ pmole/μl cell water·second). The curves drawn in each panel are the theoretical ones for zero-*trans* influx at these parameter values with S_1 at the concentration indicated. The broken lines show the expected equilibrium levels based on the measured intracellular water space.

2. The apparent Michaelis–Menten constants of the kinases with respect to nucleosides are, as a rule, much lower than those for transport (compare Tables II and III); they are more comparable to, but larger than, the Michaelis–Menten constants for uptake.

3. The Michaelis–Menten constants with respect to ATP are such that under *normal* cellular ATP levels (2–4 m*M*), the kinases are virtually saturated with ATP. Thus, their kinetic behavior *in situ* may often be approximated by unimolecular rate equations. But it should not be taken for granted that this approximation is appropriate; for example, large influx of phosphorylatable substrate

may critically deplete the cellular ATP pool. There is some indication (Goldenberg and Stein, 1978), indeed, that the affinity of uridine kinase for ATP may be modulated with changing physiological conditions.

4. Maximal kinase velocities are deliberately omitted from Table III, since generalizations in this regard would be misleading. Of course, kinase activity differs with cell type; but, more important to the present discussion, they are regulated within the same cell type in response to changing physiological conditions. Some, such as thymidine, uridine, and deoxycytidine kinases, are subject to feedback control by one or more nucleoside triphosphates; most are subject to control by a variation in enzyme level. For example, thymidine kinase of some cell lines virtually disappears in phases of the cell cycle other than S-phase (Cunningham and Remo, 1973; Roller *et al.,* 1974; Hopwood *et al.,* 1975) and, in asynchronous populations, decreases sharply as logarithmic growth of a culture begins to retard (Plagemann *et al.,* 1975).

The kinase component of nucleoside uptake is thus contingent not only on the inherent enzymatic characteristics exhibited by purified enzymes, but on the growth phase of cells employed in a given uptake study and on the physiological state of the cells, especially with respect to nucleoside triphosphate levels. Clearly, our assumption of a simple, unimolecular, Michaelian model for the simulations of uptake kinetics (see Section II,C) is an oversimplification; nonetheless we regard the simulations as instructive and useful in predicting the kinetic behavior of nutrient uptake.

A further shortcoming of our uptake model is its failure to provide for additional, nonphosphorylative metabolism. The magnitude of this problem must be determined for each nucleoside substrate. Thymidine, for example, is metabolized in mammalian cells only by thymidine kinase, which, in turn, is easily eliminated by selection for bromodeoxyuridine-resistant mutants (Thompson and Baker, 1973). Uridine is subject to phosphorylation and also to phosphorolysis, one product of which, uracil, is free to exit the cell. (If [5-^{3}H]uridine is used as substrate, uracil is the only permeable radioactive product.) Uridine phosphorylase is not easily deselected genetically, nor are there specific inhibitors of it. Under many circumstances, however, the extent of phosphorolysis is negligible in comparison to phosphorylation, and the simple transport–phosphorylation model describes uridine uptake reasonably well (Rozengurt *et al.,* 1977; Plagemann *et al.,* 1978b). Adenosine uptake is more complex, largely because of a very active adenosine deaminase present in many cell lines. In wild-type P388 mouse leukemia cells (Lum *et al.,* 1979), the main route of [^{3}H]adenosine incorporation into nonpermeable metabolites involves entry, deamination to inosine (some of which exits and eventually reenters), phosphorolysis to hypoxanthine (which also exits and reenters), and phosphoribosylation to IMP. Our simple uptake model is useless in the face of this

TABLE III

MICHAELIS–MENTEN CONSTANTS OF NUCLEOSIDE KINASES FROM VARIOUS SOURCES[a]

Trivial name of enzyme	Source	Substrate	K_m (M)	References
Adenosine kinase[b]	Rabbit liver	Adenosine	1.6×10^{-6}	Lindberg et al. (1967)
	Ehrlich ascites cells	Adenosine	2.8×10^{-6}	Murray (1968)
		ATP	2.2×10^{-4}	
Deoxyadenosine kinase[b]	Calf thymus	Deoxyadenosine	7.4×10^{-4}	Krygier and Momparler (1971)
		ATP	2.0×10^{-4}	
Deoxycytidine kinase[b]	Calf thymus	Deoxycytidine	2×10^{-6}	Krenitsky et al. (1976)
		Cytosine arabinoside	2.5×10^{-5}	
		Deoxyguanosine	1.8×10^{-4}	
		Deoxyadenosine	3.3×10^{-4}	
	Calf thymus	Deoxycytidine	5.3×10^{-6}	Durham and Ives (1970)
		ATP	3.5×10^{-5}	
	L1210 cells	Deoxycytidine	1.1×10^{-5}	Kessel (1968)
		ATP	2.0×10^{-4}	
	Human granulocytes	Deoxycytidine	7.8×10^{-6}	Coleman et al. (1975)
		Cytosine arabinoside	2.6×10^{-5}	
Uridine kinase	Calf thymus	Uridine	4.0×10^{-5}	Lee et al. (1974)
		Cytidine	5.0×10^{-5}	
		5-Azacytidine	2.0×10^{-4}	

	Substrate	K_m	Reference
Ehrlich ascites cells	Uridine	4.8×10^{-5}	Sköld (1960)
	Cytidine	2.3×10^{-5}	
	5-Fluorouridine	3.8×10^{-5}	
	ATP	5.0×10^{-5}	
Novikoff ascites rat tumor	Uridine	2.7×10^{-4}	Orengo (1969)
	ATP	3.1×10^{-4}	
Thymidine kinase			
Regenerating rat liver	Thymidine	3.2×10^{-6}	Kizer and Holman (1974)
	ATP	3.0×10^{-4}	
Walker carcinoma	Thymidine	3.7×10^{-6}	Bresnick and Thompson (1965)
	ATP	3.4×10^{-4}	
Novikoff ascites hepatoma	Thymidine	2.7×10^{-6}	Bresnick and Thompson (1965)
Human lymphocytes (isozyme IS)	Thymidine	6.0×10^{-6}	Munch-Peterson and Tyrsted (1977)
	ATP	1.0×10^{-3}	

[a] Enzymes were purified from the cells or tissue of origin to various extents, not necessarily to homogeneity. All of these activities are present in the cytoplasm, but some proportion of total cellular activity of uridine and thymidine kinases are associated with nuclei and/or mitochondria (see text).

[b] In calf thymus, deoxyadenosine is apparently also a substrate for deoxycytidine kinase (Krenitsky et al., 1976). Deoxyadenosine may also be a substrate for adenosine kinase (see Fox and Kelley, 1978). Inosine and guanosine are generally not phosphorylated in mammalian cells and tissues (see Friedmann et al., 1969; Henderson and Paterson, 1973; Marz et al., 1979), although direct phosphorylation has been reported in Novikoff and Ehrlich ascites tumor cells (Schaffer et al., 1973; Pierre and LePage, 1968).

complexity. Deoxycoformycin, however, is a potent and fairly specific inhibitor of adenosine deaminase (Fox and Kelley, 1978); treatment of P388 cells with deoxycoformycin blocks the major route of adenosine uptake and reduces uptake to the transport–phosphorylation paradigm.

C. Nucleoside Transport and Phosphorylation in Tandem

The characterization of nucleoside transport summarized in Section III,B has been accomplished as a consequence of a growing technical and conceptual facility to cope with the rapidity of that transport. However, a great deal of information has been published on the characteristics of metabolizable nucleoside uptake in cultured cells measured at relatively long time intervals (say, > 1 minute). The characteristics of uptake revealed in such experiments have generally been attributed to the transport system on the grounds (1) that the kinetic parameters of uptake were dissimilar to those of the corresponding nucleoside kinases (see Table III), (2) that little or no free nucleoside could be detected in cells taking up exogenous nucleoside, and (3) that uptake could be inhibited by agents known to interact with the cell membrane, but to be inert toward nucleoside kinases. [The evidence and arguments have been discussed in detail by Plagemann and Richey (1974).]

There were, however, observations not well accomodated by this hypothesis indicating that uptake reflected the properties of the nucleoside kinases. For example, Schuster and Hare (1971) noted a correlation between rates of thymidine uptake (but not of uridine uptake) and thymidine kinase activities in various hamster cell lines. Littlefield (1966) and Adams (1969a,b) observed a close parallelism between changes in thymidine uptake and thymidine kinase activity during the cell cycle of mouse fibroblasts. Mitogenic stimulation of lymphocytes results in a simultaneous increase in thymidine uptake and in thymidine kinase activity (Barlow and Ord, 1975; Barlow, 1976). In mouse L-cells, thymidine uptake parallels thymidine kinase activity in wild-type cells, where the kinase changes markedly with growth phase, but not in a subline thereof, Cl 139, carrying herpes simplex virus-coded thymidine kinase, which is not regulated with growth phase (Lin and Munion, 1974; Plagemann et al., 1976). And an analysis of isotope dilution in nucleotide pools of thymocytes exposed to radiolabeled nucleosides (Forsdyke, 1971; Sjostrom and Forsdyke, 1974) indicated a rate limitation beyond the tranport step. Moreover, the pattern of specificity of nucleoside uptake systems seemed suspiciously similar to that of the nucleoside kinases.

Our thesis is that this dilemma can be resolved by analyzing the data in terms of the kinetic behavior expected for transport and phosphorylation operating as a tandem pathway. Thus the two contradictory perspectives—that uptake reflects transport, or that it reflects kinase activities—are seen to correspond to the extreme situations possible with the tandem model, respectively, i.e., that transport is virtually the only rate determinant of uptake, or that phosphorylation is. The simulations presented in Section II reveal a complex middle ground between those simple extremes. Within the middle ground, uptake rate is biphasic: before any data analysis is undertaken, we must ascertain which rate has actually been measured, i.e., the initial attainment of steady state, or the steady-state rate of intracellular isotope accumulation. Also within this middle ground, the contribution of each component to the determination of the steady-state rate depends not only on the inherent kinetic properties and total capacities of the transport and phosphorylation machinery of a given cell type, but also on the exogenous concentration of substrate employed.

The biphasic nature of nucleoside uptake is illustrated in Fig. 9 for thymidine uptake (see also Marz *et al.*, 1977a) and in Fig. 10 for uridine uptake in Novikoff hepatoma cells (Plagemann *et al.*, 1978b). Similar biphasic uptake curves have been reported also for uridine in NIL-8 hamster cells (Heichal *et al.*, 1979), for deoxycytidine and cytidine arabinoside in Novikoff hepatoma cells (Plagemann *et al.*, 1978a), and for adenosine in P388 murine leukemia cells blocked in adenosine deaminase (Lum *et al.*, 1979). There is no doubt that these biphasic uptake curves are real examples of the simulations of Fig. 3, i.e., examples of an an uptake system comprising a high-K_m, high velocity transport and a low-K_m,

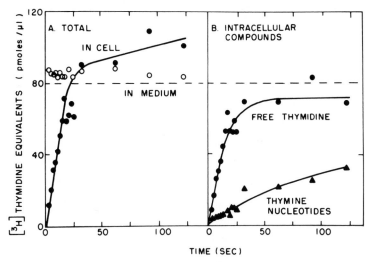

FIG. 9. The contribution of thymidine transport and phosphorylation to total uptake in Novikoff rat hepatoma cells. The incorporation of [*methyl*-^{3}H]thymidine (80 μM) was followed with time (at 24°C) by centrifuging samples of cell suspension at short intervals through silicone oil into 0.2 M trichloroacetic acid. The acid extracts were fractionated into thymidine and thymine nucleotides. Experimental details are given in Wohlhueter *et al.* (1978a), from which the figure is reprinted with permission. Panel A records the measured radioactivity remaining in the medium (O–O) and the total recovered in the cells (●–●); Panel B records the intracellular thymidine (●–●) and phosphorylated products (▲–▲). The broken lines show the nominal extracellular concentration.

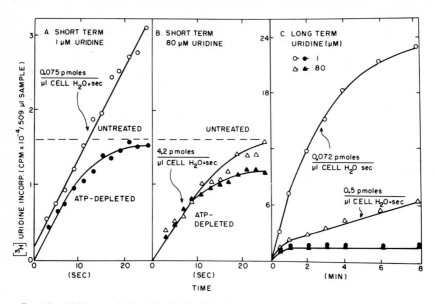

FIG. 10.   Uridine uptake into Novikoff cells in the presence and absence of uridine phosphorylation. The uptake of [5-³H]uridine (1 and 80 μM; 750 cpm/μl, irrespective of concentration) was
followed with time (at 24°C) in suspensions of wild-type Novikoff cells, either untreated (open
symbols) or treated with 5 mM KCN plus 5 mM iodoacetate to deplete cellular ATP (closed
symbols). The intracellular water space was 26.5 μl/ 509 μl sample of suspension. Other experimental details are found in Plagemann *et al.* (1978b), from which the figure is reprinted with permission.
Total intracellular radioactivity is reported for early sampling times (Panels A and B) and for later
times (Panel C) in the same experiment. The velocities indicated were estimated graphically from
linear segments of the corresponding curves. The broken line indicates the concentration of radioactivity equivalent to that in the medium.

low velocity kinase operating in tandem. Figure 9 resolves the overall uptake into
its component parts: the rapid accumulation within the cell of free thymidine,
attaining a steady state in about 25 seconds, followed by the slower accumulation
of thymine nucleotides.

The dependence of the initial rate of uptake on exogenous substrate concentration corresponds to the Michaelian kinetics in a zero-trans influx experiment, a correspondence demonstrated directly for uridine uptake in Novikoff
cells by measuring uridine uptake in cells with normal or depleted ATP pools
[see Fig. 10 and Plagemann *et al.*, 1978b].

The slope of the second, linear phase of uptake is, in general, a complex
function of the kinetic parameters for transport and phosphorylation [see Eq.
(5)]. But in view of the disparity in these parameters (compare Tables II and III),
the slope depends much more heavily on those of phosphorylation. In fact,
virtual saturation of the nucleoside kinases can be achieved at exogenous concentrations of nucleoside that are still in the first-order range with respect to

transport. In such a situation, it is a straightforward matter to determine the maximal velocity of the kinase *in situ*. An examination of the literature on the kinetics of nucleoside uptake (see Plagemann and Richey, 1974, for summary) shows that just this situation typically obtains. It follows that the maximum velocity apparent in such uptake experiments is that of the kinase *in situ,* and that, where this velocity is less than maximal velocity of the same kinase in cell-free extracts, the discrepancy is attributable to the difference in milieu.

A similar argument can be made for disparities between K_m of uptake and K_m for the corresponding kinase in cell-free extracts. In general, the reported Michaelis constants for uptake are smaller than those for kinases (compare Tables II and III). But since intracellular concentration of free nucleoside must be less than or equal to the exogenous concentration, an apparent K_m^{uptake}, based on the relationship between uptake velocity and *exogenous* substrate concentration, would be greater than or equal to the actual K_m of the kinase. That the coupled transport–phosphorylation model cannot easily account for underestimations of the K_m for phosphorylation suggests to us that the effective K_m with respect to nucleoside *in situ* is less than that observed in cell-free extracts.

It is difficult to express quantitatively the difference in actual K_m of a kinase and the apparent K_m of uptake because the relationship between steady-state uptake velocity and exogenous concentration of substrate is not, strictly, that of a rectangular hyperbola, so that K_m is not strictly defined. Thus, the apparent K_m of uptake depends, in part, on how a true hyperbola was fitted to the data points at hand. A graphical appreciation of the problem is evident in Fig. 5.

Given accurate values for the kinetic parameters of transport, one should, in fact, be able to assess the kinetic properties of a kinase effective *in situ* by fitting Eq. (5) to uptake data at various exogenous concentrations, or by Lineweaver–Burk analysis of uptake rate versus measured, intracellular, steady-state substrate concentrations. We have successfully applied the former analysis to deoxyglucose uptake (see Graff *et al.,* 1978, and Section V,C), and Goldenberg and Stein (1978) have applied the latter technique to uridine kinase of quiescent cells stimulated by serum. They detect an apparent change in the K_m of uridine kinase *in situ* with respect to ATP, a change that was not obvious in cell-free extracts, and that was therefore attributed to some change in the cellular milieu.

D. Does Nonmediated Permeation Contribute to the Uptake of Nucleosides?

The kinetic description of nucleoside uptake has frequently required introduction of a term linear with respect to substrate concentration, in addition to the standard hyperbolic term, thus:

$$v_0 = \frac{VS}{K_m^{uptake} + S} + kS \tag{10}$$

The linear term has generally been taken to indicate a second mode of nucleoside entry, namely nonmediated permeation (equivalent to "simple diffusion"; see, for example, Lassen, 1967; Hawkins and Berlin, 1969; Schuster and Hare, 1971; Plagemann, 1971a; Roos and Pfleger, 1972; Stein and Rozengurt, 1975; Sixma *et al.*, 1976; Cass and Paterson, 1977). While the nonmediated route has been ascribed little physiological significance, at substrate concentrations in the range of zero-trans K_m (100–400 μM, see Table II), it becomes, typically, the larger term in Eq. (10).

As demonstrated in Fig. 8 (and discussed in Section III,A), however, nucleoside *transport* can be adequately described by the rate equation for a saturable transporter without recourse to a nonmediated component, even at concentrations in excess of 1 mM. Additional evidence for the insignificance of nonmediated component is the fact that the first-order rate constant for cytosine entry into Novikoff cells, which involves apparently only nonmediated permeation (Graff *et al.*, 1977), is about $^1/_{100}$ that of thymidine in its first-order range with respect to transport [Wohlhueter *et al.*, 1976].

The apparent nonsaturable component of uptake kinetics is, we believe, an artifact of uptake measurements, for which we offer the following rationalization. It will be appreciated from the simulation of Fig. 3 or from the data of Fig. 10, that at a given time, long with respect to the time of attainment of steady state, cell-associated radioactivity comprises two classes of compounds: (1) phosphorylated compounds, the accumulation of which progresses measurably at this time scale, and which show an approximately Michaelis–Menten dependence on exogenous substrate concentration, and (2) free intracellular nucleoside, which, at the same time scale appears to be constant, and approximately proportional to exogenous substrate concentration. If one equates, erroneously, total accumulation per given time as a measure of uptake velocity, and analyzes this "velocity" as a function of exogenous substrate concentration, one perceives one component that is apparently saturable and another that is apparently unsaturable.

We conclude that careful analysis of nucleoside transport indicates only negligible rates of nonmediated permeation, even at millimolar concentrations.

E. Physiological Regulation of Nucleoside Uptake

The thrust of our argument thus far is that the kinetic characteristics of nucleoside transport and of nucleoside phosphorylation are such that, when acting in tandem, it is the latter process that governs the rate of nucleoside incorporation in the long run. Expressed quantitatively, for the example of thymidine in Novikoff cells (where $K_{12}^{zt} \cong 225\ \mu M$, $V_{12}^{zt} \cong 27$ pmole/μl cell water·second), the coefficient of rate determination by transport [see Eq. (6)] is only about 0.02. Transport is a near-equilibrium "reaction," and consequently, may be expected to be an unsuitable site for regulation of uptake.

This prediction is borne out by several recent studies on the mode of regulation of nucleoside uptake. In asynchronous cultures of Novikoff rat hepatoma cells (Marz *et al.*, 1978), the transport capacity for thymidine and uridine remain constant regardless of growth phase of the culture, whereas thymidine and uridine uptake decrease markedly as cultures enter stationary phase. The changes in uptake parallel those of, respectively, thymidine or uridine kinase activity. However, the turnover of the nucleoside carrier protein, as judged by decay following treatment with puromycin or actinomycin D, is slow with respect to that of thymidine kinase. Conversely, quiescent 3T3 cells, stimulated to divide by addition of serum or insulin, increase uridine uptake by virtue of an increase in uridine kinase activity *in situ* (Rozengurt *et al.*, 1977). Similar conclusions have been reached (Heichal *et al.*, 1979) also for uridine uptake in stimulated NIL-8 hamster fibroblasts and their murine sarcoma virus-transformed counterparts.

Earlier observations [see Section III,C] of a correlation between nucleoside kinase activity and nucleoside uptake are easily accommodated by our model, with kinase activity taken as the chief determinant of the rate (and specificity) of nucleoside uptake. What sets the more recent observations apart from these earlier observations is that they were conducted with techniques capable of kinetically resolving the transport and phosphorylation components of uptake, and thus demonstrated, directly and simultaneously, modulation of the kinase component and constancy of the transport component.

Although modulation of nucleoside uptake at the kinase level has been rigorously demonstrated in only the few cases cited, there seems little doubt that kinase limitation pertains to nucleoside uptake in mammalian cells generally. This conclusion follows from an argument from analogy on the general disparity between Michaelis–Menten constants for transport and uptake (Table III). Transport kinetics similar to those given in Table II have been reported with at least seven nucleoside substrates in 10 cell lines; uptake kinetics comparable to those in Table II have been even more widely observed (see Wohlhueter and Plagemann, 1980, for extensive tabulations).

The primary physiological consequence of the kinetic properties of the nucleoside transporter and nucleoside kinases is that the latter are the key sites for regulation of nucleoside uptake. Particularly the uptake of uridine has been associated with the physiological status of cultured cells. When untransformed cells become quiescent because of high cell densities or of serum starvation, uridine uptake decreases markedly (Weber and Rubin, 1971; Hare, 1972a,b; Hale *et al.*, 1975; Eilam and Bibi, 1977). Conversely, stimulation of quiescent cells with serum, or of lymphocytes with plant lectins, causes a progressive rise in uridine uptake commencing within minutes after exposure to mitogen (Weber and Rubin, 1971; Pariser and Cunningham, 1971; Hare, 1972a,b; Kram *et al.*,

1973; Kram and Tomkins, 1973; Rozengurt and Stein, 1977). Mitogenic stimulation of thymidine uptake, in contrast, occurs first after a lag of 10 or more hours, and correlates with the entry of cells into S-phase (Peters and Hausen, 1971; Cunningham and Remo, 1973; Barlow and Ord, 1975; Barlow, 1976).

Knowledge of the broad specificity of the nucleoside carrier (see Section III,A), the comparison of the kinetic properties or uridine transport and uptake (see Section III,C), and the direct demonstration by Rozengurt et al. (1977, 1978) and Heichal et al. (1979) that it is only the second phase of the uridine uptake curve whose slope is increased upon stimulation of serum-starved cells, constitute firm evidence that the physiological regulation of nucleoside uptake is exerted at a posttransport level of the uptake pathway.

The posttransport regulation of thymidine uptake is manifested in an increase in the amount of thymidine kinase present in the cell (Littlefield, 1966; Adams, 1969a,b; Schuster and Hare, 1971). The posttransport regulation of uridine uptake is more apt to involve some allosteric affect on uridine kinase activity, since the increase in uptake is not blocked by inhibitors of protein synthesis (Jimenez de Asua and Rozengurt, 1974; Stein and Rozengurt, 1975), nor is an increase in kinase activity discernable in cell-free extracts from stimulated cells (Hare, 1972b; Rozengurt et al., 1978; Goldenberg and Stein, 1978).

We wish to point out, also, two consequences of potential pharmacological import. In Novikoff hepatoma cells made dependent on exogenous thymidine (and a purine) for growth by treatment with the folate antagonist methotrexate, maximal growth rate, equal to that in the absence of methotrexate, is attained at 6 μM exogenous thymidine (Marz et al., 1977b). This concentration is well below the $K_{1/2}^{zt}$ for thymidine transport (225 μM), but is saturating for thymidine uptake ($K_m^{uptake} = 0.5 \ \mu M$). Thus under normal conditions transport is clearly not rate-limiting for growth of these cells. Dipyridamole ("Persantin"), however, inhibits nucleoside transport, and renders it the rate-limiting step in uptake. Under this artificial condition, the growth of cells is limited by the rate of permeation of an "essential" nutrient. A related phenomenon has been observed by Warnick et al. (1972) in cultured lymphoblastoid cells. Treatment of these cells with a nucleoside transport inhibitor p-nitrobenzylthioinosine decreased the permeation rate of several cytotoxic drugs and thus provided protection against their antiproliferation effects.

The relative rates of transport and phosphorylation in the uptake of nucleosides explains, at least in part, the failure to select for transport deficient mutants upon treatment with cytotoxic nucleosides. Resistance to nucleoside analogs that become active antimetabolites only upon phosphorylation is usually associated with diminished activity of that kinase that phosphorylates the drug (Thompson and Baker, 1973; Harrap, 1976). In our experience, resistance to increasing levels of drug accrues stepwise, with concurrent diminution of kinase activity. A partial diminution of the superabundant transport capacity would carry little advantage;

even an absolute deficiency of transport carrier might not suffice to confer drug resistance, because of residual entry by nonmediated routes.

F. Effect of Transport Inhibitors on Uptake

It is evident from the foregoing discussion that there exist substances that inhibit the transport of nucleosides into cultured cells. One class of such substances is the nitrobenzylthiopurine nucleosides, which are remarkably potent (K_i of 1 nM or less) and apparently specific (Paterson and Oliver, 1971; Pickard et al., 1973; Cass et al., 1974; Eilam and Cabantchik, 1976, 1977; Wohlhueter et al., 1978b) inhibitors of nucleoside transport. Other compounds, such as dipyridamole and cytochalasin B, inhibit nucleoside transport and hexose and nucleobase transport as well. In fact, many lipophilic substances seem to exert some inhibition of transport, perhaps because of a general perturbation of the lipid bilayer or of lipophilic membrane proteins.

Now, the kinetics of transport inhibition by reversible inhibitors that bind to carrier proteins at their substrate binding sites or at others sites is a complicated business. Rate equations have been developed (Devés and Krupka, 1978a,b,c) to describe certain cases of such inhibition. But there are too little data currently available that are amenable to such analysis to support much speculation on the possible mechanisms of transport inhibition.

We do, however, want to give some practical consideration to the influence of inhibitors of nucleoside *transport* on rates of *uptake*. The question is treated theoretically in Section II,D,5, and it is obvious that the situation simulated in Fig. 6, for example, closely approximates the actual situation with nucleosides. The fact that transport and phosphorylation of nucleosides operate as tandem processes introduces an extra dimension of complexity to the interpretation of the kinetics of inhibition of uptake. Thus, the apparently competitive kinetic patterns of inhibition of nucleoside uptake reported, for example, for cytochalasin B (Plagemann and Estensen, 1972; Plagemann and Erbe, 1974a), for dipyridamole (Plagemann and Roth, 1969; Plagemann, 1971a), for colcemid (Plagemann and Erbe, 1974a), for theophylline and prostaglandin (Plagemann and Sheppard, 1974), and for nitrobenzylthiopurine nucleosides (Cass and Paterson, 1977; Eilam and Cabantchik, 1977; Paterson et al., 1977a,b) cannot be construed to indicate that these inhibitors are competitive in the mechanistic sense that they bind in mutually exclusive competition with nucleosides to the nucleoside transport carrier.

IV. Uptake of Nucleobases

Exogenous purine bases are utilized efficiently by cultured cells. The salvage pathways comprise mediated permeation through the cell membrane followed by phosphoribosylation, which yields directly purine riboside monophosphates. Though the uptake of purines has not been studied so extensively as that of nucleosides, it is clear that they conform to the paradigm discussed theoretically in Section II, viz. they enter cells by a saturable, nonconcentrative transport system and are converted in a single and irreversible enzymatic step to nonpermeable products.

Pyrimidine bases, on the other hand, are utilized little, if at all, by cells free of mycoplasma contamination (Levine, 1974; Schneider *et al.*, 1974; Long *et al.*, 1977), even though pyrimidines gain entry into the cells. For example, uracil is transported into some mammalian cells (Plagemann *et al.*, 1978b), but the only available route for its anabolism is phosphorolytic condensation with ribose-1-phosphate to yield uridine. Uridine phosphorylase, though present in many mammalian cells (Henderson and Paterson, 1973; Plagemann, 1971b, and unpublished observations), functions sluggishly, at best, in the anabolic direction. Thymine also appears to be transported, but is not incorporated into nucleotides (Wohlhueter *et al.*, 1978b). Orotic acid is, of course, phosphoribosylated in the normal course of pyrimidine *de novo* synthesis, but exogenous supplies are apparently not utilized (Pasternak *et al.*, 1961; Skehel *et al.*, 1967; Plagemann, 1971b), presumably because its charged carboxyl group prevents entry. Cytosine permeates Novikoff rat hepatoma cells only by a nonmediated route and is not metabolized by these cells, nor has its uptake been reported for other cell lines (Graff *et al.*, 1977). Therefore, because pyrimidine bases are poorly utilized, the following discussions are confined to purines.

A. Transport of Purines

Table IV summarizes the kinetic properties observed for purine transport as measured by rapid kinetic techniques in cells incapable of metabolizing the transport substrate, as well as those for steady-state purine uptake in metabolically active cells. Michaelis–Menten constants and maximal velocities of transport are generally in the same order of magnitude, but somewhat higher, than those for nucleosides. Adenine is exceptional in that its affinity for the carrier seems to be drastically less than those of guanine or hypoxanthine (except in Chinese hamster ovary cells). Mutual inhibition among these substrates implicates two different carriers: one specific for hypoxanthine and guanine, the other for adenine (Marz *et al.*, 1979). Differential inhibition by nitrobenzylthioinosine indicates that both nucleobase carriers are distinct from the nucleoside transporter, at least in Chinese hamster ovary cells (Wohlhueter *et al.*, 1978b) and P388 leukemia cells (unpublished observations by the authors). On the other hand, nucleosides inhibit hypoxanthine transport quite strongly, while hypoxanthine inhibits nucleoside transport more weakly. In short, a rigorous assignment of specificity classes is not yet possible. (Consult Plagemann and Wohlhueter, 1980, for more extensive treatment.)

TABLE IV

KINETIC CHARACTERISTICS OF PURINE NUCLEOBASE TRANSPORT AND UPTAKE IN VARIOUS CELL TYPES[a]

Substrate	Cell type	Transport		Uptake		References[c]
		K_{12}^{z1} (μM)	V_{12}^{z1} (pmole/μl cell water·second)	K_m^{uptake} (μM)	$V^{uptake[b]}$ (pmole/μl cell water·second)	
Adenine	N1S1-67, Novikoff hepatoma	2800	102	35	2.1	1
	Chinese hamster ovary	2300	48	—	—	
	Human platelets	—	—	0.16	0.1	2
Guanine	N1S1-67, Novikoff hepatoma	430	31	8	1.1	1
Hypoxanthine	N1S1-67, Novikoff hepatoma	279	38	7	1.1	1
	MH$_1$, C$_1$, rat hepatoma	—	—	5	0.6	3
	Chinese hamster ovary	1460	8	—	—	
	L-cells, mouse	207	64	—	—	
	P388 murine leukemia	455	48	—	—	
	HeLa, human carcinoma	303	86	—	—	

[a] Transport was measured at 23° to 25°C by rapid sampling techniques in cells incapable of metabolizing the transport substrate either by virtue of enzyme deficiency or by depletion of phosphoribosyl pyrophosphate. Uptake was measured at 37°C in metabolizing cells as the second-phase, steady-state velocity of accumulation of radioactivity within the cell (see text, Section II). Where a range of values was available from the original literature, the median of this range is tabulated.

[b] Velocity values were converted to these dimensions assuming 1.3 μl/10⁶ cells.

[c] Reference for all transport data is: Marz *et al.* (1979). References for uptake data are: (1) Zylka and Plagemann (1975); (2) Sixma *et al.* (1973); (3) Dybing (1974).

Still, the transport of a given base is rapid: for example, the $V/K_{1/2}^{zt}$ ratio for hypoxanthine in Novikoff cells (see Table IV) corresponds to a rate constant of 0.14 sec^{-1}, thus in the first-order range of concentration, 90% of transmembrane equilibrium is attained in 17 seconds. And the Michaelis–Menten constant for transport is two orders of magnitude greater than that for phosphoribosylation: a situation reminiscent of the nucleosides.

B. PHOSPHORIBOSYLATION OF PURINES

As with nucleosides, the apparent Michaelis–Menten constant for uptake into intact cells (see Table IV) is more comparable to that of the respective phosphoribosyltransferase as measured in cell-free extracts (compare Tables IV and V) than to that of transport as measured in the absence of substrate metabolism.

To the extent that cosubstrate PRPP saturates the transferase, the latter can be

TABLE V

MICHAELIS–MENTEN CONSTANTS OF THE PURINE PHOSPHORIBOSYLTRANSFERASES
FROM VARIOUS TISSUES AND CELLS[a]

Trivial name	Tissue/cell type	Substrate	K_m (M)	References[b]
Hypoxanthine–Guanine Phosphoribosyltransferase	Human erythrocytes	Hypoxanthine	9.9×10^{-6}	1
		Guanine	4.0×10^{-6}	
		PRPP[c]	2.4×10^{-4}	
	Chinese hamster brain	Hypoxanthine	5.2×10^{-7}	2
		Guanine	1.1×10^{-6}	
		PRPP	5.3×10^{-6}	
	Morris hepatoma 3924A	Hypoxanthine	4.0×10^{-6}	3
		PRPP	5.0×10^{-6}	
	N1S1-67, Novikoff rat hepatoma cells	Hypoxanthine	1.9×10^{-6}	4
		Guanine	1.3×10^{-6}	
Adenine Phosphoribosyltransferase	Rat liver	Adenine	9×10^{-7}	5
		PRPP	5×10^{-6}	
	Ehrlich ascites carcinoma cells	Adenine	9×10^{-7}	6
		PRPP	5×10^{-6}	
	N1S1-67, Novikoff rat hepatoma cells	Adenine	1.3×10^{-6}	4

[a] The first entries for each enzyme are based on studies with partially purified enzymes. The other entries are based on studies using cell sap as a source of enzyme. In all cases, the enzymes are regarded as soluble, cytoplasmic enzymes.

[b] References are: (1) Henderson et al. (1968); (2) Olsen and Milman (1978); (3) Wohlhueter (1975); (4) Zylka and Plagemann (1975); (5) Groth et al. (1978); (6) Hori and Henderson (1966).

[c] 5-Phosphoribosyl-1-pyrophosphate, most likely as the monomagnesium complex.

regarded as unimolecular with respect to base. But this assumption must be treated with caution. Under normal physiological conditions, cellular pools of PRPP are on the order of 1 mM (Henderson and Khoo, 1965; Lalanne and Henderson, 1974), and are, thus, sufficient to saturate the enzyme. Replenishment of PRPP is apparently limited by the activity of PRPP synthetase, however, such that an increased demand on PRPP (for example, to phosphoribosylate exogenously supplied purines) quickly depletes the pool.

Purine nucleoside phosphorylase catalyzes the phosphorolytic condensation of ribose-1-phosphate and hypoxanthine. It represents a potential alternate route of hypoxanthine metabolism. But, as with uridine phosphorylase, it seems to function only slowly in this anabolic capacity; in Fig. 11, for example, it can be seen that the formation of inosine from hypoxanthine in Novikoff rat hepatoma cells is negligible in comparison to the rate of phosphoribosylation.

Mutant sublines deficient in hypoxanthine, guanine, or adenine phosphoribosyltransferase are readily obtained by selection for resistance to cytotoxic purines, for example, to 8-azaguanine, 6-thioguanine, or 2,6-diaminopurine (Thompson and Baker, 1973), respectively. The utility of such enzyme-deficient sublines in assaying for transport of hypoxanthine, guanine, or adenine is discussed in Section II,E. Treatment of cells with 5 mM KCN plus 5 mM iodoacetate, which effectively depletes cellular ATP (Plagemann et al., 1976), also depletes PRPP as judged by the failure of such cells to phosphoribosylate purines (Marz et al., 1979). PRPP-depletion is, thus, a useful means to dissociate chemically the two components of purine uptake, viz. transport and phosphoribosylation.

C. TRANSPORT AND PHOSPHORIBOSYLATION IN TANDEM

Figure 11 resolves hypoxanthine uptake in Novikoff hepatoma cells into its component parts: the intracellular accumulation of free hypoxanthine and its phosphorylated products. At 160 μM exogenous hypoxanthine, influx greatly exceeds the rate of phosphoribosylation, resulting in a biphasic uptake curve, resembling the pattern predicted by simulating the kinetic behavior of sequential transport and enzymatic conversion (see Fig. 3).

A comparison of the data for 0.5 μM and 160 μM exogenous hypoxanthine illustrates a point made in Section II,D,3: that rate determination in an uptake system is dependent upon substrate concentration, as well as on the kinetic characteristics of transport and catalysis. For hypoxanthine, $V^{\text{transport}} >> V^{\text{phosphoribosylation}}$. Thus at concentrations of hypoxanthine that saturate the phosphoribosyltransferase in situ, transport velocity exceeds phosphoribosylation, the uptake curve is biphasic, and steady-state S_2 of hypoxanthine approaches S_1. But the rate constant effective in the first-order range of a Michaelis–Menten process is given by $V_{\text{max}}/K_{\text{m}}$. This quotient is greater for the phosphoribosyltransferase than for transport, so that at a given hypoxanthine S_1 in the first-order

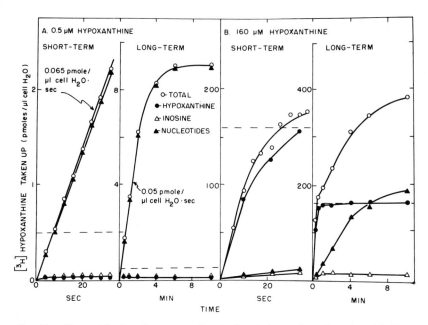

FIG. 11.   The contribution of transport, phosphoribosylation, and condensation with ribose-1-phosphate to the total uptake of hypoxanthine. Wild-type Novikoff cells were incubated at 24°C with 0.5 μM (Panel A) or 160 μM (Panel B) [³H]hypoxanthine for various times (abscissa). Then cells were rapidly separated from medium by centrifugation into silicone oil ("total uptake") or through silicone oil into 0.2 M trichloroacetic acid. In the latter case, the acid extracts were analyzed chromatographically for hypoxanthine, inosine, and phosphorylated products. The velocities noted in the figure were estimated graphically from linear segments of the curve indicated. Experimental details are given in Marz *et al.* (1979), from which the figure is reproduced with permission.

range with respect to both processes (as in Fig. 11A, where it is 0.5 μM), the rate of phosphoribosylation (which, of course, can never exceed that of entry) becomes nearly equal to that of transport. Thus for a given low S_2, reaction with the transferase exceeds exit, the steady-state S_2 is well below S_1, and influx becomes the predominate rate determinant. Even in such a situation, however, the apparent K_m of uptake reflects that of the enzyme: as hypoxanthine S_1 is elevated, the phosphoribosyltransferase becomes saturated and its maximum rate becomes limiting for long-term uptake.

D. Physiological Regulation of Purine Uptake

The uptake of hypoxanthine (and of adenine) decreases as asynchronous cultures of Novikoff rat hepatoma cells enter the stationary phase (Zylka and Plagemann, 1975), reminiscent of the change in nucleoside uptake. In an attempt to elucidate the mechanism of this decrease in uptake, we have assayed both hypoxantine transport in PRPP-depleted cells and hypoxanthine phosphoribosyl-transferase activity in cell-free extracts prepared therefrom, as a function of the growth phase of the culture (Wohlhueter and Plagemann, unpublished observations). Neither quantity changed with growth phase to an extent sufficient to explain the uptake pattern. Our tentative conclusion is that the supply of PRPP, which, in turn, may reflect the activity of PRPP synthetase, decreases as cells enter the stationary phase, thereby decreasing the rate of hypoxanthine uptake.

V. Uptake of Hexoses

Tissues specialized for the absorption of D-glucose (intestinal mucosa and renal tubules) possess an active transport system for glucose. In most other tissues, even in the retina and in the working muscle, which derive their energy almost exclusively from glycolysis, glucose enters the cell by a carrier-mediated, nonconcentrative transport system. Such a mechanism pertains also to animal cells in culture.

The regulation of glycolysis is a vast topic, quite beyond the scope of this article, though certainly not irrelevant to a consideration of hexose uptake (the reader is referred to comprehensive treatments by Newsholme and Smart, 1973; and Krebs, 1972). Our purpose here is to focus on the initial steps in the utilization of exogenously supplied hexose (glucose, in particular), and we limit ourselves to the mention of some features of glycolysis that contribute directly to an understanding of the role of the early steps.

In the absence of other fuels, the rate of utilization of glucose by a cell culture depends primarily (1) on the energy requirement of the culture and (2) on the energy yield from the glucose consumed. The first factor relates obviously to the growth rate of the culture (i.e., to its biosynthetic demands), but also more subtly, to the efficiency with which ATP-requiring processes are carried out (Racker, 1972, 1976). The energy requirements of the cell are communicated to the glycolytic apparatus in two ways: (1) the balance between ATP production and utilization establishes the "energy charge", which controls the activities of phosphofructokinase and pyruvate kinase, and (2) the rate of production of P_i and ADP, which are obligatory cosubstrates for glyceraldehyde-3-phosphate dehydrogenase and pyruvate kinase, pace stoichiometrically the reactions catalyzed by these enzymes.

The efficiency with which energy is extracted from glucose is the nub of the long-standing debate over the relative prominence of glycolysis and respiration in normal versus malignant cells (see essays by Warburg, 1955; Potter, 1963; Weinhouse, 1976; Perdue, 1979). Alleged differences in lactate production by transformed and nontransformed cells for the moment aside, it seems to be a fair generalization that a given cell culture presented with an abundance of glucose (typical in culture media) utilizes it profligately, and returns it to the medium as lactate. The same culture presented with limiting amounts of glucose, utilizes it prudently with little or no production of lactate (Graff *et al.*, 1965; Renner *et al.*, 1972; Rheinwald and Green, 1974; Schwartz *et al.*, 1975). In fact, several investigators have successfully contrived to meter glucose slowly to cells (in a way that does not lead to excretion of lactate) without impairment of growth (Graff *et al.*, 1965; Rheinwald and Green, 1974; Naiditch and Cunningham, 1977), or of the maintenence of ATP pools (Liss and Aiken, 1972), or of the ability to respond to mitogenic stimulation (Fodge and Rubin, 1975; Naiditch and Cunningham, 1977).

Investigation of the initial steps of glucose utilization is hampered by the rapid distribution of isotope, originating in glucose, into numerous metabolite pools, and its discharge from the cell as isotopic lactate and/or CO_2. These problems may be circumvented by the use of 2-deoxy-D-glucose, which is transported by the hexose carrier and phosphorylated by hexokinase (see Sections V,B and V,C), but not further metabolized to any significant degree (Yushok, 1964; Renner *et al.*, 1972; Weber, 1973). 2-Deoxy-D-glucose, thus, is a substrate that meets well the requirements of our simple model of transport followed by phosphorylation, the kinetic behavior of which was described in Section II.

Another useful substrate is 3-O-methyl-D-glucose, which likewise is transported by the glucose carrier but is not metabolized at all (Crane *et al.*, 1957; Regen and Morgan, 1974). 3-O-Methylglucose has been widely used as a substrate to assess the characteristics of hexose transport without the complicating interference of metabolic conversion.

A. TRANSPORT OF HEXOSES

Much of the literature on the "transport" of hexoses is based on studies of the uptake of metabolizable substrates, particularly glucose and 2-deoxyglucose. A rigorous interpretation of such studies, we believe, must take cognizance of the fact that transport and phosphorylation constitute, at the very simplest, a two-component "pathway," the kinetic behavior of which is predictable from that of its component parts, but is not necessarily a direct reflection of the kinetic behavior of either.

An analysis of the uptake of metabolizable substrates in terms of tandem pathway kinetics is undertaken in Sections V,D and E. We consider in this section transport per se, as measured in the absence of substrate metabolism.

This condition is met most frequently by using the nonmetabolizable substrate 3-O-methyl-D-glucose. In our experience, the influx of 3-O-methylglucose into cultured cells is rapid, and we have had to apply rapid kinetic techniques to measure it with any precision (Graff *et al.*, 1978). For example, in five mammalian cell lines, the first-order rate constant describing uptake of 3-O-methylglucose at sub-K_m concentrations ranged from 0.011 to 0.060 sec^{-1} (at 24°C; cf. Table VI). In practical terms this means (1) that one-half of the equilibrium concentration is attained intracellularly in from 12 to 63 seconds ($t_{\frac{1}{2}}$, mean = 37 seconds); (2) that by $t_{1/2}$, the *momentary* velocity of influx is already only one-half of the initial velocity; (3) that the slope of the straight line between the origin and intracellular accumulation at $t_{1/2}$ underestimates the initial velocity by about 30%; and (4) that the extent of underestimation changes (decreases) as substrate concentration increases in the range of the Michaelis–Menten constant for transport. In view of this experience, we tend to treat reports of 3-O-methylglucose transport kinetics with some skepticism, expecially when based on single time points and without supporting time courses.

The Michaelis–Menten parameters for the influx of 3-O-methylglucose in several cell lines are tabulated in Table VI. The variation in K_{12}^{zt} may be attributed in part to the temperature dependence of this parameter, but stems also from the technical difficulties involved in obtaining accurate initial velocities of substrate entry. We cannot reconcile the wide discrepancies in maximum velocities reported by different laboratories for chick embryo cells.

There is a consensus that hexose transport is saturable and nonconcentrative, and that 3-O-methyl-D-glucose, 2-deoxy-D-glucose, D-glucose, and D-galactose (but not D-fructose nor any L enantiomer) are all substrates for the same carrier (Weber, 1973; Christopher, 1977; Graff *et al.*, 1978).

A significant contribution of nonsaturable permeation of hexoses has been postulated (Renner *et al.*, 1972; Kletzien and Perdue, 1974a; Hatanaka, 1976; Germinario *et al.*, 1978), a conclusion that, insofar as it is based on studies with deoxyglucose, is open in principle to the objections raised in regard to nonmediated permeation of nucleosides (see Section III,D). The error of including steady-state, intracellular substrate pools in "velocity" computations almost certainly stands behind the conclusion of Hatanaka (1976) that several sugars in the D-configuration enter mouse embryo fibroblasts by nonmediated routes at 50 or more times the rate of L-glucose. Eilam and Vinkler (1976), however, have avoided these objections by employing equilibrium exchange experiments, using 3-O-methylglucose in BHK-21 cells. They detect an apparently nonsaturable component, with a first-order rate constant of 0.02 to 0.08 min^{-1}, which is the only mode of entry in quiescent cells. At sparser cell densities, a saturable mechanism predominated, with a first-order rate constant (when $S << K^{ee} = 1.3$ mM) of 0.27 to 0.44 min^{-1}. Based on median values, these data would indicate a minimal contribution of nonmediated permeation of about 14% in BHK cells, and an increase in that contribution with cell density and substrate concentration. Another approach to the problem has been to compare the permeation of D-hexoses to that of L-glucose or mannitol, whose permeation is nonsaturable, and occurs, presumably, at rates equivalent to those for the nonsaturable route for D-hexoses. This method, too, has led to widely disparate estimates of

TABLE VI
KINETIC PROPERTIES OF 3-O-METHYL-d-GLUCOSE INFLUX IN ANIMAL CELLS[a]

Cell type	Temperature (°C)	K_{12}^{zt} (μM)	Zero-trans Influx V_{12}^{zt} (pmole/μl cell water · second)	V_{12}^{zt}/K_{12}^{zt} (sec^{-1})	References
Erythrocytes, pig neonatal	22	23,000	450	0.020	Zeidler et al. (1976)
Thymocytes, rat	37	4,400	2.1	0.00048	Yasmeen et al. (1977)
Thymocytes, rat	37	7,700	11	0.0014	Whitesell et al. (1977b)
N1S1-67, Novikoff rat hepatoma	24	1,300	26	0.020	Graff et al. (1978)
L-cells, mouse	24	3,700	47	0.032	Graff et al. (1978)
P388, murine leukemia	24	3,100	100	0.032	Graff et al. (1978)
Chinese hamster ovary	24	2,100	23	0.011	Graff et al. (1978)
HeLa, human carcinoma	24	1,100	66	0.060	Graff et al. (1978)
3T3, mouse fibroblasts	37	5,600	87[b]	0.016	Romano and Colby (1975)
Chick embryo cells	24	3,500	3.2[c]	0.00091	Weber (1973)
Chick embryo cells	37	4,300	141[c]	0.033	Kletzien and Perdue (1974a)
Chick embryo cells	24	2,500	14[c]	0.0056	Christopher et al. (1976a)

[a] Data for cultured cells were taken for growing cells, and in some cases, are corrected for apparent, nonsaturable components. 3T3 and chick embryo were assayed as monolayers; the other cell types were in suspension. Median values are tabulated, if the original article specified a range of values.

[b] Based on 2.3 μl cell water/mg protein (Romano, 1976).

[c] Based on 4.3 μl cell water/mg protein (Weber, 1973).

the contribution of nonmediated permeation. Germinario *et al.* (1978) attribute about half the entry of deoxyglucose (at 1 mM) into cultured human fibroblasts to nonmediated permeation. Weber (1973) found that mannitol entered chick embryo cells at about 10% the rate of 3-O-methyl-D-glucose (at 2 mM). Other investigators (Illiano and Cutrecasas, 1971; Czech *et al.*, 1974; Kletzien and Perdue, 1974a; Salter and Cook, 1976; Graff *et al.*, 1978) have found the entry of L-glucose to be insignificant in comparison to mediated entry. And for the substrates listed in Table VII also, no nonsaturable component was obvious in plots of v versus S_1. It seems most prudent at this point not to generalize, but to urge investigators to consider the problem on a case-by-case basis.

A comparison of the kinetic parameters of zero-trans influx obtained with four substrates of the hexose carrier of Novikoff rat hepatoma cells are shown in Table VII. These parameters reflect operation of only the transport system, since influx was measured in cells depleted of ATP and thus incapable of initiating the metabolism of any of these substrates.

A discussion of the regulation of hexose transport is deferred to Section V,E, following an examination of the kinetic behavior of hexose uptake.

B. Phosphorylation of Hexoses

Mammalian tissues contain four hexokinase isozymes (Gonzalez *et al.*, 1964; Grossbart and Schimke, 1966; Hanson and Fromm, 1965, 1967; Kosow and Rose, 1968; Colowitz, 1973). One of these (Type IV = "glucokinase") predominates in liver, and its kinetic properties are trimmed to cope with the postprandial influx of glucose from the portal system. Nonhepatic tissues in adult and fetal rats (Knox, 1976), hepatomas, and apparently most cultured cell lines

TABLE VII
Kinetic Parameters of Zero-Trans Influx of Various Hexoses in Novikoff Hepatoma Cells[a]

Substrate	K_{12}^{zt} (μM)	V_{12}^{zt} (pmole/sec·μl)	V_{12}^{zt}/K_{12}^{zt} (sec^{-1})
3-O-Methyl-D-glucose	2,100	29	0.014
2-Deoxy-D-glucose	870	19	0.022
D-Glucose	5,800	33	0.0057
D-Galactose	24,000	84	0.0035

[a] Cells were grown in suspension culture, and depleted of ATP by treatment with 5 mM KCN plus 5 mM iodoacetate, so that none of the substrates were phosphorylated. Transport assays were conducted at 24°C by rapid sampling techniques detailed in Graff *et al.* (1978), from which the data are taken.

possess one or more of the other isozymes (Type I, II, and III = "hexokinases"), characterized by lower Michaelis–Menten constants with respect to glucose than Type IV, and by a sensitivity to inhibition by glucose-6-phosphate. The hexokinases have broad specificity for phosphoryl group acceptor, including the D enantiomers of glucose, fructose, mannose, glucosamine, and 2-deoxyglucose (but excluding galactose) (Sols and Crane, 1954). The kinetic characteristics of hexokinases from various sources are summarized in Table VIII. The majority of hexokinase activity is cytoplasmic, but a portion is also associated with particulate fractions of cell homogenates, apparently the mitochondria (Kosow and Rose, 1968; Bustamante and Pedersen, 1977).

Phosphofructokinase is a key enzyme regulating glycolytic flux in response to the energy status and citrate levels of the cell. Hexokinase, however, is also a regulatory enzyme, concerned directly with the rate of utilization of glucose for nonglycolytic purposes, such as the generation of NADPH and the production of glycogen (Krebs, 1972). The regulation of hexokinase is coordinated with that of phosphofructokinase, whose substrate, glucose-6-phosphate, is a noncompetitive (with respect to glucose) inhibitor of hexokinase.

TABLE VIII

MICHAELIS–MENTEN CONSTANTS FOR HEXOKINASE ISOZYMES FROM VARIOUS SOURCES[a]

Tissue or cell line/ substrate	Isozyme type	K_m (M)	$\dfrac{V_{\text{substrate}}}{V_{\text{glucose}}}$	References
Rat muscle	II			Grossbard and Schimke (1966)
D-Glucose		2.3×10^{-4}	1	
2-Deoxy-D-glucose		8.6×10^{-4}	—	
Fructose		3.0×10^{-3}	1.2	
ATP		7.8×10^{-4}	—	
Rat liver				Gonzalez et al. (1964)
D-Glucose	I	4.4×10^{-5}	1	
ATP		4.2×10^{-4}	—	
D-Glucose	II	1.3×10^{-4}	1	
ATP		7.0×10^{-4}	—	
D-Glucose	III	6.0×10^{-6}	1	
D-Glucose	IV	1.8×10^{-2}	1	
Novikoff rat hepatoma				Renner et al. (1972)
D-Glucose		2.4×10^{-4}	1	
2-Deoxy-D-glucose		$6 \ \times 10^{-4}$	0.3	
Chick embryo fibroblasts				Kletzien and Perdue (1974a)
2-Deoxy-D-glucose		$8 \ \times 10^{-4}$	1	
D-Glucose		$(1.6 \times 10^{-4})^b$		

[a] Isozyme type is assigned on the basis of chromatographic behavior, where known. The kinetics of hexokinases from culture were obtained in crude, cell-free extracts. In some cases the median of a range of values are given.

[b] This is the median value for the K_i for glucose with deoxyglucose as substrate.

There is an apparent correlation between hexokinase activity, as measured in cell-free extracts, and cellular growth rate. For example, the activity of hexokinase increases with growth rate of the Morris hepatomas (Burk *et al.*, 1967), and in cultured hepatoma cells as compared to normal (rat) tissue (Renner *et al.*, 1972; Bustamante and Pedersen, 1977; Knox, 1976). But the actual rates of glucose utilization by tissues and cultured cells is often well below the maximal kinase capacity apparent in cell-free extracts (Kletzien and Perdue, 1974a; Germinario *et al.*, 1978). That is, the rate of catalysis *in situ* as a function of intracellular substrate concentration cannot, in general, be expected to conform to the Michaelis–Menten relationship observed in cell-free preparations. For a given set of intracellular conditions, however, there are, presumably, an apparent K_m and V that adequately relate reaction rate to local substrate concentration. The kinetic characteristics effective *in situ* are, of course, the ones relevant to our understanding of the contribution of hexokinase action to the uptake of hexoses; there is little known about them in cultured cells.

We noted above the utility of 2-deoxyglucose in assessing the relationship between transport and phosphorylation of hexoses. There are two problems connected with its use that should be mentioned: (1) since it is not metabolized further than 2-deoxyglucose-6-phosphate, it drains ATP and traps intracellular phosphate. With relatively high loads (e.g., 2 mM), it may, within a few minutes, deplete ATP sufficiently to diminish hexokinase activity; and (2) in analogy to glucose-6-phosphate, it may be an effective allosteric inhibitor of hexokinase. Its inhibitory activity has, to our knowledge, not been examined.

C. Hexose Transport and Phosphorylation in Tandem

The hexose transport system and hexokinase have comparable Michaelis–Menten parameters with respect to hexoses (compare Tables VI, VII, and VIII). This fact and our theoretical considerations in Section II lead us a priori to the judgments (1) that the time course of uptake should be biphasic and establish a finite, steady-state, intracellular pool of hexose; (2) that neither the transport step nor the phosphorylation step will play an exclusively rate-limiting role in uptake; but, rather, (3) that each step will contribute to the uptake rate determination with a roughly equal predominance and will not be strongly dependent on the external substrate concentration.

To a first approximation we may expect the uptake of deoxyglucose to behave like the hypothetical system simulated in Fig. 3. That these expectations are met in reality is demonstrated in Figs. 12 and 13 for Novikoff rat hepatoma cells. The first of these figures shows the biphasic kinetics of total incorporation of exogenous, radiolabeled deoxyglucose into the cells. The first phase persists for 15 seconds or less; by 30 seconds a steady-state, intracellular deoxyglucose concentration has been established, and the rate of total uptake beyond this time reflects

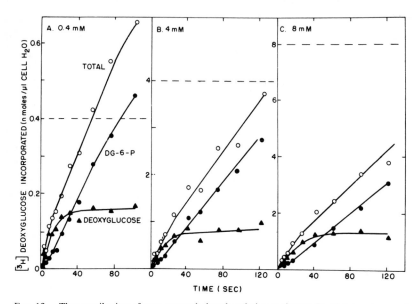

FIG. 12. The contribution of transport and phosphorylation to the uptake of deoxyglucose. Suspensions of Novikoff rat hepatoma cells were incubated at 25°C with 0.4 mM (Panel A), 4 mM (Panel B), or 8 mM (Panel C) [³H]deoxyglucose in glucose-free medium. At the times indicated on the abscissa, samples of the cell suspension were centrifuged through silicone oil into 0.2 M trichloroacetic acid. One aliquot of the acid extract was used for estimation of total radioactivity (O–O), another was chromatographed to separate deoxyglucose (▲–▲) and deoxyglucose-6-phosphate ("DG-6-P," ●–●). The broken lines indicate the extracellular concentration of substrate. Details are given in Graff *et al.* (1978), from which the figure is reproduced with permission.

only the rate of accumulation of deoxyglucose-6-phosphate. The initial slope (corresponding to the rate of zero-trans entry) exceeded the slope of the second, longer lived segment (corresponding to phosphorylation rate) by 20 to 40% (depending inversely on S_1).

The theoretical expectations regarding the relationship between steady-state S_2 and S_1 of deoxyglucose are confirmed in Fig. 13. In this experiment, Eq. (5) was fitted by a least squares regression program to the measured concentrations of deoxyglucose in the cell corresponding to several, experimentally given, exogenous concentrations. Most consistent with these data is a hexokinase operating *in situ* with $K_m = 0.56$ mM [identical to that measured in extracts of Novikoff cells (see Table VIII)] and $V_{max} = 20$ pmole/μl cell water·second, comparable to that measured for deoxyglucose uptake (Renner *et al.*, 1972). The inset in Fig. 13 demonstrates that, for this set of kinetic parameters, the coefficient of rate determination of transport is about 0.5 throughout the whole range of deoxyglucose concentrations, as is that for phosphorylation. If we assume that about the same relationship holds for glucose, we would conclude that glycolytic control of hexokinase activity, by way of modulation of the glucose-6-phosphate pool, for example, would alter uptake rates about equally as effectively as a modulation of transport velocity, by virtue of the production of new carrier protein, for example.

We stress these points because they bear importantly on a longstanding debate among investigators of hexose utilization by cultured cells. One school of thought holds that, regardless of the rate of hexose entry into the cell, regulation of flux in the glycolytic pathway conforms to the classic models of glycolytic control, exerted at the enzymatic level. The evidence mustered in support of this point of view involves analysis of metabolite pools established at various rates of transport. Though transport of 3-O-methylglucose is increased 2-fold by Rous sarcoma virus-induced transformation of chick embryo cells (Singh *et al.*,

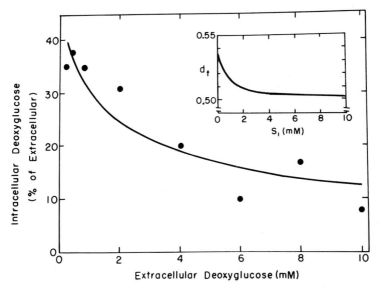

FIG. 13. Steady-state intracellular substrate concentrations as a function of extracellular concentration in cells taking up deoxyglucose. Novikoff rat hepatoma cells were exposed at 25°C to [^{3}H]deoxyglucose at various concentrations, as indicated on the abscissa, and the time courses of intracellular accumulation of deoxyglucose and of deoxyglucose-6-phosphate were monitored (examples are given in Fig. 12) as described by Graff *et al.* (1978), from which this figure is reproduced with permission. The steady-state levels of deoxyglucose apparent in these time courses are here plotted as percentage of extracellular concentration versus extracellular concentration. The curve is obtained by fitting Eq. (5) to these data, whereby K^t ($= 650 \, \mu M$) and V^t ($= 20$ pmole/μl cell water · second) were fixed at values determined independently for deoxyglucose transport (see also Table VII). The best fitting Michaelis–Menten parameters for the operation of hexokinase *in situ* were $K^e = 560 \pm 220 \, \mu M$ and $V^e = 20.2 \pm 4.8$ pmole/μl cell water · second. The inset shows the coefficient of rate determination for transport (d_t), computed according to Eq. (6), for an uptake system with these kinetic parameters, over the same range of exogenous substrate concentrations S_1.

1974b), a comparison of glycolytic metabolites with glucose as substrate (1 mM) shows a crossover plot indicative of increased flux at the levels of hexokinase, phosphofructokinase, and pyruvate kinase (but not at transport), corroborated by measurably increased activities of these enzymes (Singh *et al.*, 1974a). Increased phosphofructokinase and glucose-6-phosphate dehydrogenase activities have been observed in dense cultures of virally transformed 3T3 cells as compared to the parental line, although hexokinase activities rose similarly with culture density in both normal and transformed cells (Gregory *et al.*, 1976). In C6 astrocytoma cells (Passonneau, 1976) addition of 5.5 mM glucose to depleted medium caused the intracellular pools of glucose, glucose-6-phosphate, and UDP-glucose to swell. But, though a further increase in exogenous glucose (to 55 mM) caused further enlargement of the glucose pool, it did not affect the metabolites distal to hexokinase. And, as Romano and Colby (1973; Colby and Romano, 1975; Romano, 1976) have cogently argued, the existence of an intracellular deoxyglucose concentration, comparable to that in the medium, indicates directly the "near-equilibrium" character of the transport "reaction" and the manifest excess of transport over phosphorylation rate. (The argument is valid provided that the near-equilibrium steady state is not simply a consequence of ATP depletion, as noted in Section V,B; the data of Romano and Colby, 1973, may fall short of this provision).

The opposing view holds that transport is the "rate-limiting" step of hexose utilization. Its proponents argue that manipulations of the rate of permeation have been observed to alter the rate of utilization of glucose (or the rate of accumulation of deoxyglucose-6-phosphate), and, in some cases, the pool size of glycolytic intermediates (Bissell *et al.*, 1973; Bissell, 1976). The rate of hexose transport (as measured with nonmetabolizable hexoses) has been manipulated in a variety of cell types by several means: (1) by treatment with transport inhibitors such as cytochalasin B (Kletzien and Perdue, 1974c; Bissell *et al.*, 1973); (2) by reversible phenotypic transformation induced by temperature-sensitive viruses (Kletzien and Perdue, 1974b; Lang and Weber, 1978; Dubrow *et al.*, 1978); (3) by nutritional induction of carriers (Kalckar and Ullrey, 1973; Salter and Cook, 1976; Christopher *et al.*, 1976a,b); (4) by treatment with serum, hormones, or mitogens (reviewed by Perdue, 1979); and (5) by sundry changes in the ionic composition of the transport medium (Amos *et al.*, 1976; Bader *et al.*, 1976, 1978; Bowen-Pope and Rubin, 1977; Kaminskas, 1978).

The common feature of these studies is that, whatever the treatment, alterations in hexose transport rate are reflected in the rate of hexose utilization, an observation that is construed to mean that transport is "rate-limiting" for hexose uptake.

While it is not possible to reanalyze these studies retrospectively in terms of the kinetic behavior of transport and phosphorylation operating as a tandem pathway, it is clear that the two positions are not mutually exclusive. Given the similarity of kinetic properties of hexose transport and phosphorylation, one must expect on theoretical grounds that both processes should contribute to determination of the flux in the pathway, and that the predominance of each component in determining overall flux will shift continuously as the kinetic characteristics of either component are experimentally or physiologically altered.

D. NUTRITIONAL AND HORMONAL REGULATION OF HEXOSE UPTAKE

The regulation of hexose uptake is a broad topic, many aspects of which have been the subject of reviews: Newsholme and Smart (1973) and Krebs (1972) have addressed the fundamental aspects of glycolysis and its regulation; Morgan and Whitfield (1973) have considered the uptake of hexoses in eukaryotic cells generally, including mammalian and avian erthrocytes; Czech (1977) and Clausen (1975) have reviewed current concepts of insulin action; Wallach (1975) and Perdue (1979) have summarized the changes recorded in hexose uptake upon nutritional and mitogenic stimulation of cultured cells.

Much of the original literature is phenomenological in character, in that it describes changes in rates of overall uptake in response to various experimental manipulations, whereby the specific component or components of the uptake pathway responsible for the change is, by the nature of the experimental design, not discernable. In this section we endeavor to summarize in broad strokes these widely observed phenomena, and then to focus specifically on those experiments that do assess the roles of transport and phosphorylation in altering the rates of uptake.

1. *Nutritional Influences*

The rate of uptake of hexoses into at least three types of cells has been demonstrated to be contingent upon the carbohydrate source (or lack thereof) in the culture medium (Zielke *et al.*, 1976, 1978; Demetrakopoulos and Amos, 1976; Demetrakopoulos *et al.*, 1977). Chick embryo cells (Martineau *et al.*, 1972; Kletzien and Perdue, 1974a; Musliner *et al.*, 1977), NIL hamster kidney fibroblasts (Ullrey *et al.*, 1975; Christopher *et al.*, 1976b), and human skin cells (Salter and Cook, 1976) show a marked (as much as 10-fold) increase in the uptake of glucose following 24 hour incubation in a medium free of sugar or containing "derepressing" sugars such as fructose or xylose. How much of this increased uptake is attributable to increased transport rates, and how much to glycolytic flux? One approach to the answer is to utilize nonmetabolizable hexoses to assay transport rates. Although (for reasons discussed in Section V,A) graphical estimations of initial rates of 3-O-methylglucose tend to be underestimations, there is little doubt that glucose starvation enhances the rate at which 3-O-methylglucose is transported into some cells: increased rates of 3-O-methylglucose entry have been reported by Salter and Cook (1976) using human skin cells, by Christopher *et al.* (1976c) using NIL hamster cells, and by Musliner *et al.* (1977) using chick embryo cells. The effect on transport was corroborated by Christopher (1977), also, by measuring galactose entry into NIL cells whose galactose kinase had been selectively poisoned with N-ethylmaleimide. As discussed in Section V,C, the theoretical analysis of the kinetic relationship between hexose transport and hexokinase suggests that the transport step would be a plausible target for the regulation of hexose uptake; in fact regulation at the transport level seems to exist.

But it seems equally clear that changes in the transport apparatus are not the only ones consequent to glucose starvation. Several unresolved problems still stand in the way of a comprehensive elucidation of the phenomenon. These are well illustrated by the study of Musliner *et al.* (1977). For example, the extent of "derepression" of uptake is different for galactose, 2-deoxyglucose, glucose, and glucosamine, though all are regarded as substrates of the same transport system, but subject to different metabolic fates. Furthermore, glucose uptake in "derepressed" cells can be reduced by about 75% by refeeding glucose for as little as 15 minutes. The reduction, however, apparently does not involve modulation of the transport system, since the transport of 3-O-methylglucose is not affected.

The implication of these observations is that the physiological adaptation to glucose starvation involves metabolic loci in addition to the transport system. These loci have not been identified, nor are we in a position to distinguish quantitatively the contribution each regulatory locus might make to the overall change in uptake rate. Such a distinction will require a detailed analysis of metabolite pools and isotope fluxes, whereby permeation and the intracellular pool of free hexose must be regarded as reactions and intermediates, respectively, in the glycolytic pathway.

Correlated with the regulability of hexose carrier activity is the apparent turnover of that carrier. The basal level of galatose uptake in NIL cells in glucose medium decays to less than 20% in 24 hours when protein synthesis is blocked by cycloheximide (Christopher *et al.*, 1976a). Conversely, the increase in the rate of hexose uptake upon incubation in sugar-free medium is blocked by inhibitors of protein synthesis in hamster and chick cells (Kletzien and Perdue, 1974c; Christopher *et al.*, 1976b,c; Christopher, 1977; Musliner *et al.*, 1978). The decay of "derepressed" carrier, in the absence of continued protein synthesis is much slower than that of carrier at the basal (glucose-fed) level, prompting Christopher (1977) to postulate modulation of carrier degradation as well as of carrier synthesis.

But while such studies are suggestive of the metabolic instability of the hexose carrier, a note of precaution is in order: (1) These conclusions are based on studies of the uptake of metabolized substrates, with the attendant ambiguities of interpretation, and (2) they take uptake rate at a single substrate concentration (often $< K_m^{\text{uptake}}$) as a gauge of the amount of carrier present. Quantitative conclusions regarding carrier turnover and its role in the regulation of hexose uptake must await further, more incisive, experiments.

2. Hormonal Regulation

It is well recognized that quiescent cells stimulated to growth increase their uptake of glucose; it could hardly be otherwise. That the stimulation of uptake is observed also with 2-deoxyglucose as substrate implies that the response to stimulation involves, if only in part, the first two reactions of the uptake pathway, transport and hexokinase.

As the density of chick embryo cells in monolayer culture increases, the rate of deoxyglucose uptake decreases. Sefton and Rubin (1971) refed dense cultures with 6% serum, which caused about a 2-fold increase in deoxyglucose uptake by 75 minutes (the earliest time studied), an increase that rose to 7-fold and persisted for several hours. The uptake of thymidine, uridine, and α-aminoisobutyric acid were relatively small at the early times, though that of thymidine increased markedly after a 4 hour lag, presumably as cells began to enter S-phase. These observations on deoxyglucose uptake have been extended to mammalian cells (Plagemann, 1973; Bose and Zlotnick, 1973; Jimenez de Asua and Rozengurt, 1974; Oshiro and DiPaolo, 1974; Bradley and Culp, 1974; Moolten et al., 1978) and carried out with greater time resolution with 3T3 and chick embryo cells (Kletzien and Perdue, 1974a). The increase in deoxyglucose uptake begins with no perceptible time lag, and proceeds in two phases: an initial rise, followed by a period of constancy, and a final rise commencing at about 2 hours (Jimenez de Asua and Rozengurt, 1974; Kletzien and Perdue, 1974a).

The implied mechanistic complexity is borne out by the facts that the initial phase is reversible upon removal of serum and insensitive to treatment with cycloheximide, while the latter phase is blocked by addition of cycloheximide.

Kletzien and Perdue (1974c) regard both changes as membrane events. Their argument is based (1) on their observation (Kletzien and Perdue, 1974a) that total intracellular radioactivity in growing and in density-inhibited cells was accounted for, virtually completely, by phosphorylated derivatives, and (2) that cytochalasin B, a known inhibitor of transport, inhibits uptake: hence transport must be the predominant rate determinant. Furthermore, they argue, serum does not increase the activity of hexokinase as measured in cell-free extracts of chick embryo cells (Kletzien and Perdue, 1973). Smith and Temin (1974) come to the same conclusion on the grounds that parallel changes are seen in both deoxyglucose and 3-O-methylglucose uptake in chick embryo cells stimulated by serum (or by a partially purified "multiplication-stimulating activity"). And Dubrow et al. (1978) have reported biphasic increase in 3-O-methylglucose transport following serum stimulation of 3T3 cells. Still, there are problems with these interpretations. The cellular levels of free deoxyglucose is subject to three sources of error all favoring underestimation (see Section II,D,3). That other investigators have detected finite pools of free hexose (Romano, 1976; Passonneau, 1976; Naiditch and Cunningham, 1977; Graff *et al.*, 1978) (albeit in other cell types) gives cause for skepticism in the present case. Furthermore, the stimulation witnessed by Smith and Temin (1974) was not inhibited by cycloheximide, and though it coincides temporally with the cycloheximide-sensitive second phase, it must be considered, functionally, part of the first phase of increase. That inhibition of transport velocity by cytochalasin B should be reflected in a diminished rate of uptake is predictable, even for the situation where transport is not rate-limiting (see Section II,C,5). The case, we believe, is not tightly closed.

This is not to assert that transport changes play *no* role in the stimulation of hexose uptake by serum. Indeed, the studies of Smith and Temin (1974), Kletzien and Perdue (1974c), Eilam and Vinkler (1976), Gregory and Bose (1977), and Dubrow *et al.* (1978) with 3-O-methylglucose transport clearly establish modulation at the transport level, which (see Section V,C.) very plausibly does exert an influence on uptake rates.

Eilam and Vinkler (1976) provide insight into how the modulation of transport is effected. Using an equilibrium exchange protocol, which facilitates detection of multiple kinetic components, they have characterized two components: one, more rapid, characteristic of growing cultures of BHK cells; the other, less rapid, characteristic of quiescent cells. As an asynchronous culture grew to higher densities, the proportion of the more rapid component decreased from 100% to zero. For a single cell, the modulatory event would appear to be an instantaneous, all-or-none event; the gradual shift seen typically in uptake experiments results from population averaging.

Mitogenic lectins have also been shown to stimulate the uptake of 3-O-methylglucose in rat thymocytes (concanavalin A) (Whitesell *et al.*, 1977a; Yasmeen *et al.*, 1977; Hume and Weidemann, 1978) and in cultured bovine lymphocytes (phytohemagglutinin) (Peters and Hausen, 1971), compelling the conclusion that these mitogens impinge at the transport level of uptake. The enhancement of 3-O-methylglucose transport is detectable within a few minutes after treatment, is not blocked by cycloheximide, but is antagonized when the intracellular cAMP pool is enlarged by treatment with methylisobutylxanthine or prostaglandin E_2 (Yasmeen *et al.*, 1977). But the response to mitogens also involves stimulation of glycolysis: the pools of glycolytic intermediates of stimulated rat thymocytes (Culvenor and Weidemann, 1976) indicate activation at the levels of transport, hexokinase, and phosphofructokinase.

In many respects the effects of serum or lectins on hexose uptake duplicate the action of insulin. Insulin accelerates the rate of entry or exit of 3-O-methylglucose from intact tissues such as the diaphragm or epididymal fat (see, for example, Bihler, 1968; Clausen, 1969; Bihler and Sawh, 1973), which, in view of this compound's metabolic inertness, is attributable to accelerated permeation across the membrane. This stimulatory effect of insulin on 3-O-methylglucose transport has been observed in suspensions of cells dispersed from epididymal and perimetrical fat tissues by enzymatic and mechanical means (Vinten *et al.*, 1976; Czech, 1976a,b). In contrast to the slow flux of 3-O-methylglucose in intact tissues, the kinetic properties ($V^{ee} = 70$–200 pmole/μl cell water·second; $K^{ee} = 5$ mM) for equilibrium exchange of 3-O-methylglucose in fat cells were similar to those for influx in cultured cells (see Table VI); upon insulin stimulation, V^{ee} increases to 1700 pmole/μl cell water·second, without change in K^{ee} (Vinten *et al.*, 1976).

Secondary cultures of chick embryo fibroblasts have been extensively used to assess the influence of serum factors on hexose uptake, and such cells respond also to insulin (see, for example, Vaheri *et al.*, 1972; Smith and Temin, 1974; Raizada and Perdue, 1976). There have, however, been relatively few reports of insulin-stimulated hexose uptake in cells of established lines: the maximal velocity of deoxyglucose uptake in density-inhibited 3T3 cells is stimulated by insulin (Bradley and Culp, 1974), while in a glial tumor cell line (Passonneau, 1976), insulin enlarges the intracellular pools of glucose, glucose-6-phosphate, and glycogen. Moreover, the derepressive effect on glucose uptake of hexose starvation in chick embryo cells is potentiated by insulin (Musliner *et al.*, 1977), and, especially in sparse cultures, may be entirely dependent on the presence of this hormone.

Treatment of chick embryo cells with neuraminidase (Vaheri *et al.*, 1972) or trypsin (Sefton and Rubin, 1971) also mimics the effects of serum on deoxyglucose uptake. Raizada and Perdue (1976) have demonstrated that treatment with these enzymes unmasks insulin binding sites on the cell surfaces, while stimulation of serum-starved cells induces the formation of new receptors. Thus, it appears that a diversity of stimulatory agents converge mechanistically in that they all give rise to additional insulin receptors.

All of these studies of insulin stimulation of hexose uptake, however, have been based on relatively longterm assays of uptake of metabolizable sugars. A clearcut involvement of insulin at the transport level is not demonstrated.

E. Malignant Transformation and Hexose Uptake

Debate on the Warburg (1926, 1955) hypothesis has focused over the years on three central problems: (1) the validity of Warburg's contention that tumor tissues possess (and rely upon) a greater capacity for aerobic glycolysis than do normal tissues (Potter, 1963; Burk *et al.*, 1967; Wenner, 1975); (2) the catalytic and regulatory properties of the glycolytic pathways in normal and neoplastic tissue that might account for the alleged differences (Racker, 1972; Soulinna *et al.*, 1975; Weinhouse, 1976); and (3) the possible contribution of enhanced glycolysis to malignancy and invasiveness (Racker, 1976).

With increasing awareness of the involvement of the cell membrane in malignant transformation (reviewed by Wallach, 1975), however, the role of hexose transport has been brought into the debate. The technical advantages to transport measurement in cultured cells and the availability of nontransformed and virally-transformed versions of the same cell type have established cultured cells as a logical choice for such investigations. Increased conversion of glucose to lactate has been observed in chick embryo cells infected with Rous sarcoma virus in comparison with noninfected cells (Rubin, 1960; Morgan and Ganapathy, 1963; Bissell et al., 1972, 1973). With growing mammalian cells, however, failure to detect significant increases have been reported for chemically and polyoma virus-transformed BHK cells (Moolten et al., 1977; Broadfoot et al., 1964) and NRK cells (Gregory and Bose, 1977). In any case, the significance of changes in lactate production as a consequence of viral transformation must be considered in the context of the strong dependence of this parameter on the nature of the hexose offered as substrate, on the available concentration of exogenous glucose, and on the physiological state of a culture. Even when observed, the mere fact of increased lactate production provides no information on which of the many conceivable molecular mechanisms might have given rise to it.

Of direct relevance to our present inquiry are several reports that the transport of 3-O-methylglucose into chick embryo cells is enhanced by infection with transforming viruses (Venuta and Rubin, 1973; Weber, 1973; Kletzien and Perdue, 1974b; Lang and Weber, 1978). Technical problems in coping with the rapidity of 3-O-methylglucose transport notwithstanding, these studies leave no doubt that the velocity of hexose transport is increased 2- to 3-fold as a result of viral transformation. Upon kinetic analysis, the difference has proved to be in the maximum velocity of influx, while the Michaelis–Menten constant remained unchanged.

Deoxyglucose uptake into chick embryo cells is also stimulated by transformation (Venuta and Rubin, 1973; Weber, 1973; Bissell et al., 1973; Kletzien and Perdue, 1974b; Bader et al., 1976), and, in some cases, the stimulation parallels that of 3-O-methylglucose uptake. Thus Weber (1973) and Lang and Weber (1978) found comparable increases in the maximal velocities of 3-O-methylglucose and deoxyglucose (ca. 5-fold), and thus considered the increase in transport to account entirely for the increase in uptake. Kletzien and Perdue (1974b) drew similar conclusions from experiments on cells infected with a temperature-sensitive mutant of Rous sarcoma virus (Ts-68) and shifted from a permissive to a nonpermissive temperature. And Bissell et al. (1973) pointed out that the increased uptake of deoxyglucose, in turn, parallels the increase in lactate formation from glucose by chick embryo cells. But the data of Lang and Weber (1978) indicate a more complex, concerted regulation of both transport and hexokinase. Upon transformation, the steady-state intracellular concentration of deoxyglucose increased from 20 to 30% of extracellular concentration, while the steady-state rate of phosphorylation increased about 7-fold: clearly not possible for a Michaelian enzyme operating before and after transformation with the same, effective kinetic parameters. In fact, judging from the pool analysis in growing cells, the kinetic relationship between transport and phosphorylation in chick embryo cells would seem to resemble that in Novikoff cells (see Figs. 12 and 13), in which both components contribute about equally to determination of uptake rate. Furthermore, it is likely that phosphofructokinase is a further determinant, as indicated by the manifest increase in flux at the phosphofructokinase level (Bissell *et al.*, 1973). Observations that the hexokinase activity, as measured in cell-free extracts, does not increase upon transformation of chick cells is not germaine to this point: that activity is nearly an order of magnitude higher than hexokinase flux observed *in situ*.

Analogous studies on 3-*O*-methylglucose transport in mammalian cells leave a rather more ambiguous picture. Hatanaka and colleagues (Hatanaka *et al.*, 1969; Graff *et al.*, 1973; Hatanaka, 1976) reported marked increases in 3-*O*-methylglucose transport (at nonsaturating concentrations) into 3T3 cells that were infected with murine sarcoma virus. These investigators claimed that the difference in rate stems from a shift in affinity of the transport system for hexoses (Hatanaka *et al.*, 1969), or even a shift to a nonsaturable mode of entry as the predominate one (Hatanaka, 1976). But it appears that these studies suffer from serious underestimation of initial rates of entry, and from the inclusion of near-equilibrium, intracellular substrate levels into apparent velocities, as discussed in Section V,A. Eckhart and Weber (1974) reported an 8-fold stimulation of 3-*O*-methylglucose in polyoma virus-transformed 3T3 cells; Dubrow *et al.* (1978), a 12-fold stimulation in SV40-transformed 3T3 cells; and Gregory and Bose (1977) observed a 4-fold increase in 3-*O*-methylglucose influx when NRK cells were transformed by Kirsten sarcoma virus.

On the other side of the ledger, Romano and Colby (1975) and Romano (1976) failed to detect any kinetic differences in 3-*O*-methylglucose transport between 3T3 and SV40-transformed 3T3 cells. Nor did Lever (1976) detect a difference in glucose transport into vesicles prepared from such cells: vesicles that exhibit stereospecific transport of D-glucose but phosphorylate very little of it. And Graff *et al.* (1978) noted that untransformed and Rous sarcoma virus-transformed vole cells behaved similarly with respect to 3-*O*-methylglucose entry.

In view of these conflicting results and of the technical problems attending the use of 3-*O*-methylglucose (which may be responsible in part for them) we regard the question of a possible effect of transformation on hexose transport in mammalian cells as not completely resolved.

The uptake of deoxyglucose into mammalian cells (as distinct from the transport of 3-*O*-methylglucose) is accelerated in response to viral transformation, especially when compared in dense cultures: in 3T3 cells (Bose and Zlotnick, 1973; Bradley and Culp, 1974; Moolten *et al.*, 1977, in NRK cells (May *et al.*,

1973; Gregory and Bose, 1977), in mouse embryo cells (Plagemann, 1973), and in BHK cells (Isselbacher, 1972). This fact establishes that some regulatory infuence(s), consequent to transformation, is (are) localized to the first two steps of hexose utilization. But again, it does not distinguish to what extent the transport and/or phosphorylation step is responsible for the increase in overall flux. Nor does increased deoxglucose uptake seem to be a necessary adjunct of transformation, judging from a report by Moolten et al. (1977) of a chemically transformed BHK strain whose uptake of deoxyglucose and production of lactate are less than those of the parental strain.

The galactose uptake system casts the ambiguity into another perspective. Recall that galactose enters cells via the same hexose carrier that transports 3-O-methylglucose, deoxyglucose, and glucose, but that [apparently because of the limited capacity of UDPgalactose 4-epimerase (Kalckar et al., 1973)] its carbon is metered much more slowly into the glycolytic pathway than is glucose carbon. Confluent polyoma virus-transformed NIL hamster cells (grown on galactose) take up galactose 2.5 times more rapidly than do their nontransformed counterparts (Ullrey et al., 1975). Steady-state, intracellular galactose in these cells equaled 89 and 74% of extracellular galactose, respectively. Thus with galactose, to a much greater extent than with deoxyglucose (and presumably glucose), transport constitutes a near-equilibrium reaction, and modulation of it alone could be neither a necessary nor sufficient cause of increased galactose uptake.

A rigorous, quantitative statement of the contributions of hexose transport and phosphorylation to an increased glycolytic flux in transformed cells in culture remains, thus, elusive. All things considered, we favor the hypothesis that both transport and metabolic apparatuses are modulated by transformation and function as codeterminants of glycolytic rate, i.e., that there is no single, simple off-on switch. And we are optimistic that a rigorous statement is feasible; the substrates, methods, cell types, and theoretical framework are available for a step-by-step analysis of isotopic flux and of steady-state intermediate pools. From this information, the rates of transport and phosphorylation in hexose uptake can be quantitatively assessed.

VI. Uptake of Vitamins and Choline

The mode of nutrient uptake under consideration in this article—mediated permeation of the cell membrane and entrapment within the cell as phosphorylated products—is plausible for a number of water soluble vitamins. Possible candidates for this mode of uptake, based on the existence of phosphorylated coenzyme forms of the vitamins, include riboflavin (Horwitt and Witting, 1972), pantothenic acid (Brown, 1971a), pyridoxine (Snell and Haskell, 1971), thiamine (Brown, 1971b), and choline (Plagemann, 1971c).

Our perusal of the vitaminological literature reveals that the intestinal absorption of vitamins has been a subject of frequent investigation (see, for example, the review by Rindi and Ventura, 1972). However, little work appears to have been done on vitamin transport into cultured cells or into single cells dispersed from solid tissue. It is this latter work on which we focus in this section.

Thiamine is accumulated in jejunal mucosa to a concentration approximately equal to that of the bathing medium, whereas thiamine pyrophosphate accumulated to about three times this concentration (Rindi and Ferrari, 1977). This is indicative of nonconcentrative permeation, either mediated or not. Chen (1978) has described thiamine uptake into collagen-dispersed hepatocytes. Under aerobic conditions (and in the presence of Na^+), he found a 5-fold concentration of radioactivity within the cells as compared to medium. While all the cell-associated radioactivity cochromatographed with thiamine, less than a third of it exited the cells upon incubation in thiamine-free medium. In the absence of O_2 or Na^+, cellular accumulations approximated the extracellular concentration. Chen interpreted his results as evidence of active transport, but there are ambiguities in the study that, to our mind, leave the question of the mode of thiamine uptake open. In any case the apparent K_m of aerobic uptake was 0.31 mM.

Various forms of *pyridoxine* accumulate in brain and cerebrospinal fluid to levels $\geqq 25$ times the concentration of the vitamin in plasma. Spector and Greenwald (1978) have measured total levels of radioactivity accumulated in rabbit brain slices, intact choroid plexus, and erythrocytes, and fractionated the cell-associated radioactivity into free and phosphorylated material. In all cases, accumulation in excess of that in medium was attributable to phosphorylated forms. Whether the permeation step is mediated or nonmediated is not established. In intestine, permeation appears to be nonmediated (Middleton, 1977).

Pyridoxal increases the extent to which amino acids are concentrated from medium into Ehrlich ascites cells (Pal and Christensen, 1961). Accumulation of pyridoxal within these cells, however, is limited to about 1.4 times medium concentration; the excess is attributable to intracellular binding and, apparently, not to phosphorylation.

Choline is taken up by cultured and ascites tumor cells, phosphorylated, and incorporated into membranes as phosphatidylcholine. Uptake has been characterized kinetically for cultured Novikoff rat hepatoma cells ($K_m^{\text{uptake}} = 4$–$7 \, \mu M$; Plagemann, 1971c), for cultured N_{18} and S_{21} neuroblastoma cells ($K_m^{\text{uptake}} = 2.6$ and $0.1 \, \mu M$, respectively; Massarelli *et al.*, 1974), and for Ehrlich-Lettre ascites tumor cells ($K_m^{\text{uptake}} = 59 \, \mu M$; Haeffner, 1975). As uptake studies (in the sense employed throughout this article) they relate the steady-state rate of appearance of cell-associated radioactivity to the exogenous choline concentration, without clearly delineating the contributions of transport and phosphorylation to the kinetics of uptake.

The Michaelis–Menten constant for exchange (6.5 μM; Martin, 1968) and efflux ($\cong$ 25 μM; Edwards, 1973) of choline in human erythrocytes (which, apparently, do not phosphorylate choline) are roughly comparable to the K_m^{uptake} in cultured cells, thus lending credence to, but certainly not proving, the notion that the kinetics of choline uptake reflect those of the transport system, i.e., that transport is the primary rate determinant of choline uptake.

VII. Concluding Discussion

Three main classes of nutrients—nucleosides, nucleobases, and hexoses (and possibly a fourth, the water-soluble vitamins)—are taken up into cultured cells as a result of the tandem action of a nonconcentrative transport system and an intracellular enzyme that introduces into the transported substrate an anionic group, and thus confers impermeability.

For several substrates representing the first three of these nutrient classes, both members of the uptake pathway have been characterized kinetically: the transport systems with cells and substrates that, by virtue of enzyme deficiency, ATP depletion, or chemical design, are metabolically inert; the enzymes, as purified proteins, in idealized milieu. We have endeavored in this article to reintegrate these members, to inquire how they function in relation to one another in the intact cell.

The question is put theoretically: Given knowledge of the kinetic characteristics of the component parts, can we predict how they might work together? A simple analysis of their behavior as a pathway shows that transport and phosphorylation play interdependent roles in determining overall flux in the uptake pathway. It follows that an adequate appreciation of changes in flux must take this interdependence into consideration. The consequences of changes in substrate concentration, in temperature, and in the effective kinetic parameters of transport and phosphorylation on rates of uptake demonstrate the complexity of this interdependence, and the errors of interpretation that can accrue if the complexity is overlooked.

Flux in a tandem pathway follows a biphasic time course. The first phase corresponds to the attainment of steady-state levels of free, intracellular substrate, and its slope at zero time measures transport. The second phase represents steady-state flux through the pathway, that is, the rate of accumulation of impermeable product. Such biphasic behavior has been reported for the uptake of nucleosides, nucleobases, and hexoses in experiments that have been designed adequately to discern the first phase (which, with all of these substrates, is of short duration). Conventional sampling schedules (say, at intervals > 30 seconds), employed with phosphorylatable substrates, almost invariably detects only the second phase of uptake, though measurements made under these experimental conditions have often been construed as measurements of "transport".

The complexity of even this two-component pathway demands precise and careful usage of the term "rate-limiting." In its simplest sense, this term applies only in the extremes: when (for a given exogenous substrate concentration) the velocity of substrate entry greatly exceeds that of phosphorylation, or when the velocity of entry is much less than the velocity of phosphorylation *would be* if the phosphorylating enzyme were exposed to that substrate concentration prevailing beyond the membrane barrier. Between the extremes, *both* components may codetermine overall flux, and the predominance of each can be stated quantitatively.

For nucleosides and nucleobases, the Michaelis–Menten parameters of transport and phosphorylation (or phosphoribosylation) are such that the first extreme is generally approximated: the phosphorylation component is the predominant rate determinant and the apparent site of physiological regulation of nutrient uptake. But even this generalization must be qualified. The predominance of one component or the other in the determination of uptake rate is contingent on substate concentration, as well as on the kinetic parameters of the transport and phosphorylation systems. At very low concentration of substrate (in a range that may be quite relevant physiologically; if not to cells in culture, then certainly to cells in tissues) it is the ratio of first-order rate constants for the two components (given by V_{max}/K_m) that is crucial to the determination of the steady state.

For hexoses, the kinetic characteristics of transport and phosphorylation place the uptake pathway squarely between the extremes. It is theoretically predictable and experimentally demonstrable that both components contribute comparably to determination of the rate of uptake. Thus, both components qualify as appropriate loci for physiological regulation. And, in fact, it seems likely that both transport and phosphorylation of hexoses are modulated when cultured cells are hexose-starved, serum-starved, or virally transformed. Many published experiments have been designed to detect a change in *one* component—transport or phosphorylation—that is then held to be responsible for the change in uptake rate. What is needed now are experiments designed to ascertain how much (and how) each component is modified to effect an observed change in uptake.

ACKNOWLEDGMENTS

The authors are indebted to a number of individuals, to whom they express their gratitude. To Drs. Richard Marz and Jon Graff, for their contributions as colleagues during the period in which many of the ideas expressed here were thrashed out intramurally; to John Erbe, for his adeptness with computers, and for much skillful technical help with our experimental work; to Drs. Harold Amos, William Christopher, John Cook, Ruth Koren, James Perdue, and Wilfred Stein, who supplied us with preprints of their work; and to Timothy Leonard and Cheryl Thull, for their expert graphical and clerical help. Financial support came from USPHS research grants GM24468 and AM23001 and training grant CA 09138 (R.M.W.).

REFERENCES

Adams, R. L. P. (1969a). *Exp. Cell Res.* **56**, 49.
Adams, R. L. P. (1969b). *Exp. Cell Res.* **56**, 55.
Amos, H., Christopher, C. W., and Musliner, T. A. (1976). *J. Cell. Physiol.* **89**, 669.
Anderson, E. P. (1973). *In* "The Enzymes" (P. Boyer, ed.), 3rd Ed. Chap. 2. Academic Press, New York.
Bader, J. P., Lew, M. A., and Brown, N. R. (1976). *Arch. Biochem. Biophys.* **175**, 196.
Bader, J. P., Sege, R., and Brown, N. R. (1978). *J. Cell. Physiol.* **95**, 179.
Barlow, S. D. (1976). *Biochem. J.* **154**, 395.
Barlow, S. D., and Ord, M. G. (1975). *Biochem. J.* **148**, 295.
Berlin, R. D., and Oliver, J. M. (1975). *Int. Rev. Cytol.* **42**, 287.
Bhargava, P. M. (1977). *J. Theor. Biol.* **68**, 101.
Bihler, I. (1968). *Biochim. Biophys. Acta* **163**, 401.
Bihler, I., and Sawh, P. C. (1973). *Can. J. Physiol. Pharmacol.* **51**, 371.
Bissell, M. J. (1976). *J. Cell. Physiol.* **89**, 701.
Bissell, M. J., Hatie, C., and Rubin, H. (1972). *J. Natl. Cancer Inst.* **49**, 555.
Bissell, M. J., White, R. C., Hatie, C., and Bassham, J. A. (1973). *Proc. Natl. Acad. Sci. U.S.A.* **70**, 2951.
Bose, S. K., and Zlotnik, B. J. (1973). *Proc. Natl. Acad. Sci. U.S.A.* **70**, 2374.
Bowen-Pope, D. F., and Rubin, H. (1977). *Proc. Natl. Acad. Sci. U.S.A.* **74**, 1585.
Bradley, W. E. C., and Culp, L. A. (1974). *Exp. Cell Res.* **84**, 335.
Bresnick, E., and Thompson, U. B. (1965). *J. Biol. Chem.* **240**, 3967.
Broadfoot, M., Walker, P., Paul, J., MacPherson, I., and Stoker, M. (1964). *Nature (London)* **204**, 79.
Brown, G. M. (1971a). *In* "Comprehensive Biochemistry" (M. Florkin and E. H. Stotz, eds.), Vol. 21, p. 73. Elsevier, Amsterdam.
Brown, G. M. (1971b). *In* "Comprehensive Biochemistry" (M. Florkin and E. H. Stotz, eds.), Vol. 21, p. 1. Elsevier, Amsterdam.
Burk, D., Woods, M., and Hunter, J., (1967). *J. Natl. Cancer Inst.* **38**, 839.
Bustamante, E., and Pedersen, P. L. (1977). *Proc. Natl. Acad. Sci. U.S.A.* **74**, 3735.
Cabantchik, Z. I., and Ginsburg, H. (1977). *J. Gen. Physiol.* **69**, 75.
Cass, C. E., and Paterson, A. R. P. (1972). *J. Biol. Chem.* **247**, 3314.
Cass, C. E., and Paterson, A. R. P. (1973). *Biochim. Biophys. Acta* **291**, 734.
Cass, C. E., and Paterson, A. R. P. (1977). *Exp. Cell Res.* **105**, 427.
Cass, C. E., Gaudette, L. A., and Paterson, A. R. P. (1974). *Biochim. Biophys. Acta* **345**, 1.
Chen, C.-P. (1978). *J. Nutr. Sci. Vitaminol.* **24**, 351.
Christensen, H. N. (1975). *In* "Biological Transport" (2nd ed.). Benjamin, Reading, Massachusetts.
Christopher, C. W. (1977). *J. Supramol. Struct.* **6**, 485.
Christopher, C. W., Kohlbacher, M. S., and Amos, H. (1976a). *Biochem. J.* **158**, 439.

Christopher, C. W., Ullrey, D., Colby, W., and Kalckar, H. M. (1976b). *Proc. Natl. Acad. Sci. U.S.A.* **73**, 2429.
Christopher, C. W., Colby, W. W., and Ullrey, D. (1976c). *J. Cell. Physiol.* **89**, 683.
Čihak, A., and Rada, B. (1976). *Neoplasma* **23**, 233.
Clausen, T. (1969). *Biochim. Biophys. Acta* **183**, 625.
Clausen, T. (1975). *Curr. Top. Membr. Transp.* **6**, 169.
Cleland, W. W. (1963a). *Biochim. Biophys. Acta* **67**, 104.
Cleland, W. W. (1963b). *Biochim. Biophys. Acta* **67**, 173.
Colby, C., and Romano, A. H. (1975). *J. Cell. Physiol.* **85**, 15.
Coleman, C. N., Stoller, R. G., Drake, J. C., and Chabner, B. A. (1975). *Blood* **46**, 791.
Colowitz, S. P. (1973). *In* "The Enzymes" (P. Boyer, ed.), 3rd ed., Chap. 1. Academic Press, New York.
Crane, R. K., Field, R. A., and Cori, C. F. (1957). *J. Biol. Chem.* **224**, 649.
Culvenor, J. G., and Weidemann, M. J. (1976). *Biochim. Biophys. Acta* **437**, 354.
Cunningham, D. D., and Pardee, A. B. (1969). *Proc. Natl. Acad. Sci. U.S.A.* **64**, 1049.
Cunningham, D. D., and Remo, R. A. (1973). *J. Biol. Chem.* **248**, 6282.
Czech, M. P. (1976a). *J. Biol. Chem.* **251**, 1164.
Czech, M. P. (1976b). *J. Cell. Physiol.* **89**, 661.
Czech, M. P. (1977). *Annu. Rev. Biochem.* **46**, 359.
Czech, M. P., Lawrence, J. C., and Lynn, W. S. (1974). *J. Biol. Chem.* **249**, 5421.
Demetrakopoulos, G. E., and Amos, H. (1976). *Biochem. Biophys. Res. Commun.* **72**, 1169.
Demetrakopoulos, G. E., Gonzalez, F., Colofiore, J., and Amos, H. (1977). *Exp. Cell Res.* **106**, 167.
Devés, R., and Krupka, R. M. (1978a). *Biochim. Biophys. Acta* **510**, 186.
Devés, R., and Krupka, R. M. (1978b). *Biochim. Biophys. Acta* **510**, 339.
Devés, R., and Krupka, R. M. (1978c). *Biochim. Biophys. Acta* **513**, 156.
Dubrow, R., Pardee, A. B., and Pollack, R. (1978). *J. Cell. Physiol.* **95**, 203.
Durham, J. P., and Ives, D. H. (1970). *J. Biol. Chem.* **245**, 2276.
Dybing, E. (1974). *Biochem. Pharmacol.* **23**, 395.
Eckhart, W., and Weber, M. (1974). *Virology* **61**, 223.
Edwards, P. A. W. (1973). *Biochim. Biophys. Acta* **311**, 123.
Eilam, Y., and Bibi, O. (1977). *Biochim. Biophys. Acta* **467**, 51.
Eilam, Y., and Cabantchik, Z. I. (1976). *J. Cell. Physiol.* **89**, 831.
Eilam, Y., and Cabantchik, Z. I. (1977). *J. Cell. Physiol.* **92**, 185.
Eilam, Y., and Stein, W. D. (1974). *In* "Methods in Membrane Biology" (E. D. Korn, ed.), Vol. 2, Chap. 5. Plennum, New York.
Eilam, Y., and Vinkler, C. (1976). *Biochim. Biophys. Acta* **433**, 393.
Fodge, D. W., and Rubin, H. (1975). *J. Cell. Physiol.* **86**, 453.
Forsdyke, D. R. (1971). *Biochem. J.* **125**, 721.
Fox, I. H., and Kelley, W. N. (1978). *Annu. Rev. Biochem.* **47**, 655.
Friedmann, T., Seegmiller, J. E., and Subak-Sharpe, J. H. (1969). *Exp. Cell Res.* **56**, 425.
Germinario, R. J., Oliveira, M., and Leung, H. (1978). *Can. J. Biochem.* **56**, 80.
Goldenberg, G. J., and Stein, W. D. (1978). *Nature (London)* **274**, 475.
Gonzalez, C., Ureta, T., Sanchez, R., and Niemeyer, H. (1964). *Biochem. Biophys. Res. Commun.* **16**, 347.
Graff, J. C., Hanson, D. J., and Hatanaka, M. (1973). *Int. J. Cancer* **12**, 602.
Graff, J. C., Wohlhueter, R. M., and Plagemann, P. G. W. (1977). *J. Biol. Chem.* **252**, 4185.
Graff, J. C., Wohlhueter, R. M., and Plagemann, P. G. W. (1978). *J. Cell. Physiol.* **96**, 171.
Graff, S., Moser, H., Kastner, D., Graff, A. M., and Tannenbaum, M. (1965). *J. Natl. Cancer Inst.* **34**, 511.

Gregory, S. H., and Bose, S. K. (1977). *Exp. Cell Res.* **110**, 387.

Gregory, S. H., Kumari, H. L., Lakshmi, M. J., and Bose, S. K. (1976). *Arch. Biochem. Biophys.* **175**, 644.

Grossbard, L., and Schimke, R. T. (1966). *J. Biol. Chem.* **241**, 3546.

Groth, D. P., Young, L. G., and Kenimer, J. G. (1978). *Methods Enzymol.* **51**, 574.

Haeffner, E. W. (1975). *Eur. J. Biochem.* **51**, 219.

Hale, A. H., Winkelhake, J. L., and Weber, M. J. (1975). *J. Cell Biol.* **64**, 398.

Hanson, T. L., and Fromm, H. J. (1965). *J. Biol. Chem.* **240**, 4133.

Hanson, T. L., and Fromm, H. J. (1967). *J. Biol. Chem.* **242**, 501.

Hare, J. D. (1972a). *Biochim. Biophys. Acta* **282**, 401.

Hare, J. D. (1972b). *Biochim. Biophys. Acta* **255**, 905.

Harrap, K. R. (1976). *In* "Scientific Foundations of Oncology" (T. Symmington and R. C. Carter, eds.). Year Book Medical Publishers, Chicago, Illinois.

Hatanaka, M. (1974). *Biochim. Biophys. Acta* **355**, 77.

Hatanaka, M. (1976). *J. Cell. Physiol.* **89**, 745.

Hatanaka, M., Huebner, R. J., and Gilden, R. V. (1969). *J. Natl. Cancer Inst.* **43**, 1091.

Hauschka, P. V., Everhart, L. P., and Rubin, H. (1972). *Proc. Natl. Acad. Sci. U.S.A.* **69**, 3542.

Hawkins, R. A., and Berlin, R. D. (1969). *Biochim. Biophys. Acta* **173**, 324.

Heichal, O., Ish-Shalom, D., Koren, R., and Stein, W. D. (1979). *Biochim. Biophys. Acta* **551**, 169.

Henderson, J. F., and Khoo, M. K. Y. (1965). *J. Biol. Chem.* **240**, 2358.

Henderson, J. F., and Paterson, A. R. P. (1973). "Nucleotide Metabolism: an Introduction." Academic Press, New York.

Henderson, J. F., Brox, L. W., Kelley, W. N., Rosenbloom, R. M., and Seegmiller, J. E. (1968). *J. Biol. Chem.* **243**, 2514.

Holley, R. W. (1972). *Proc. Natl. Acad. Sci. U.S.A.* **69**, 2840.

Hopwood, L. E., Dewey, W. C., and Hejny, W. (1975). *Exp. Cell Res.* **96**, 425.

Hori, M., and Henderson, J. F. (1966). *J. Biol. Chem.* **241**, 1406.

Horwitt, M. K., and Witting, L. A. (1972). *In* "The Vitamins" (W. H. Sebrell and R. S. Harris, eds.), p. 53. Academic Press, New York.

Hume, D. A., and Weidemann, M. J. (1978). *J. Cell. Physiol.* **96**, 303.

Illiano, G., and Cuatrecasas, P. (1971). *J. Biol. Chem.* **246**, 2472.

Isselbacher, K. (1972). *Proc. Natl. Acad. Sci. U.S.A.* **69**, 585.

Jimenez de Asua, L., and Rozengurt, E. (1974). *Nature (London)* **251**, 624.

Jimenez de Asua, L., Rozengurt, E., and Dulbecco, R. (1974). *Proc. Natl. Acad. Sci. U.S.A.* **71**, 96.

Kalckar, H. M., and Ullrey, D. (1973). *Proc. Natl. Acad. Sci. U.S.A.* **70**, 2502.

Kalckar, H. M., Ullrey, D., Kijomoto, S., and Hakomori, S. (1973). *Proc. Natl. Acad. Sci. U.S.A.* **70**, 839.

Kaminskas, E. (1978). *Biochem. J.* **174**, 453.

Kessel, D. (1968). *J. Biol. Chem.* **243**, 4739.

Kessel, D. (1978). *J. Biol. Chem.* **253**, 400.

Kessel, D., and Shurin, S. B. (1968). *Biochim. Biophys. Acta* **163**, 179.

Kizer, D. E., and Holman, L. (1974). *Biochim. Biophys. Acta* **350**, 193.

Kletzien, R. F., and Perdue, J. F. (1973). *J. Biol. Chem.* **248**, 711.

Kletzien, R. F., and Perdue, J. F. (1974a). *J. Biol. Chem.* **249**, 3366.

Kletzien, R. F., and Perdue, J. F. (1974b). *J. Biol. Chem.* **249**, 3375.

Kletzien, R. F., and Perdue, J. F. (1974c). *J. Biol. Chem.* **249**, 3383.

Kletzien, R. F., and Perdue, J. F. (1975). *Cell* **6**, 513.

Kletzien, R. F., and Perdue, J. F. (1976). *J. Cell. Physiol.* **89**, 723.

Knox, W. E. (1976). *In* "Enzyme Patterns in Fetal, Adult and Neoplastic Rat Tissues" (2nd ed.). Karger, Basel.

Koren, R., Shohami, E., Bibi, O., and Stein, W. D. (1978). *FEBS Lett.* **86**, 71.

Koren, R., Shohami, E., and Yeroushalmi, S. (1979). *Eur. J. Biochem.* **95**, 333.

Kosow, D. P., and Rose, I. A. (1968). *J. Biol. Chem.* **243**, 3625.

Kram, R., and Tomkins, G. M. (1973). *Proc. Natl. Acad. Sci. U.S.A.* **70**, 1659.

Kram, R., Mamont, P., and Tomkins, G. M. (1973). *Proc. Natl. Acad. Sci. U.S.A.* **70**, 1432.

Krebs, H. A. (1972). *Essays Biochem.* **8**, 1.

Krenitsky, T. A., Tuttle, J. V., Koszalka, G. W., Chen, I. S., Beachum, L. M., Rideout, J. L., and Elion, G. B. (1976). *J. Biol. Chem.* **251**, 4055.

Krygier, V., and Momparler, R. L. (1971). *J. Biol. Chem.* **246**, 2745.

Lalanne, M., and Henderson, J. F. (1974). *Anal. Biochem.* **62**, 121.

Lang, D. R., and Weber, M. J. (1978). *J. Cell. Physiol.* **94**, 315.

Lassen, U. V. (1967). *Biochim. Biophys. Acta* **135**, 146.

Lee, T., Karon, M., and Momparler, R. L. (1974). *Cancer Res.* **34**, 2482.

Lever, J. E. (1976). *J. Cell. Physiol.* **89**, 779.

Levine, E. M. (1974). *Methods Cell Biol.* **8**, 229.

Lieb, W. R., and Stein, W. D. (1974). *Biochim. Biophys. Acta* **373**, 178.

Lin, S. S., and Munyon, W. (1974). *J. Virol.* **14**, 1199.

Lindberg, B., Klenow, H., and Hausen, K. (1967). *J. Biol. Chem.* **242**, 350.

Liss, E., and Aiken, P. (1972). *Z. Krebsforsch.* **77**, 292.

Littlefield, J. W. (1966). *Biochim. Biophys. Acta* **114**, 398.

Littlefield, J. W. (1977). *In* "Molecular Biology of the Mammalian Genetic Apparatus" (P. Tsu, ed.), Chap. 16. Elsevier-North Holland Biomedical Press, Amsterdam.

Long, C. W., Del Giudice, R., Gardella, R. S., and Hatanaka, M. (1977). *In Vitro* **13**, 429.

Lum, C. T., Marz, R., Plagemann, P. G. W., and Wohlhueter, R. M. (1979). *J. Cell. Physiol.* **101**, 173.

Martin, K. (1968). *J. Gen. Physiol.* **51**, 497.

Martineau, R., Kohlbacher, M., Shaw, S. N., and Amos, H. (1972). *Proc. Natl. Acad. Sci. U.S.A.* **69**, 3407.

Marz, R., Wohlhueter, R. M., and Plagemann, P. G. W. (1977a). *J. Supramol. Struct.* **6**, 433.

Marz, R., Wohlhueter, R. M., and Plagemann, P. G. W. (1977b). *J. Membr. Biol.* **34**, 277.

Marz, R., Wohlhueter, R. M., and Plagemann, P. G. W. (1978). *J. Supramol. Struct.* **8**, 511.

Marz, R., Wohlhueter, R. M., and Plagemann, P. G. W. (1979). *J. Supramol. Struct.* **8**, 511.

Marz, R., Wohlhueter, R. M., and Plagemann, P. G. W. (1979). *J. Biol. Chem.* **254**, 2329.

Massarelli, R., Ciesielski-Treska, J., Ebel, A., and Mandel, P. (1974). *Biochem. Pharmacol.* **23**, 2857.

May, J. T., Somers, K. D., and Kit, S. (1973). *Int. J. Cancer* **11**, 377.

Meuwissen, H. J., Pickering, R. J., Polara, B., and Porter, I. H. (1975). *In* "Combined Immunodeficiency Disease and Adenosine Deaminase Deficiency: A Molecular Defect." Academic Press, New York.

Middleton, H. M. (1977). *J. Nutr.* **107**, 126.

Moolten, F. L., Moolten, D. N., and Capparell, N. J. (1977). *J. Cell. Physiol.* **93**, 147.

Morgan, H. E., and Ganapathy, S. (1963). *Proc. Soc. Exp. Biol. Med.* **113**, 312.

Morgan, H. E., and Whitfield, C. F. (1973). *Curr. Top. Membr. Transp.* **4**, 255.

Munch-Petersen, B., and Tyrsted, G. (1977). *Biochim. Biophys. Acta* **478**, 364.

Murray, A. W. (1968). *Biochem. J.* **106**, 549.

Murray, A. W. (1971). *Annu. Rev. Biochem.* **40**, 811.

Murray, A. W., Elliott, D. C., Atkinson, M. R. (1970). *Prog. Nucleic Acid Res.* **10**, 87.

Musliner, T. A., Chrousos, G. P., and Amos, J. (1977). *J. Cell. Physiol.* **91**, 155.

Naiditch, W. P., and Cunningham, D. D. (1977). *J. Cell. Physiol.* **92,** 319.

Newsholme, E. A., and Smart, C. (1973). *In* "Regulation in Metabolism" Chap. 1. Wiley, New York.

Oliver, J. M., and Paterson, A. R. P. (1971). *Can. J. Biochem.* **49,** 262.

Olsen, A. S., and Milman, G. (1978). *Methods Enzmol.* **51,** 543.

Orengo, A. (1969). *J. Biol. Chem.* **244,** 2204.

Oshiro, Y., and DiPaolo, J. A. (1974). *J. Cell. Physiol.* **83,** 193.

Pal, P. R., and Christensen, H. N. (1961). *J. Biol. Chem.* **236,** 894.

Pardee, A. B. (1971). *In Vitro* **7,** 95.

Pariser, R. J., and Cunningham, D. D. (1971). *J. Cell Biol.* **49,** 525.

Passonneau, J. V. (1976). *J. Cell. Physiol.* **89,** 693.

Pasternak, C. A., Fisher, G. A., and Handschuhmacher, R. E. (1961). *Cancer Res.* **21,** 110.

Paterson, A. R. P., and Oliver, J. M. (1971). *Can. J. Biochem.* **49,** 271.

Paterson, A. R. P., Naik, S. R., and Cass, C. E. (1977a). *Mol. Pharmacol.* **13,** 1014.

Paterson, A. R. P., Babb, L. R., Paran, J. H., and Cass, C. E. (1977b). *Mol. Pharmacol.* **13,** 1147.

Perdue, J. F. (1979). *In* "Virus-Transformed Cell Membranes" (C. Nicolau, ed.), Chap. 4. Academic Press, New York.

Peters, J. H., and Hausen, P. (1971). *Eur. J. Biochem.* **19,** 509.

Pickard, M. A., Brown, P. R., Paul, B., and Paterson, A. R. P. (1973). *Can. J. Biochem.* **51,** 666.

Pierre, K. J., and LePage, G. A. (1968). *Proc. Soc. Exp. Biol. Med.* **127,** 432.

Plagemann, P. G. W. (1971a). *Biochim. Biophys. Acta* **233,** 688.

Plagemann, P. G. W. (1971b). *J. Cell. Physiol.* **77,** 213.

Plagemann, P. G. W. (1971c). *J. Lipid Res.* **12,** 715.

Plagemann, P. G. W. (1973). *J. Cell. Physiol.* **82,** 421.

Plagemann, P. G. W., and Erbe, J. (1974a). *Cell* **2,** 71.

Plagemann, P. G. W., and Erbe, J. (1974b). *J. Cell Biol.* **55,** 161.

Plagemann, P. G. W., and Estensen, R. D. (1972). *J. Cell Biol.* **55,** 179.

Plagemann, P. G. W., and Richey, D. P. (1974). *Biochim. Biophys. Acta* **344,** 263.

Plagemann, P. G. W., and Roth, M. F. (1969). *Biochemistry* **8,** 4782.

Plagemann, P. G. W., and Sheppard, J. R. (1974). *Biochem. Biophys. Res. Commun.* **56,** 869.

Plagemann, P. G. W., and Wohlhueter, R. M. (1980). *Curr. Top. Membr. Transp.* (in press).

Plagemann, P. G. W., Richey, D. P., Zylka, J. M., and Erbe, J. (1975). *J. Cell Biol.* **64,** 29.

Plagemann, P. G. W., Marz, R., and Erbe, J. (1976). *J. Cell. Physiol.* **89,** 1.

Plagemann, P. G. W., Marz, R., and Wohlhueter, R. M. (1978a). *Cancer Res.* **38,** 978.

Plagemann, P. G. W., Marz, R., and Wohlhueter, R. M. (1978b). *J. Cell. Physiol.* **97,** 49.

Potter, V. R. (1963). *Adv. Enz. Reg.* **1,** 279.

Pritchard, J. B., Chavez-Peon, F., and Berlin, R. D. (1970). *Am. J. Physiol.* **219,** 1263.

Quinlan, D. C., and Hochstadt, J. (1974). *Proc. Natl. Acad. Sci. U.S.A.* **71,** 5000.

Quinlan, D. C., and Hochstadt, J. (1976). *J. Biol. Chem.* **251,** 344.

Racker, E. (1972). *Am. Sci.* **60,** 56.

Racker, E. (1976). *J. Cell. Physiol.* **89,** 697.

Raizada, M. K., and Perdue, J. F. (1976). *J. Biol. Chem.* **251,** 6445.

Regen, D. M., and Morgan, H. E. (1964). *Biochim. Biophys. Acta* **79,** 151.

Renner, E. D., Plagemann, P. G. W., and Bernlohr, R. W. (1972). *J. Biol. Chem.* **247,** 5765.

Rheinwald, J. G., and Green, H. (1974). *Cell* **2,** 287.

Rindi, G., and Ferrari, G. (1977). *Experimentia* **33,** 211.

Rindi, G., and Ventura, U. (1972). *Physiol. Rev.* **52,** 821.

Roller, B., Hiral, K., and Defendi, V. (1974). *J. Cell. Physiol.* **83,** 163.

Romano, A. H. (1976). *J. Cell. Physiol.* **89,** 737.

Romano, A. H., and Colby, C. (1973). *Science* **179,** 1238.

Roos, H., and Pfleger, K. (1972). *Mol. Pharmacol.* **8**, 417.

Rosenbloom, F. M., Kelley, W. N., Miller, J., Henderson, J. F., and Seegmiller, J. E. (1967). *J. Am. Med. Assoc.* **202**, 175.

Roux, J. M. (1973). *Enzyme* **15**, 361.

Rozengurt, E., and Stein, W. D. (1977). *Biochim. Biophys. Acta* **464**, 417.

Rozengurt, E., Stein, W. D., and Wigglesworth, N. M. (1977). *Nature (London)* **267**, 442.

Rozengurt, E., Mierzejewski, K., and Wigglesworth, N. (1978). *J. Cell. Physiol.* **97**, 241.

Rubin, H. (1960). *Virology* **10**, 29.

Salter, D. W., and Cook, J. S. (1976). *J. Cell. Physiol.* **89**, 143.

Schaffer, M., Hurlbert, R. B., and Orengo, A. (1973). *Cancer Res.* **33**, 2265.

Schneider, E. L., Stambridge, E. J., and Epstein, C. J. (1974). *Exp. Cell Res.* **84**, 311.

Schuster, G. S., and Hare, J. D. (1971). *In Vitro* **6**, 427.

Schwartz, J. P., Lust, W. D., Lauderdale, J. R., and Passonneau, J. V. (1975). *Mol. Cell Biol.* **9**, 67.

Sefton, B. M., and Rubin, H. (1971). *Proc. Natl. Acad. Sci. U.S.A.* **68**, 3154.

Sheppard, J. R. (1972). *Nature (London) New Biol.* **236**, 14.

Singh, M., Singh, V. N., August, J. T., and Horecker, B. L. (1974a). *Arch. Biochem. Biophys.* **165**, 240.

Singh, V. N., Singh, M., August, J. T., and Horecker, B. L. (1974b). *Proc. Natl. Acad. Sci. U.S.A.* **71**, 4129.

Sixma, J. J. Holmsen, H., and Trieschnigg, A. M. C. (1973). *Biochim. Biophys. Acta* **298**, 460.

Sixma, J. J., Lips, J. P. M., Trieschnigg, A. M. C., and Holmsen, H. (1976). *Biochim. Biophys. Acta* **443**, 33.

Sjostrom, D. A., and Forsdyke, D. R. (1974). *Biochem. J.* **138**, 253.

Skehel, J. J., Hay, A. J., Burke, D. C., and Cartwright, L. N. (1967). *Biochim. Biophys. Acta* **142**, 430.

Sköld, O. (1960). *J. Biol. Chem.* **235**, 3273.

Smith, G. L., and Temin, H. M. (1974). *J. Cell. Physiol.* **84**, 181.

Snell, E. E., and Haskell, B. E. (1971). *In* "Comprehensive Biochemistry" (M. Florkin and E. H. Stotz, eds.), Vol. 21, p. 47. Elsevier, Amsterdam.

Sols, A., and Crane, R. K. (1954). *J. Biol. Chem.* **210**, 581.

Sols, A., and Marco, R. (1970). *Curr. Top. Cell. Reg.* **2**, 227.

Soulinna, E.-M., Buchsbaum, R. N., and Racker, E. (1975). *Cancer Res.* **35**, 1865.

Spector, R., and Greenwald, L. L. (1978). *J. Biol. Chem.* **253**, 2373.

Stein, W. D. (1967). *In* "The Movement of Molecules Across Cell Membranes." Academic Press, New York.

Stein, W. D., and Rozengurt, E. (1975). *Biochim. Biophys. Acta* **419**, 112.

Thompson, L. H., and Baker, R. M. (1973). *Methods Cell Biol.* **6**, 210.

Ullrey, D., Gammon, M. T., and Kalckar, H. M. (1975). *Arch. Biochem. Biophys.* **167**, 410.

Vaheri, A., Ruoslahti, E., and Nordling, S. (1972). *Nature (London) New Biol.* **238**, 211.

Venuta, S., and Rubin, H. (1973). *Proc. Natl. Acad. Sci. U.S.A.* **70**, 653.

Vinten, J., Gliemann, J., and Østerlind, K. (1976). *J. Biol. Chem.* **251**, 794.

Wallach, D. F. H. (1975). *In* "Membrane Molecular Biology of Neoplastic Cells," Chap. 7. Elsevier, Amsterdam.

Warburg, O. (1926). *In* "Über den Stoffwechsel der Tumoren." Springer-Verlag, Berlin and New York.

Warburg, O. (1955). *Naturwissenschaften* **14**, 401.

Warnick, C. T., Muzik, H., and Paterson, A. R. P. (1972). *Cancer Res.* **32**, 2017.

Weber, M. J. (1973). *J. Biol. Chem.* **248**, 2978.

Weber, M. J., and Rubin, H. (1971). *J. Cell. Physiol.* **77**, 157.

Weinhouse, S. (1976). Z. Krebsforsch. Klin. Onkol. **87,** 115.

Wenner, C. E. (1975). In "Cancer: A Comprehensive Treatment" (F. F. Becker, ed.), Vol. 3, Chap. 14, Plenum, New York.

Whitesell, R. R., Johnson, R. A., Tarpley, H. L., and Regen, D. M. (1977a). J. Cell Biol. **72,** 456.

Whitesell, R. R., Tarpley, H. L., and Regen, D. M. (1977b). Arch. Biochem. Biophys. **181,** 596.

Wohlhueter, R. M. (1975). Eur. J. Cancer **11,** 463.

Wohlhueter, R. M., and Plagemann, P. G. W. (1980). In "Handbook of Nutrition and Foods." CRC Press, West Palm Beach, Florida (in press).

Wohlhueter, R. M., Marz, R., Graff, J. C., and Plagemann, P. G. W. (1976). J. Cell. Physiol. **89,** 605.

Wohlhueter, R. M., Marz, R., Graff, J. C., and Plagemann, P. G. W. (1978a). Methods Cell Biol. **20,** 211.

Wohlhueter, R. M., Marz, R., and Plagemann, P. G. W. (1978b). J. Membr. Biol. **42,** 247.

Wohlhueter, R. M., Marz, R., and Plagemann, P. G. W. (1979). Biochem. Biophys. Acta **553,** 262.

Yasmeen, D., Laird, A. J., Hume, D. A., and Weidemann, M. J. (1977). Biochim. Biophys. Acta **500,** 89.

Yushok, W. D. (1964). Cancer Res. **24,** 187.

Zeidler, R. B., Lee, P., and Kim, H. D. (1976). J. Gen. Physiol. **67,** 67.

Zielke, H. R., Ozand, P. T., Tildon, J. T., Sevdalian, D. A., and Cornblath, M. (1976). Proc. Natl. Acad. Sci. U.S.A. **73,** 4110.

Zielke, H. R., Ozand, P. T., Tildon, J. T., Sevdalian, D. A., and Cornblath, M. (1978). J. Cell. Physiol. **95,** 41.

Zylka, J. M., and Plagemann, P. G. W. (1975). J. Biol. Chem. **250,** 5756.

INTERNATIONAL REVIEW OF CYTOLOGY, VOL. 64

The Contractile Apparatus of Smooth Muscle

J. Victor Small and Apolinary Sobieszek

Institute of Molecular Biology of the Austrian Academy of Sciences, Salzburg, Austria

I. Introduction 241
II. Architecture of the Contractile Apparatus 242
   A. General Morphology 242
   B. Organization of the Contractile Material: Fibrils and Contractile
      Units 242
   C. General Development of Ultrastructural Studies 245
   D. Thick Filament Architecture and Structural Polarity 249
   E. The 10 nm Skeletin Filaments 256
   F. Dense Bodies and Attachment Plaques 262
   G. The Contractile Apparatus 263
III. The Contractile Proteins 266
   A. Smooth Muscle Actomyosin 266
   B. The Thin Filaments 274
   C. Myosin 277
   D. Additional Proteins in Actomyosin 280
IV. Regulation of the Actin–Myosin Interaction via Ca^{2+} 283
   A. Myosin Phosphorylation and Ca-Dependent Regulation . . . 283
   B. The Activating Effect of Tropomyosin 292
   C. The Myosin Light Chain Kinase 293
   D. Myosin Dephosphorylation and Relaxation 295
   E. Kinase–Phosphatase Regulation 298
   References 299

I. Introduction

From its common integration into organs and tissues rather than, with a few exceptions, into distinct anatomical muscular units, vertebrate smooth muscle has invariably been considered a less favorable system for the study of contractility than either vertebrate striated muscle or even invertebrate smooth muscles. Perhaps least discouraged by these disadvantages and prompted more by the functional importance of vertabrate smooth muscle have been the muscle physiologists. Through their persistence of effort over the last half century, the excitatory properties of diverse vertebrate smooth muscles have been documented in detail (see Büllbring *et al.*, 1970; Büllbring and Shuba, 1976; Daniel and Paton, 1975). But an understanding of the process of excitation–contraction coupling in

241

smooth muscle has been continually frustrated by the notable dearth of information on the contractile machinery itself, i.e., on its protein composition and structural organization. On this subject not much more could be said than indicated by the appropriate definition of smooth muscle cells as formulated by Fischer (1944); "Cells specialized for contraction, only with their protoplasmic constituents homogeneous in the longitudinal direction."

Fortunately considerable progress has been made in the last decade toward the elucidation of the structure and biochemistry of vertebrate smooth muscle protoplasm. The advances that have been made are encouraging since they begin to point the way to an understanding of the excitation–contraction coupling pathway. This article will be devoted to an analysis of these more recent developments.

II. Architecture of the Contractile Apparatus

A. General Morphology

As was established by various nineteenth century histologists, (see McGill, 1909; Needham, 1971) smooth muscle cells, when isolated, are generally spindle-like in shape with tapered ends and possess a centrally placed, elongated nucleus. *In vivo*, where the cells occur in bundles and overlapping layers, they adapt themselves to the contours of the tissue; in some instances, they may be essentially straight, while in others, for example in the vascular wall, they may be markedly curved along their length. This adaptation extends further to their shape in cross section, which may vary within the same muscle layer from circular to polygonal to a very flattened form (McGill, 1909; Burnstock, 1970). Thus, the contractile machinery must of necessity have a design that not only conforms to these requirements for flexibility in cell shape, but tolerates the presence of the cell nucleus in a central position.

B. Organization of the Contractile Material: Fibrils and Contractile Units

In the early studies, the protoplasm of the smooth muscle cell was seen in histological sections as either devoid of structure or exhibiting coarse or fine fibrillar units. Since the investigations were then bedeviled by the variability in the degree of preservation of these fibrils, their significance in relation to the contractile apparatus was a matter of some debate (see McGill, 1909).

More recently, a reinvestigation of the structure of the contractile apparatus of smooth muscle, using the light microscope, has been prompted by the development of procedures to isolate smooth muscle cells in an intact and viable form (Bagby *et al.,* 1971; Fay and Delise, 1973; Small, 1974, 1977a; Fay *et al.,*

1976; Fay, 1977; Fisher and Bagby, 1977). From these studies two general conclusions have emerged.

1. The contractile elements are arranged in fibrillar contractile units that have their attachment sites on the plasma membrane. These units probably correspond to the fibrils observed in the early histological investigations.

2. In the extended cell, the contractile elements and their corresponding units lie approximately parallel to the cell axis, whereas, during shortening, they become progressively more obliquely arranged. At the shortest length of the cell, the units may subtend angles between 25° (Small, 1974), and 45° (Fisher and Bagby, 1977) with respect to the cell axis.

The presence of contractile units was inferred from two lines of evidence. First, for living isolated cells from toad stomach, electrical stimulation caused the coextensive formation of closely spaced blebs over the cell surface and the blebs were interspersed by small undistorted regions. Following subsequent cell shortening, the blebs were withdrawn and the cells assumed their original, smooth appearance. The transitory formation of the blebs could be explained by the attachment of "force generating units" (Fay and Delise, 1973) over the entire cell area, with the proposed attachment sites lying in the zones between the blebs. The presence of densely staining plaques on the plasmalemma in these zones, as seen in the electron microscope, was also consistent with the existence of such attachment sites.

Second, more direct evidence for the existence of fibrillar contractile units came from experiments with detergent-extracted isolated cells (Small, 1974, 1977a). By the use of conditions for detergent treatment by which the extraction of contractile proteins was minimal, "cell models" were obtained that retained the ability to contract in the presence of Ca^{2+} and MgATP (Fig. 1a–f). In the absence of ATP, the cells (from chicken gizzard and from guinea pig taenia coli and vas deferens) were rigid and in a state corresponding to rigor in striated muscle. In this rigor state, the cytoplasm showed the presence of distinct fibrillar units arranged in an oblique fashion with respect to the cell axis (Fig. 1g,h). The degree of obliquity of the fibrils was related to cell length, the fibrils being more obliquely arranged at the shorter cell lengths. The onset of contraction in the presence of MgATP was marked by the dissolution of the fibrils and the adoption by the cytoplasm of a homogeneous appearance that persisted during shortening.

Evidence consistent with changes in orientation of the contractile elements in smooth muscle on shortening was first obtained from birefrigence measurements on living muscle (Fischer, 1944). Contrary to the situation with striated muscle, smooth muscles, both invertebrate (Bozler and Cottrell, 1937) and vertebrate (Fischer, 1946), showed a clear dependence of birefringence on muscle length;

244   J. VICTOR SMALL AND APOLINARY SOBIESZEK

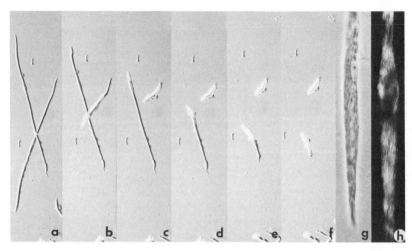

FIG. 1. Isolated smooth muscle cell models obtained by extraction with Triton X-100. (a–f) Sequence of frames showing the contraction of the isolated and Triton-extracted cells in response to the addition of MgATP and Ca ions (for precise composition of the medium, see Small, 1977a). Times after the initiation of contraction are, in seconds: (a) 0; (b) 30; (c) 60; (d) 90; (e) 120; and (f) 150. ×96. (g,h) Triton-extracted smooth muscle cell models showing cytoplasmic fibrils. These are obliquely arranged with respect to the cell axis. This state of the cells corresponds essentially to the state of rigor in striated muscle. (g) Cell from chicken gizzard; phase contrast, ×360. (From Small, 1974, with permission.) (h) Cell from guinea pig taenia coli; polarization optics. ×720.

elongation was accompanied by increased birefringence. A reinvestigation of this phenomenon using living, isolated cells from toad stomach was made by Fisher and Bagby (1977). They confirmed the earlier data by demonstrating a dramatic loss of birefringence on cell shortening or in areas of localized contractions. That this loss was due to a change in orientation of the contractile elements was supported by the recovery of areas of birefringence on rotation of the shortened cells through angles of up to about 45°. This observed change in orientation was also consistent with the change in angling of the cytoplasmic fibrils noted in the Triton-extracted "cell models" (Small, 1974).[1]

In concluding this section, we should note that the existence of a contractile unit as a distinct entity of a certain dimension has yet to be defined. In cells in rigor, the diameter of the observed fibrils is somewhat variable, indicating that some lateral aggregation can take place. Also, separate and functional contractile units have not so far been isolated. Nevertheless, the concept of fibrils consisting

[1]It may be noted that the X-ray diffraction patterns from smooth muscle that have shown reflections from actin and myosin (see following section) were obtained only from extended muscle strips in which the contractile elements would be expected to be essentially parallel to the cell axis. Patterns from shortened muscles showing the filament reflections have not been reported.

of contractile elements coupled together and to the cell membrane fits well with the data currently available. This concept will be developed in further discussions of the structural organization of the contractile apparatus.

C. General Development of Ultrastructural Studies

It is perhaps not surprising that efforts to define the detailed organization of the contractile elements in smooth muscle by electron microscopy have been impeded by a variability in preservation comparable to that encountered by the early histologists. At the ultrastructural level, this variability has related primarily to the degree of preservation of the thick, myosin-containing filaments. In most of the early ultrastructural studies (see reviews by Burnstock, 1970; and Shoenberg and Needham, 1976), with some notable exceptions (Taxi, 1961; Pavlova *et al.*, 1961; Choi, 1962), thick filaments of a diameter comparable to those seen in striated muscle could not be detected. The only filamentous components that were observed (see Lane, 1965; Fawcett, 1966) were actin filaments and filaments of slightly larger diameter (around 100 Å) now generally referred to as "10 nm," "intermediate," or "skeletin" filaments. The latter filaments are discussed in more detail in Section II,E.

Kelly and Rice (1968) and Pease (1968) demonstrated significant numbers of thick filaments in smooth muscle using, respectively, glycerinated material and tissue prepared by the method of "inert dehydration." In subsequent studies, the former authors could demonstrate thick filaments only in muscle fixed in a contracted state (Kelly and Rice, 1969; Rice *et al.*, 1970) and proposed that thick filaments may form at the onset of contraction. This view was not inconsistent at that time with the negative X-ray evidence for thick filaments in smooth muscle (Elliott and Lowy, 1968) and the difficulties in isolation of thick filaments in muscle homogenates under relaxing conditions (Shoenberg, 1969; see also Section II,D). In a spate of ultrastructural investigations that followed, thick filaments were reported in different vertebrate smooth muscles fixed either in relaxation or contraction (Devine and Somlyo, 1971; Garamvölgyi *et al.*, 1971; Heumann, 1971; Uehara *et al.*, 1971; Cooke and Fay, 1972a; Campbell and Chamley, 1975). But a primary dilemma in these, as in the earlier studies, was the lack of an independent control of the preservation during chemical fixation, i.e., of the *in vivo* organization of the contractile apparatus.

The first firm evidence for the presence of thick filaments in living smooth muscle was provided by the X-ray diffraction studies of Lowy *et al.* (1970, 1973). For strips of guinea pig taenia coli relaxed for the prolonged periods required to obtain the patterns (by incubation at low temperature or in hypertonic solutions), a meridional reflection, characteristic for ordered aggregates of myosin (Huxley and Brown, 1967) could be detected. The same reflection could also be shown to be present (although much weaker) in taenia coli muscle undergoing rhythmic, phasic contraction (Lowy *et al.*, 1973; P. J. Vibert, personal communication). In an attempt to define the form of the myosin filaments, parallel electron microscope studies were undertaken utilizing the X-ray method as a control for the fixation process (Lowy and Small, 1970; Small and Squire, 1972). For muscle, relaxed as for the X-ray experiments and showing the myosin reflection following aldehyde fixation, thick filaments were observed, but were found to vary in their cross-sectional shape and to commonly show a ribbon-like form. Such ribbons were then considered to be the form of the myosin elements *in vivo* (Lowy and Small, 1970; Small and Squire, 1972; Lowy *et al.*, 1973).

While there was little doubt that the latter tissue prepared for electron microscopy had been well preserved, it was argued that the unphysiological nature of the methods employed for relaxation may themselves induce an artificial aggregation of filaments into ribbons (Somlyo *et al.*, 1971, 1973; Rice *et al.*, 1971; Jones *et al.*, 1973). This possibility was reemphasised in a further analysis of the X-ray data by Shoenberg and Haselgrove (1974), but, from the presence of only one reflection from myosin, a clear distinction between cylindrical filaments and ribbons of narrow width could still not be made.

A new approach toward resolving this problem was offered by the development of procedures to obtain smooth muscle cell models (see Section II,B) essentially analogous to the myofibrils of striated muscle. In these demembranated cells, the contractile apparatus was directly accessible to the bathing medium and was therefore not subject to uncontrolled changes in the state or aggregation of the filaments during the fixation process (see, for example, Shoenberg, 1973). For cells fixed under conditions corresponding to "rigor" and "relaxation," thick filaments, but no ribbon-like-structures, were observed in the electron microscope (Small, 1977a). In rigor, that is, in the absence of ATP, the thick filaments characteristically showed a square form in cross section (Fig. 2), while in relaxation, they appeared more cylindrical in profile (Fig. 3).[2] From the fragmentation of the isolated cells, the thick filaments could also be isolated (Fig. 5, Section II,D), and while not ribbon-like in shape, they did exhibit the ability to aggregate, in fact, with their cross-bridges precisely in register (Small, 1977a). These aggregates clearly corresponded to the ribbon structures observed in thin sections, which showed lateral striations and projections at the characteristic cross-bridge repeat around 140 Å (Small and Squire, 1972).

From these and the other electron microscope studies, it seems reasonable to conclude that the images obtained of different smooth muscle preparations showing approximately cylindrical thick filaments and optimal ratios of numbers of thick to thin filaments (cf. Fig. 3, and Devine and Somlyo, 1971) closely repre-

---

[2]It is worth noting at this juncture that the demonstration of myosin filaments in such "relaxed" model cells provided independent evidence for the presence of myosin in filament form in the state of relaxation (see also Small, 1977a).

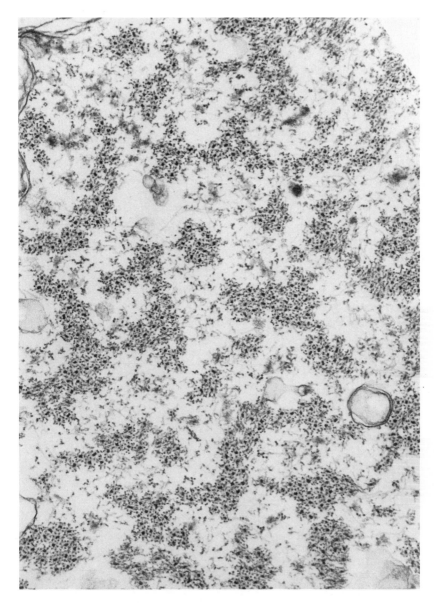

FIG. 2. Cross section of Triton-extracted smooth muscle cell from guinea pig taenia coli. This state of the cell corresponds to the rigor-like state shown in Fig. 1g and h. The actin and myosin filaments are formed into irregularly shaped groups. ×53,550.

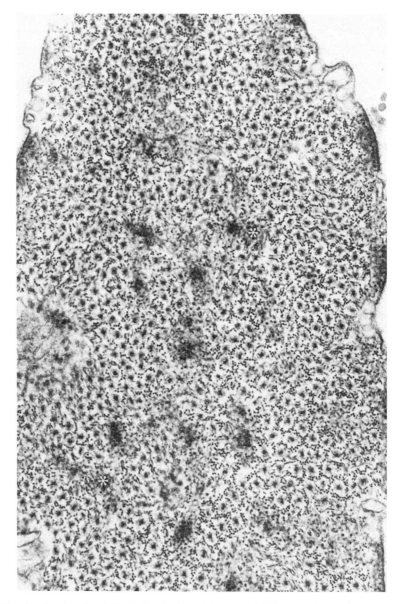

FIG. 3.   Taenia coli cell fixed, after 1 minute exposure at 4°C, in a relaxing medium containing MgATP, EGTA, 0.05% Triton X-100 (pH 6.5). Thick and thin filaments are present and uniformly distributed. Many dense bodies occur and are frequently seen to be encircled by the 10 nm skeletin filaments (asterisks). ×74,375. (From Small, 1977a, with permission.)

sent the *in vivo* situation. Further, confidence in this conclusion derives from a comparison of estimates of the actin to myosin mass ratio obtained from the filament counts (see Tregear and Squire, 1973) with the measured ratio obtained using biochemical methods. Of the latter, the most reliable estimates have come from the densitometry of sodium dodecyl sulfate (SDS)–poly-acrylamide gels of whole muscle preparations (Tregear and Squire, 1973; Small and Sobieszek, 1977a; Cohen and Murphy, 1978). While some differences in the data are evident (Section III,A), the values obtained are, within experimental error, consistent with the filament counts.

From the studies of thin section material, it appears that the thick and thin filaments are distributed more or less uniformly in the longitudinal direction. A definite segregation into skeletal muscle type "A" or "I" segments is not evident, even at the level of small groups of myofilaments such as observed in cells in rigor (Fig. 2). Ashton *et al.* (1975), in a study of semithick sections by intermediate high voltage electron microscopy, claimed to have detected some degree of longitudinal grouping of thick filaments, but, by the nature of this type of investigation, the data is so far too limited to allow firm conclusions to be drawn. More studies by high voltage microscopy using thicker sections or isolated cells should help to shed more light on the three-dimensional organization of the myofilaments.

Taking the observed presence of thick and thin filaments, together with the invariance in spacing of the meridional X-ray reflections from actin and myosin in relaxation and contraction, it may be assumed, as indeed was early proposed by Hanson and Huxley (1955), that the process of contraction in smooth muscle involves a relative sliding of the contractile elements. At the same time, some significant differences between the two muscle systems are apparent. Not only are there obvious differences in the filament distributions, but also in the molecular organization and mode of coupling of the actin and myosin filaments. The key to an elucidation of the organization of the contractile elements, of course, lies in defining the molecular architecture of the thick myosin filaments. Progress toward this end is discussed in the following section.

D. THICK FILAMENT ARCHITECTURE AND STRUCTURAL POLARITY

1. Difficulties in Isolation of Thick Filaments

Efforts to isolate the thick filaments of smooth muscle by homogenization of fresh or glycerinated tissue have met with no or only limited success (Hanson and Lowy, 1964; Panner and Honig, 1967; Kelly and Rice, 1968; Nonomura, 1968; Shoenberg, 1969; Cooke and Chase, 1971; Sobieszek, 1972). Factors contributing to this result are undoubtedly: (1) the high content of connective tissue in smooth muscle that necessitates relatively brutal treatment to effect cell fragmentation (Hanson and Lowy, 1964; Shoenberg, 1969), as well as, (2) the presence of a network of 10 nm filaments that closely embraces the contractile elements (see Section II,E). A more important barrier than this, however, is presented by the special solubility properties of smooth muscle myosin and actomyosin that is discussed further in Section III,A. Under the relaxation conditions used for filament isolation from fresh skeletel muscle (Huxley, 1963; Hanson *et al.,* 1971; Hinssen *et al.,* 1978), smooth muscle actomyosin is soluble, and homogenates obtained under such conditions show only thin filaments in the electron microscope (Hanson and Lowy, 1964; Panner and Honig, 1967; Shoenberg, 1969; Sobieszek, 1972). Slight modification of the composition of the extraction medium, by the addition of divalent cations or adjustment of the pH, gives rise to the appearance of very short myosin filaments of no more than 0.5 μm in length (Shoenberg, 1969; Nonomura, 1968). But whether or not these poorly formed aggregates represent fragments of the *in vivo* filaments, or filaments assembled from solubilized myosin, has not been established. It is clear, however, that as ATP is hydrolyzed during the ageing of these homogenates, the spontaneous self assembly of synthetic myosin filaments takes place (Shoenberg, 1969; Sobieszek, 1972; Sobieszek and Small, 1973b) so that progressively more and longer thick filaments may be detected.

A likely explanation for the fragility of the smooth muscle myosin filaments has come from studies of the α-helical subfragments of the myosin molecule (Huriaux, 1972; Hamoir, 1973). Huriaux has shown that smooth muscle light meromyosin (LMM), unlike the same subfragment from skeletal muscle, is soluble at low ionic strength and shows a higher net charge at neutral pH. This interesting result was subsequently confirmed by Sobieszek (1977a), who further correlated the lower solubility with a marked reluctance toward the formation of paracrystals so readily obtained with skeletal muscle LMM (Huxley, 1963). It was also found that smooth muscle LMM paracrystals did not show the 430 Å repeat characteristic of LMM paracrystals from skeletal muscle, but only weaker striations at 144 and 72 Å (Fig. 5h,i; Sobieszek, 1977a). Despite these special properties of LMM, the rod part of smooth muscle myosin (Kendrick-Jones *et al.,* 1971; Sobieszek, 1977a), as well as the whole molecule, do not exhibit the same high solubility. In fact, for purified myosins from smooth and skeletal muscle, the solubilizing effect of increasing ionic strength and ATP concentration are remarkably similar (Hamoir, 1973; Hinssen *et al.,* 1978). It is thus apparent that during the initial stages of homogenization and extraction of actomyosin (as well as under some conditions of fixation for electron microscopy) the solubility properties of the LMM part of the molecule have a predominating effect. Subsequently, in later stages of actomyosin and myosin purification, the myosin must adopt a conformation in which it is less soluble and which tends to favor filament reformation. Presumably, the insoluble region of the molecule becomes in some way unmasked either by the loss of an associated component

(Hamoir, 1973) or by a conformational change within the molecule itself, for example, by a movement of the globular heads away from the terminal (S-2) region of the rod.

2. Synthetic Myosin Filaments

In the light of the difficulties to isolate the native filaments, efforts have been mostly directed at characterizing the mode of assembly of purified smooth muscle myosin, on the assumption that this should be related to the structural assembly *in vivo* (Hanson and Lowy, 1964; Kaminer, 1969; Sobieszek, 1972, 1977a; Sobieszek and Small, 1973b; Wachsberger and Pepe, 1974; Cooke, 1975; Pollard, 1975; Craig and Megerman, 1977; Hinssen *et al.*, 1978). Following the considerable experience already gained with skeletal muscle myosin (Huxley, 1963; Lowey, 1971), these studies have also included, as indicated above, investigations of the α-helical subfragments of myosin obtained via proteolytic cleavage of the whole molecule with trypsin, chymotrypsin, or papain (Kendrick-Jones *et al.*, 1971; Sobieszek, 1977a).

It will be convenient to consider first the self assembly properties of the intact myosin molecule. From the various studies that have been made, it has been generally concluded that smooth muscle myosin assemblies only into short filaments of no more than 0.6 μm in length (Kaminer, 1969; Wachsberger and Pepe, 1974; Cooke, 1975; Pollard, 1975; Kaminer *et al.*, 1976) and that the much longer filaments that have been observed (Fig. 5c,d; Sobieszek, 1972, 1977a; Sobieszek and Small, 1973a) represent an anomalous or different form of assembly of myosin (Kaminer *et al.*, 1976; Pollard, 1975; Craig and Megerman, 1977). A more detailed analysis of the synthetic filaments (Hinssen *et al.*, 1978) has led to a clarification of the situation and, at the same time, has pointed to interesting features of myosin assembly that may have more general significance.

Simply stated, the short filaments that have been most commonly observed and that show a central bare region may be considered merely as early stages in the assembly of the longer filaments. The structural features of synthetic filaments formed from gizzard myosin at different stages of growth are shown in Fig. 4. For the shortest filaments, around 0.4 μm in length, two different forms may be recognized (Fig. 4a), one form showing a rectangular and the other a rhomboidal-shaped bare zone. In both cases, rough projections corresponding to the myosin heads occur at each end of the filament. From a cursory inspection of the micrographs (Fig. 4a; see also Hinssen *et al.*, 1978) it is readily apparent that the two forms are simply two different projections of the same type of filament, the one separated by an axial rotation of 90° from the other. From the projection in which the bare zone appears rectangular (Fig. 4a, "r"), the complement of myosin heads at one end faces mainly toward the support film, while at the other, the bridges point away from the film. This situation gives the false impression of an asymmetrical distribution of cross-bridges between the two ends. It is impor-

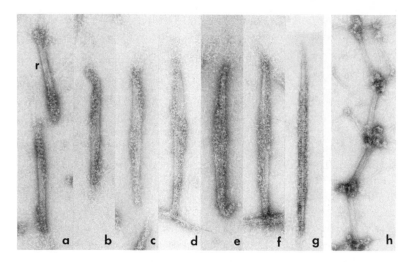

FIG. 4.   Synthetic filaments formed from gizzard myosin showing different stages in filament growth. (a) Short synthetic filaments seen in two different projections, one showing a rectangular bare zone (r) and the other a rhomboidal-shaped bare zone (see text). (b–g) Filaments at progressive stages of growth showing the gradual separation of the bare zone into bare edges at the filament ends and the population of the center of the filaments with myosin cross-bridges. (h) Filaments incapable of further growth, formed from myosin pretreated at pH 3.0 (see text). (a–f), ×72,000; (g) ×57,600; (h) ×60,000. (From Hinssen et al., 1978.)

tant to note that only under optimal conditions of preservation and staining can such fine differences in filament appearance be detected. We contend that filaments of the same type, although rather less well preserved, have constituted the major population of filaments commonly reported.

As the filaments grow, a shearing of the bare zone occurs in such a way as to produce bare edges at the filament tips (Figs. 4 and 5c,d). Concomitantly, the central part of the filament becomes populated with myosin projections whose regular organization gives rise to the characteristic cross-bridge repeat period of myosin, around 140 Å.[3] This type of assembly of myosin may be explained by the staggered packing of a single type of building unit consisting of an antiparallel myosin dimer (Sobieszek, 1972, 1977a). From optical diffraction analysis of the synthetic filaments, Sobieszek (1977a) has obtained evidence for an organization of cross-bridges on the filament surface based on a six-stranded helix of repeat 720 Å, residue translation 288 Å, and pitch 3 × 1440 Å. A further development of this model, which incorporates the mixed polarity of the myosin molecules along the filament length as an inherent consequence of the packing

[3]That the projections correspond to the enzymatic head region of the molecule is confirmed by the smooth appearance of filaments assembled from the myosin rod and light meromyosin (Fig. 5e–g; Sobieszek, 1977a, and Section III,C).

scheme, is illustrated in Fig. 6 (Hinssen *et al.*, 1978). The model shows that an organization of myosin molecules may be devised that is fully consistent with a sliding filament mechanism of contraction but that features a unique type of alternating polarity of the constituent molecules. The interaction of actin filaments with such a myosin filament would be expected to result in the movement of adjacent actins in opposite directions.

Significantly, studies with some nonmuscle myosins, from blood platelets, slime mold, and amoeba, indicate that these myosins assemble *in vitro* in a manner similar to that seen with smooth muscle myosin (Hinssen *et al.*, 1978). Purified skeletal muscle myosin also assembles into long filaments lacking a bare zone (Moos, 1973; Moos *et al.*, 1975; Hinssen *et al.*, 1978), but the absence of terminal bare edges on these filaments suggests that the detailed molecular packing scheme may be different from that of the smooth muscle and nonmuscle myosin filaments.[4]

In concluding the discussion of synthetic filaments formed from smooth muscle myosin, brief note should be made of the assembly under certain circumstances, of very short (about 0.3 μm) symmetrically bipolar aggregates (Fig. 4h). These filaments, which are incapable of further growth, show very bulbous ends and a marked tendency to aggregate into meshworks (Sobieszek, 1972, 1977a; Wachsberger and Pepe, 1974; Cooke, 1975). Their mode of assembly is possibly related to the packing of myosin rods into the segment structures described by Kendrick-Jones *et al.* (1971).

3. The Native Thick Filaments

Having considered the formation of synthetic myosin filaments at some length, we must return to the primary problem of the organization of myosin *in vivo*.

Following the development of procedures to obtain the smooth muscle cell models that we have already described, the chances became enhanced of isolating the native thick filaments in an intact state. For these demembranated cells, conditions could be defined under which the cells could first be "relaxed" in MgATP in the presence of EGTA, and then caused to contract by the replacement of EGTA with Ca^{2+} (Small, 1977a). This showed, as indicated previously, that the myosin filaments did not become disaggregated under the relaxation conditions employed and that such conditions were appropriate for attempts to isolate the contractile elements. Accordingly, the gentle fragmentation of demembranated cells under relaxing conditions was found to release both thick and thin filaments (Small, 1977a). The length of the thick filaments was very variable, and occasionally very long filaments up to about 8 μm in length were

---

[4]From the studies of Hinssen *et al.* (1978) it is also apparent that the loosely packed filaments of aorta myosin, reported by Craig and Megermann (1977) in 0.2 to 0.3 *M* KCl and showing a "side polarity," are only poorly formed filaments of the type already described.

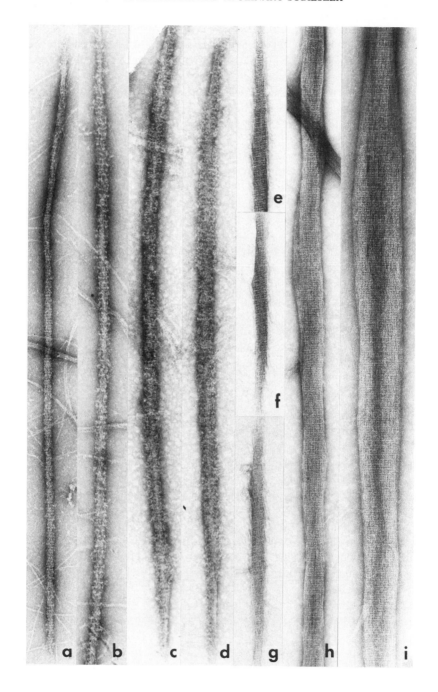

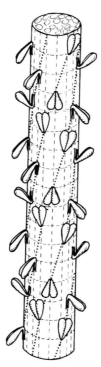

FIG. 6.   Possible mode of arrangement of cross-bridges on the myosin filament surface. The lattice arrangement is based on a six-stranded helix (dotted lines) with a residue translation of 288 Å and pitch 3 × 1440 Å. Cross-bridges lying on the same strand have alternating polarity. This arrangement gives rise to a total of 16 longitudinal rows of bridges with a common polarity within each row and with alternating polarities between rows. (From Hinssen *et al.*, 1978.)

encountered. Filaments greater than 2 μm in length were common (Fig. 5a and b; Small, 1977a). The most important parameter influencing the integrity of the filaments was the pH; cells relaxed at pH values above about 6.7 were then incapable of subsequent contraction with Ca^{2+}, and thick filaments were absent in the homogenates. This latter result was consistent with the original findings of

FIG. 5.   (a,b) Native myosin filaments isolated from smooth muscle cells from guinea pig taenia coli under relaxing conditions. (From Small, 1977a; and Hinssen *et al.*, 1978.) (c,d) Long synthetic filaments formed from gizzard myosin. Both these filaments and the native filaments showed a continuous distribution of cross-bridges along their length; their ordered arrangement gives rise to the visible repeat period of around 144 Å. (e–g) Filaments formed from the rod part of smooth muscle myosin. Comparison of these filaments with those formed from smooth muscle LMM (h,i) indicates that the frayed edges arise from the divergence of the terminal (S-2) part of the rod away from the filament surface. The fine repeat period is about 72 Å. (h,i) Paracrystals formed from smooth muscle LMM showing the characteristic repeat period of around 72 Å.

Kelly and Rice (1968) with glycerinated tissue and would further be expected from the much lower extractability of smooth muscle actomyosin from fresh muscle at lower pH.

The significant finding with the native thick filaments was that their general structural features were the same as those of the long filaments assembled from purified myosin. They showed a continuous distribution of projections along their length that gave rise to a strong repeat period of 140 Å, and in some cases, the filament tips showed the same bare edges. The only difference lay in the diameters of the filaments; the native filaments were slightly narrower than the synthetic, suggesting for the former a rather more closely packed organization of myosin molecules. The results obtained with the cell homogenates thus indicated that the general scheme of packing of myosin molecules derived for the synthetic filaments may be applicable to the form of the thick filament *in vivo*.

E. THE 10 nm SKELETIN FILAMENTS

1. *Function*

From a number of studies by electron microscopy, a class of filaments around 10nm in diameter was recognized to be present in a varieyt of cell types including smooth muscle (for earlier studies see Fawcett, 1966; Ishikawa, 1974). In addition to specific names, such as glial filaments, tonofilaments, and neurofilaments, these filaments have been commonly referred to as intermediate filaments or 10 nm filaments (Ishikawa *et al.*, 1968, 1969). Primarily from the finding of commonly occurring autoantibodies directed against the 10 nm filaments (Kurki *et al.*, 1977; Osborn *et al.*, 1977; Gordon *et al.*, 1978; Hynes and Destree, 1978), it has become clear that these filaments are as ubiquitous as actin microfilaments and microtubules.

Since the 10 nm filaments occur in relative abundance in smooth muscle, this tissue has served as a useful source in studies of filament function and composition. Cooke and Chase (1971) established that the 10 nm filaments differ markedly in their solubility properties compared to actin and myosin and are completely insoluble in solutions of high ionic strength. In muscle strips exhaustively extracted in solutions of high and low salt, the only filaments to be detected in ultrathin sections are the 10 nm filaments (Cooke, 1976; Small and Sobieszek, 1977a). From the similar extraction of isolated smooth muscle cells, pale "ghost

FIG. 7.   Smooth muscle ghost cells obtained by the extraction of isolated taenia coli cells at high and low ionic strength (see text). (a) Light micrograph showing retention of cell shape and of the nucleus after these extraction procedures. ×800. (b) Peripheral part of taenia coli ghost cell as seen after negative staining. Essentially only the 10 nm skeletin filaments may be recognized. ×35,000. Inset shows SDS gel of a corresponding taenia coli ghost cell preparation. The two predominant bands arise from actin (lower band) and the 10 nm filament protein, skeletin (upper band).

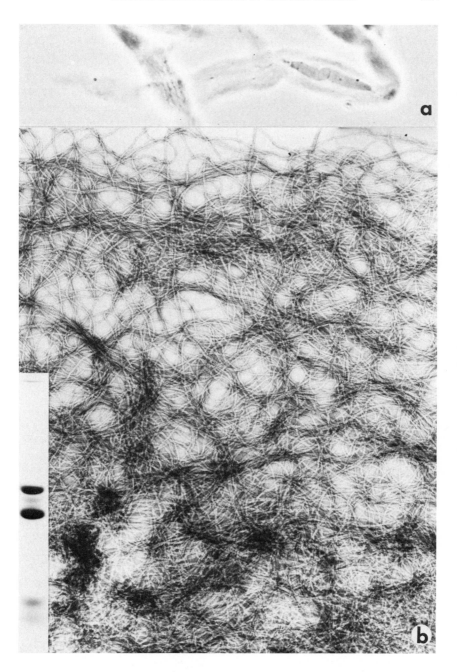

cells'' are obtained that retain the original spindle-shaped form of the cell (Fig. 7a) and that, after negative staining, are seen in the electron microscope to consist predominantly of a continuous, and entangled network of 10 nm filaments (Fig. 7b; Small and Sobieszek, 1977a). These observations, together with the noted insertion of the 10 nm filaments in the plasmalemma attachment plaques (see Section II,F) and their redistribution toward the cell center during stretch (Cooke and Fay, 1972b), suggests that these filaments play a cytoskeletal role in the smooth muscle cell.

That the 10 nm filaments of smooth muscle are not an integral component of the contractile apparatus was suggested by two observations. First, the filaments could be removed from isolated and demembranated smooth muscle cells by mild proteolysis without affecting the general organization of the myofilaments or the ability of the cells to shorten in the presence of ATP (Small and Sobieszek, 1977a). Second, the Ca-sensitivity and actin-activated ATPase activity of purified actomyosin from smooth muscle, which lacks the 10 nm filaments, was essentially the same as for homogenized muscle fragments or ''myofibrils'' in which the 10 nm filaments are present.

From the studies of cell types other than smooth muscle, it is apparent that the 10 nm filaments may generally perform a structural role (Brecher, 1975; Lehto *et al.*, 1978; Small and Celis, 1978; Eriksson and Thornell, 1979). In cow heart Purkinje fibers, the 10 nm filaments are the predominant filament type and have been shown, by the use of high salt extraction procedures, to function in the maintenance of cell form (Eriksson and Thornell, 1979). Furthermore, in cultured cells, the 10 nm filaments form a compact network that encircles the cell nucleus and retains the nucleus after the cytoskeletal frameworks of actin and tubulin have been extracted (Lehto *et al.*, 1978; Small and Celis, 1978). This has been taken to suggest that the 10 nm filaments provide structural support for the cell nucleus and possibly for other cell organelles. The 10 nm filaments of smooth muscle appear also to be involved in the support of the cell nucleus. This is suggested by the ready loss of the nucleus from isolated cells in which the 10 nm filaments have been degraded by proteolysis (unpublished observations).

While the distribution of 10 nm filaments in smooth muscle appears fairly stable, a considerable reorganization occurs in other cells in response to drugs that disrupt microtubules: namely, colcemid and colchicine. The disappearance of microtubules is normally followed by the perinuclear accumulation of the 10 nm filaments into contorted bands (Croop and Holtzer, 1975) or ''caps'' (Goldman and Follett, 1970). From this it appears that an interaction between 10 nm filaments and microtubules is necessary in these systems for maintaining the integrity of the cell cytoskeleton. In smooth muscle, relatively few microtubules are present, and the integrity of the 10 nm filament net is apparently maintained in a more stable configuration by the cytoplasmic dense bodies (Cooke, 1976; Small and Sobieszek, 1977a).

2. Composition: Skeletin

For smooth muscle residue enriched in the 10 nm filaments by extraction in solutions of high and low ionic strength (see Section II,E,1), two major bands are seen after polyacrylamide gel electrophoresis in the presence of sodium dodecyl sulfate (SDS). These occur in positions characteristic for actin (42,000 MW) and for a polypeptide of 50,000–55,000 MW. The extraction of isolated cells is, as might be expected, more complete, and essentially only these two bands may be recognized in the SDS gels (Fig. 7b). While it was noteworthy that actin filaments could not be observed in the electron microscope in thin sections of extracted muscle and were rather sparse in the ghost cells after negative staining, the latter cells showed a marked increase in density on the addition of myosin subfragment-1. This suggested that the residual actin present may exist as short disorganized filaments or in a low polymer form. Additional evidence for the presence of actin in the salt-extracted residue was provided by its reaction with anti-actin antibodies (Lazarides and Hubbard, 1979).

The expectation that the 55,000 MW material arose from the 10 nm filaments was confirmed by two further observations (Small and Sobieszek, 1977a). First, it was observed that for isolated cells from which the 10 nm filaments were removed by proteolysis (see Section II,E,1) the 55,000 band was absent from the SDS gels. Second, the 55,000 MW material could be specifically extracted from the muscle residue in acetic acid and subsequently reformed into filaments averaging around 10 nm in diameter. The 55,000 MW material obtained by the fractionation of urea extracts was also capable of filament formation (Cooke, 1976). Hubbard and Lazarides (1979) have recently republished our purification procedure for the 10 nm filament protein (Small and Sobieszek, 1977a) and confirmed our general findings. They attached some significance, however, to the contamination of the final protein preparation with variable amounts of actin, the implication being that actin may be involved in filament assembly (Hubbard and Lazarides, 1979). This conclusion is difficult to reconcile, however, with the finding that preparations of the 10 nm filament protein lacking any contaminating actin form filaments more closely resembling the native filaments (Fig. 8) than those polymerized with actin present (Hubbard and Lazarides, 1979).

In parallel studies on cultured cells (BKH-21), Starger and co-workers (Starger and Goldman, 1977; Starger et al., 1978) obtained a 10 nm filament preparation from the isolation of the birefringent caps formed during the early stages of cell spreading; this preparation showed two bands of 55,000 and 54,000 MW on SDS gels. These authors also noted a dissociation of the filaments in weak phosphate buffer at neutral pH, under which conditions the smooth muscle protein is also partly soluble (unpublished observations). The protein of glial filaments (Eng et al., 1971; Dahl and Bignami, 1976) and of the 10 nm filaments found in abundance in cow heart Purkinje fibers (Eriksson and Thornell, 1979; Stigbrand et al., 1979) also exhibit molecular weights in the region of 51,000–

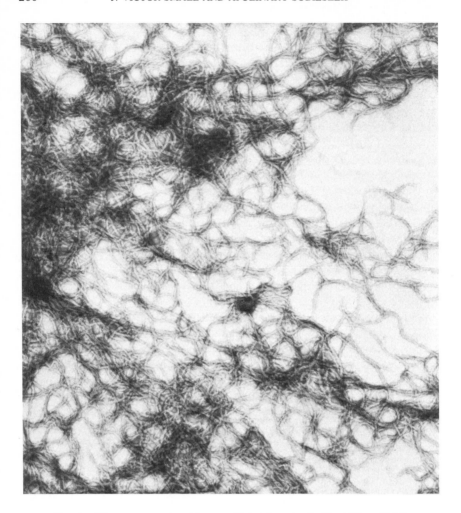

FIG. 8.   Filaments reconstituted from purified skeletin (as for Fig. 15C). ×37,500.

55,000, and these proteins, as well as the 10 nm filament protein from the BHK-21 cells (Starger *et al.*, 1978), show amino acid compositions identical to that obtained for the smooth muscle 10 nm filament protein. From other studies of cultured cell cytoskeleton preparations, the band arising from the 10 nm filament on SDS gels has been found to occur in positions corresponding to molecular weights ranging from 50,000 to 57,000 (Gordon *et al.*, 1978; Hynes and Destree, 1978; Lehto *et al.*, 1978; Franke *et al.*, 1978a).[5]

[5]Although neurofilaments and tonofilaments fall into the same 10 nm class of filaments on morphological grounds, the protein composition of these filaments differs from that already described. Neurofilaments are composed of several polypeptides ranging in molecular weight from

While these studies indicate a close relationship between the 10 nm filaments of different tissues, the slight differences in the number and molecular weight of polypeptide components as well as more pronounced differences in immunological cross reactivity (Blose *et al.*, 1977; Kurki *et al.*, 1977; Franke *et al.*, 1978b; Gordon *et al.*, 1978; Hynes and Destree, 1978; Campbell *et al.*, 1979; Toh *et al.*, 1979) have been taken to suggest the existence of a number of subclasses of the filament protein. It is worth noting that even for smooth muscle some differences have already been detected in the polypeptides obtained from different muscles: after two-dimensional gel electrophoresis, the filament protein from mammalian stomach runs as a single spot, whereas the gizzard protein runs as a doublet with isoelectric points different from that of the mammalian protein (Izant and Lazarides, 1977; P. G. Whalen, private communication).

For smooth muscle, two names have been ascribed to the 10 nm filament protein: "skeletin" (Small and Sobieszek, 1977a) and "desmin" (Lazarides and Hubbard, 1976). The first derives from the indication that the 10 nm filaments play a cytoskeletal role in smooth muscle as well as in other cells, as discussed previously. The second was suggested from the finding that antibodies to the 55,000 MW band from SDS gels of smooth muscle stained the Z-line of skeletal and cardiac muscle myofibrils, this staining being restricted to the peripheral regions of the Z-disc (Lazarides and Granger, 1978). (The latter observation will be discussed further in Section II,F). Since the presence of the 10 nm filament protein in the Z-disc of the skeletal muscle myofibril constitutes a very specialized situation, we consider "desmin" a less suitable name than "skeletin." On the same basis, Eriksson and Thornell (1979) have recently suggested that the 10 nm filaments may be more appropriately referred to as "skeletin filaments." This nomenclature will accordingly be adopted in the discussions that follow.

The mode of assembly of skeletin into filaments is currently unresolved. In cross section, the native filaments appear hollow (Rice *et al.*, 1970; Uehara *et al.*, 1971), and in appropriate orientations show four subunits of about 35 Å in diameter (Small and Squire, 1972; Eriksson and Thornell, 1979). After negative staining, the native filaments are characteristically very smooth and show no clear indication of substructure, although from optical diffraction patterns a repeat period of about 30 nm has been detected (Small and Sobieszek, 1977a). Further progress toward defining the molecular architecture of the filaments has been hampered in part by the lack of order in the synthetic filaments formed *in vitro*. In contrast to the polypeptides of α-keratin, which assembles into filaments of very constant diameter (Steinert and Gullino, 1976), smooth muscle skeletin forms, as we have indicated, branched filaments of less uniform width (see Small

---

60,000 to 200,000 (Gilbert *et al.*, 1975; Benitz *et al.*, 1976; Lasek and Hoffmann, 1976; Liem and Shelanski, 1978) and tonofilaments of epidermis and cultured epithelial cells show from six to eight bands of 47,000–68,000 MW (Steinert and Gullino, 1976; Franke *et al.*, 1978b).

and Sobieszek, 1977a; Hubbard and Lazarides, 1979; Fig. 8). Whether or not this is attributable to the choice of inappropriate conditions, or points to the requirement of minor components for filaments assembly (e.g., Starger *et al.*, 1978), remains to be established.

F. Dense Bodies and Attachment Plaques

A constant and established feature of electron micrographs of smooth muscle cells is the presence of intracellular, electron-dense regions, about $0.1 \mu m$ across in transverse sections and variously referred to as dense areas, dark bodies, or dense bodies (see review by Burnstock, 1970). In certain invertebrate muscles, areas of similar density and morphology were described before they were described in vertebrate smooth muscle, and in these former muscles were shown to be attachment sites for the thin filaments (Hanson and Lowy, 1961; Rosenbluth, 1973; Szent-Gyorgyi *et al.*, 1971; Sobieszek, 1973). From this standpoint, it has been generally supposed (Hanson and Lowy, 1960; Heumann, 1971, 1973; Ashton *et al.*, 1975) that the dense areas of vertebrate smooth muscle may likewise be intracellular sites of anchorage for the thin filaments, functionally comparable to the Z-lines of the striated muscle myofibril. However, and perhaps unexpectedly, any direct association or insertion of thin filaments into the dense bodies of vertebrate smooth muscle has, despite claims to the contrary (Ashton *et al.*, 1975; Heumann, 1971), persistently evaded demonstration. Instead, it has consistently been recognized that the dense bodies of smooth muscle have a close association with the skeletin filaments that we have already discussed (Section II,E); in transverse section, these filaments are commonly seen roughly encircling the dense bodies (Fig. 3).

In attempts to isolate the dense bodies from smooth muscle, Cooke and Chase (1971) took advantage of the noted insolubility of these structures in solutions of high salt concentrations (Prosser *et al.*, 1960). In sections and homogenates of the extracted residue, they identified not only the dense bodies, but also the skeletin filaments that remained in tight association with and radiated from the ends of the dense bodies. (As we have mentioned earlier, this observation of the same insolubility of the skeletin filaments in high salt proved useful in establishing ways of purifying the filament protein.) In a further study, Cooke and Fay (1972b) showed that a regrouping of dense bodies toward the cell axis, noted in stretched muscle strips of guinea pig taenia coli, was associated with a similar redistribution of the skeletin filaments. The interpretation of this result was that the latter filaments form a network within the smooth muscle cell, which is drawn together into nodes at the dense bodies and has its major connections at the cell ends. As already indicated (Section II,E,1), the continuity of the skeletin filament net throughout the length of the smooth muscle cell was most clearly shown from electron microscopy of the isolated smooth muscle cell ghosts.

In addition to the dense bodies, densely staining zones, first described by Pease and Molinari (1960), are found beneath the cell membrane of the smooth muscle cell. These zones appear as a discontinuous layer of variable thickness, interrupted by regions of the plasmalemma often showing pinocytotic vesicles. The feature that distinguishes these plasmalemma zones from the dense bodies already described is their clear association with the thin actin filaments, as well as with the skeletin filaments (Dewey and Barr, 1968; Cooke, 1976; Small, 1977a,b,c; Ashton *et al.*, 1975). In cross sections, both filament types may be observed embedded in or adjacent to the matrix of these plasmalemma zones. For the skeletin filaments, this association does not seem to depend on the axial position along the cell length (Small, 1977b,c), indicating that the attachment sites are not confined to the cell ends (Cooke and Fay, 1972b). In well preserved material, the filaments within the plasmalemma zones are seen to be aligned parallel to the plasmalemma and separated from it by about 100–200 Å. From their association with both the actin filaments and the skeletin filaments, it is not unreasonable to suppose that the plasmalemma dense areas constitute sites of attachment of the filaments to the cell surface (see also Rosenbluth, 1965). They will be referred to accordingly as "attachment plaques."

In a study that has been published only in part, Schollmeyer *et al.* (1976) have described the staining of both the plasmalemma attachment plaques and the dense bodies with an antibody to skeletal muscle α-actinin. If this interesting finding is confirmed, it would suggest that α-actinin, in addition to cross-linking and anchoring actin filaments (see Section III,D,2), may be capable of similar associations with the skeletin filaments. This would indeed not be surprising in the light of the demonstration that α-actinin and skeletin coexist in the Z-band of the skeletal muscle myofibril (Lazarides and Granger, 1978).

G. The Contractile Apparatus

1. *A Tentative Model*

As is clear from the foregoing discussion, the structural data so far available does not allow us to piece together the contractile apparatus in its entirety. Some pertinent features of the smooth muscle system have, however, been established and will here be brought together in a tentative model of the contractile machinery (Fig. 9).

In this model, two distinct systems of filaments are envisaged: (1) The actin and myosin filaments constitute the contractile apparatus proper, and (2) the skeletin filaments, together with the dense bodies, form a structural framework interwoven between the groups of contractile elements. An organization of the contractile elements into fibrils, or contractile units, is suggested by the data obtained from isolated cells.

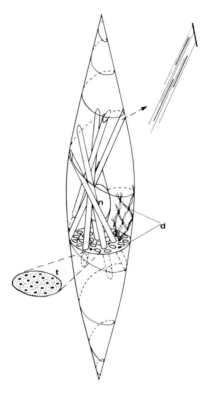

Fig. 9.   Tentative model of the contractile apparatus of the smooth muscle cell. (From Small, 1977a). The contractile units are composed only of the thick and thin filaments (t) and are anchored at their ends to the plasmalemma attachment plaques (see text). The arrangement of the attachment sites on a helix for the family of units shown is hypothetical. The dense body–skeletin filament network (d) forms a structural framework between the contractile units and is also attached to the cell surface. (n) corresponds to the cell nucleus with the associated organelle-containing regions at the nuclear poles.

We have already noted the apparent absence of a true sacromere-like organiza-tion of the actin and myosin filaments. Material analogous to the M- and Z-line substance of skeletal muscle is not found interconnecting the thick and the thin filaments of smooth muscle into well defined arrays. As is shown by the aggrega-tion of the thick filaments into ribbon-like structures under certain conditions, it is clear that the myosin filaments have considerable freedom of movement in the lateral direction. The length of the myosin filaments is presumed to vary from around 2 μm to several microns in length; the thin filaments are assumed to be several times longer than the thick filaments.

α-Actinin, and perhaps filamin (see Section III, D,1), are envisioned as being involved in the attachment of the thin filaments and the 10 nm filaments to the plasmalemma attachment plaques.

2. The Polarity Problem

The mode of coupling between the thick and thin filaments must of necessity be determined by the specific polarity of the thick filaments, which we have already discussed in Section II,D. For such thick filaments (Fig. 6), a single actin filament may interact with myosin molecules distributed along the entire filament length and may slide unhindered as long as the polarity of the actin molecules themselves remain unchanged. One complication of this scheme (Fig. 10) concerns the apparent need for a reversal of polarity of the thin filaments, presumably at their midpoints. Whether this occurs and how this may be achieved is at this stage a matter of speculation. However, two possibilities may be considered that could satisfy the requirement for matching polarities of actin and myosin and a net contraction of a coupled set of filaments. The first is that the actin filaments occur as antiparallel pairs that could be bound together at their ends by, for example, α-actinin or filamin (Fig. 10a,b). This would presumably require some degree of overlap of the ends of the thin filaments. The second possibility is that the myosin cross-bridges can interact with actin in two distinct ways, depending on the relative polarity of the thin filament (Fig. 10c). Thus, when the polarity of a thin filament relative to a row of myosin molecules is the same as in the muscle sarcomere, interaction of myosin and actin would give rise to the development of tension or to a relative sliding. When, however, the polarity of actin is reversed, relative to the same set of molecules, the binding of myosin to actin can take

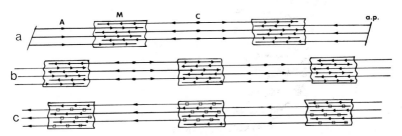

FIG. 10. Possible coupling schemes for the thick and thin filaments based on the "row polarity" model for the thick filaments (see text and Fig. 6). A, Actin filament; M, myosin filament; C, connecting proteins (e.g. α-actinin or filamin); a.p., membrane attachment plaque. Elongated arrows indicate the polarity of actin filaments, i.e., the direction of the arrowhead complex that would be formed with myosin subfragment-1. Short arrows designate the position and polarity of myosin cross-bridges on the surface of the myosin filaments. The sense of the arrows corresponds to that of the rigor complexes that would be formed between the myosin cross-bridges and actin. When the polarities of a cross-bridge row and the adjacent actin are matched, interaction and relative sliding can take place. For bipolar pairs of actin filaments (a and b) all cross-bridge rows can participate in the sliding process. In c, the case is considered for which the actin filaments have no reversal of polarity at their midpoints. Instead, it is presumed that the nonmatching of myosin and actin polarities leads, on activation, to the binding of myosin to actin and to immobilization of the actin filament. The immobilizing cross-bridges are designated by open squares. For one particular coupled unit, this last scheme would require actin filaments of the same polarity throughout the unit.

place, but no tension development or movement is possible. In the latter situation, the myosin molecules would act merely to cross-link and immobilize actin. If such a scheme is operative, it is not difficult to imagine that half of the actin filaments interacting with each thick filament will at any one time be immobilized, while the other half will be so organized as to be capable of undergoing a normal sliding interaction. In this way a net contraction of the system could readily be produced without invoking a specialized organization of actin in antiparallel pairs. At the same time, however, some provision must be introduced to restrict filament matching to those combinations giving rise to active shortening of the unit, rather than to active extension.

III. The Contractile Proteins

A. SMOOTH MUSCLE ACTOMYOSIN

1. *Early Studies*

The various properties of smooth muscle actomyosin established in studies before 1970 have been admirably reviewed by Needham (1971). Rather than attempting a similarly extensive review of this period, we shall confine ourselves to a summary of the primary developments that took place.

Although Mehl reported, as early as 1938, the extraction of "myosin," i.e., an actomyosin-like protein from beef intestine, it was Csapó, (1948, 1950a,b) who first attempted a systematic analysis of the actomyosin of smooth muscle. Concentrating his studies on uterus, Csapó was able to show that, after extraction of muscle in 0.5 M KCl, an actomyosin could be obtained with properties similar to that of skeletal muscle actomyosin. The smooth muscle actomyosin showed a similar drop in viscosity on addition of ATP (a standard test at that time) and a similar sedimentation pattern in the ultracentrifuge (Csapó *et al.,* 1950). In the absence of ATP, a single sedimenting peak was observed whereas with ATP present a new, slower peak appeared in a position characteristic for myosin. In the same manner as skeletal muscle actomyosin, the actomyosin extracted from smooth muscle could also be formed into the contractile threads of Szent-Gyorgyi (Csapó, 1950b), although it was noted that these threads gradually lost their activity during incubation in aqueous solution; this decay was ascribed to the loss of an unknown factor "X."

In 1961, Laszt and Hamoir described a new type of actomyosin preparation, which was obtained from cow carotids by extraction at low ionic strength in the presence of ATP. Under the same conditions, actomyosin from skeletal muscle cannot be extracted. This apparently new type of actomyosin, which they called tonoactomyosin to distinguish it from actomyosin extracted in high salt, was not only extractable at low ionic strength, but could not be precipitated as long as

ATP was present. Otherwise, its general properties were similar to skeletal muscle actomyosin. In subsequent studies (Filo *et al.*, 1963; Schirmer, 1965), it was shown that all the actomyosin of smooth muscle could be extracted either at low or at high ionic strength, indicating that tonoactomyosin did not, in fact, represent a second actomyosin species. Nevertheless, the observation of Laszt and Hamoir pointed to an important difference between the properties of skeletal and smooth muscle actomyosins. Thus, the smooth muscle actomyosin could be dissociated (i.e., relaxed) with the ATP anion alone, whereas skeletal muscle actomyosin required the additional presence of Mg^{2+}. As we shall see later, this difference could be correlated with the differences in the Ca-dependent regulatory mechanisms between the two muscle types.

The difficulty in obtaining a homogeneous actomyosin from smooth muscle, as emphasized, for example, by Needham and Williams (1963), generally impeded further progress in this period. Coupled to this was the inability to obtain an actomyosin that showed any Ca sensitivity with regard to its actin-activated ATPase activity. From physiological studies and from studies of other contractile systems, it was to be expected that Ca would play a primary role in regulating contraction in smooth muscle.

The first report of an actomyosin showing any sensitivity to Ca^{2+} ions was made by Sparrow *et al.* (1970) working with arterial muscle. In addition to using moderately high concentrations of ATP for actomyosin extraction, these authors effected actomyosin precipitation by dialysis and not by the usual precipitation. The main problem with these studies, as with those of earlier workers, was the low ATPase activity of the purified actomyosin (Table I). Although this activity was Ca sensitive, it was in the range of the residual activity shown by myosin alone. This situation was far from satisfactory, since the residual activity of myosin is rather dependent on ion concentration (Sobieszek, 1977a), and the modification of conditions could readily produce a Ca sensitivity independent of actin activation. Such an effect would explain the reported change in Ca sensitivity of carotid actomyosin from about 33 to 90% for a change in ionic strength from $\mu \sim 0.07$ to $\mu \sim 0.16$ (Maxwell *et al.*, 1971). The inability to distinguish clearly between the Ca sensitivity of the residual activity of myosin and the true actin-activated ATPase activity of myosin pointed to a lack of suitable methods for actomyosin purification, or to the loss of essential factors during the preparation procedures, or to both. This problem cannot be overemphasized, since reports continue to appear in which conclusions are drawn on the basis of insignificant levels of MgATPase activities (Frederiksen, 1976; Ikebe *et al.*, 1977, 1978).

2. The Improvement of Preparative Methods: Composition of Ca-Sensitive Actomyosin

A considerable advance toward the development of suitable methods for the preparation of smooth muscle actomyosin was made by Sobieszek and co-

TABLE I

ACTIN-ACTIVATED ATPASE ACTIVITIES OF DIFFERENT SMOOTH MUSCLE ACTOMYOSINS

Tissue	Temperature (°C)	Ca (nmole·mg⁻¹·min⁻¹)	EGTA	References
Vascular	25	16	1.8	Sparrow et al. (1970)
	37	15	2.7	Maxwell et al. (1970)
	?	15	2.0	Sparrow and van Bockxmeer (1972)
	37	58	3.5	Litten et al. (1977)
	37	35	3	Di Salvo et al. (1978)
Tracheal	37	40	8	Bose and Stephens (1977)
Vas deferens[a]	37	158	39	Chacko et al. (1977)
Stomach	25	204	22	Small and Sobieszek (1977)
Gizzard	25	18	3.8	Driska and Hartshorne (1975)
	25	56	8.0	Sobieszek and Bremel (1975)
	25	65	3	Dabrowska et al. (1977)
	25	224	14	Sobieszek and Small (1976)
	37	472	47	Sobieszek and Small (1976)

[a] Reconstituted actomyosin.

workers (Sobieszek and Bremel, 1975; Sobieszek and Small, 1976). From a systematic study of extraction conditions and the composition of extracts obtained from chicken gizzard, procedures were established for the purification of Ca-sensitive actomyosin showing an activity in the order of 10 times that reported earlier (Table I). The essence of the procedure was the extraction of actomyosin at low ionic strength in the presence of 10 mM ATP and chelating agents (EDTA and EGTA), followed by precipitation with divalent cations (15–25 mM MgCl$_2$ or CaCl$_2$). Figure 11 indicates in more detail the methods of preparation employed for actomyosin and the separate contractile proteins.

An important factor in the preparation procedure was the use of fresh muscle and the rapidity of actomyosin purification; the measurement of the activity of the purified actomyosin could be made within 24 to 48 hours after death. The pronounced effect of ageing on the measured activity is so important as to warrant further mention. As shown in Fig. 12, the activity of actomyosin measured 3 days after initial purification is 2- to 3- fold lower than for fresh actomyosin. This decay on ageing would explain in part the lower activities reported by others as well as the 3- to 4-fold lower activities reported for gizzard actomyosin prepared from frozen muscle (Sherry et al., 1978). Other experiments, described later (see Section IV,C), indicated that the decay in activity, as well as low initial activities, could be ascribed mainly to the decay in activity or the loss during purification of the myosin light chain kinase.

Further comment should be made concerning the conditions giving optimal MgATPase activities of actomyosin. From the early studies of Filo et al. (1963),

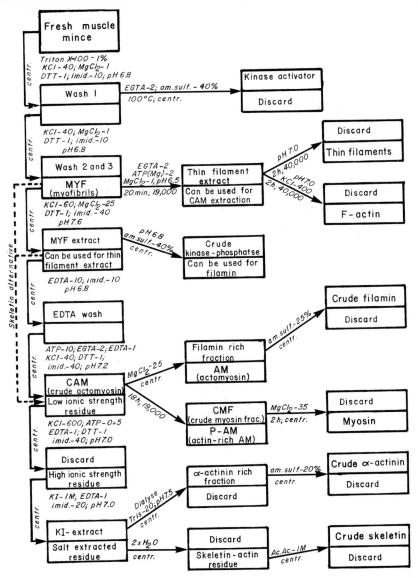

FIG. 11. Schematic diagram illustrating processing steps for the different smooth muscle preparations. Upper and lower part of the rectangles correspond to supernatants and pellets, respectively. Numbers used in the composition of the media correspond to mM concentrations. Intermediate centrifugations indicated by "centr." correspond to 12,000 r.p.m. (23,000 g) for 30 minutes, unless otherwise indicated. DTT, Dithiothreitol; EDTA, ethylenediaminetetraacetic acid; EGTA, ethyleneglycol-bis (2-aminoethyl ether)-N,N,N',N'-tetraacetic acid; Ac.Ac., acetic acid; am. sulf., with given percentage values, indicates percentage of saturated ammonium sulfate used for fractionation; imid., imidazole; Tris, Tris base.

it has generally been presumed that high Mg^{2+} concentrations, in the order of 10 mM, are required for optimal activity (see Harthshorne and Aksoy, 1977). In a study of arterial actomyosin, Russell (1973) attributed the apparent requirement for Mg^{2+} to its effectiveness in maintaining actomyosin in a precipitated state, but his highest activities, like those in earlier studies were extremely low. For chicken gizzard and hog stomach actomyosin, optimal activities are obtained in the presence of 1–3 mM free Mg^{2+} and 20–60 mM KCl. The dependence of actomyosin activity on these parameters is shown in Fig. 13. The inhibition observed at high $MgCl_2$ concentrations can be attributed to its contribution in increasing the ionic strength, since elevating the concentration of KCl or adding equivalent amounts of NaCl, in the presence of 1–3 M free Mg^{2+} has the same inhibitory effect. Under the appropriate ionic conditions, the optimal pH falls within the range of pH 6–8, with a peak around neutrality (Sobieszek and Small, 1976; Small and Sobieszek, 1977b).

The utilization of polyacrylamide gel electrophoresis in the presence of SDS assisted considerably in both the characterization of smooth muscle actomyosin and the development of procedures for its purification. In two independent studies on chicken gizzard (Driska and Hartshorne, 1975; Sobieszek and Bremel, 1975), it was established that, in contrast to skeletal muscle actomyosin, smooth muscle actomyosin contains only three major proteins, namely actin, myosin, and tropomyosin (Fig. 14D). For this actomyosin, which showed a high Ca

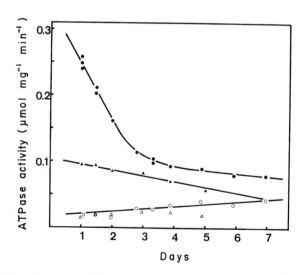

Fig. 12.   Effect of aging   on the actin-activated ATPase activities of myofibrils (▲-▲, Ca; △-△ EGTA) and actomyosin (●-●, Ca; ○-○, EGTA) stored on ice and assayed at 25°C. The rapid decay of activity within the first few days can be attributed primarily to the loss in activity of the myosin light chain kinase.

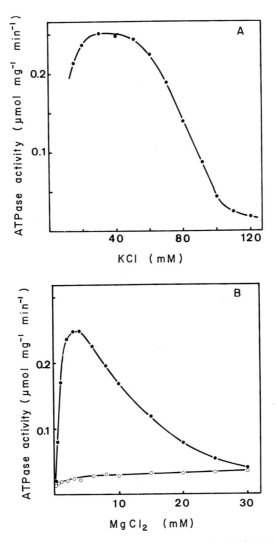

FIG. 13.   Actin-activated ATPase activities of gizzard actomyosin (CaAM; Sobieszek and Small, 1976) as a function of (A) KCl and (B) MgCl$_2$ concentrations (in the presence of: ●-●, 0.1 mM CaCl$_2$; O-O, 2 mM EGTA). Other conditions: imidazole, 20 mM, pH 7.0; 1 mM cysteine; 1 mM MgATP; 25°C; for (A), 1 mM free MgCl$_2$; for (B), 40 mM KCl. The final actomyosin concentration was in the range of 2–3 mg/ml and the assay time 30–60 seconds.

sensitivity, no components similar to those of troponin from skeletal muscle could be identified. These findings provided the first indications that the regulation of smooth muscle actomyosin via Ca^{2+} was not associated with the thin filaments, but linked to myosin (Sobieszek and Bremel, 1975; see Section IV,A).

In gizzard actomyosin, the approximate mass ratio of actin:myosin: tropomyosin were 1:2.5:0.3, respectively. This compared with a mass ratio of actin:myosin in whole muscle of about 1:1 (Table II), a difference that could be ascribed to the loss of a considerable part of the actin during the washing of the homogenized muscle, as well as the retention of an additional firmly bound fraction in the muscle residue (see Section II,E,2). Electrophoretic patterns obtained from various mammalian smooth muscle actomyosins indicate that these actomyosins have a composition essentially the same as described for gizzard (Murphy *et al.*, 1974; Chacko *et al.*, 1977; Litten *et al.*, 1977; Small and Sobieszek, 1977b; Di Salvo *et al.*, 1978). According to the method of actomyosin purification, the proportions of the contractile proteins is slightly variable (Murphy *et al.*, 1974).

Three minor components may also be identified in gels of freshly extracted

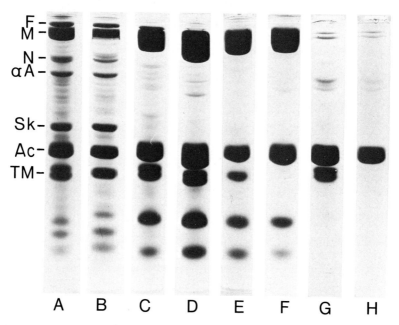

FIG. 14.   SDS–polyacrylamide gels (7.5%) of different smooth muscle preparations. See also purification scheme (Fig. 11). (A,B) Myofibrils (MYF) from chicken gizzard and hog stomach, respectively. (C) Chicken gizzard crude actomyosin (CAM). (D) Chicken gizzard actomyosin purified via precipitation with $MgCl_2$ (MgAM). (E) Hog stomach actomyosin purified by precipitation with 20 mM $CaCl_2$. (F) Tropomyosin-free chicken gizzard actomyosin. (G) Chicken gizzard thin filaments. (H) Chicken gizzard. F-actin prepared by centrifugation of thin filaments in 0.4 M KCl. Abbreviations: F, filamin (M_r, 250,000); M, myosin heavy chain (M_r, 200,000); N, 130,000 M_r component; αA, α-actinin; Sk, skeletin (M_r, 55,000); Ac, actin (M_r, 45,000); TM, tropomyosin (M_r mean, 37,000), In (C) to (F), the lower molecular weight bands correspond to the myosin light chains (see Fig. 15).

TABLE II

ESTIMATED MASS RATIOS OF CONTRACTILE PROTEINS IN SMOOTH MUSCLE[a]

Tissue	Contractile protein			References[d]
	Actin	Myosin[b,c]	Tropomyosin	
Gizzard (chicken)	1	1.1	0.35	d
Vas deferens (guinea pig)	1	1.6	0.27	a
Taenia coli (guinea pig)	1	1.02	0.26	a
	1	0.44	—	b
(glycinerated)	1	0.26	—	b
Stomach (hog)	1	0.76	0.35	a
Intestine, longitudinal (hog)	1	0.59	0.25	e
Intestine, circular (hog)	1	0.62	0.26	c
Arterial (hog)	1	0.38	0.28	c
	1	0.59	0.27	a
Uterus (hog)	1	0.90	0.28	c
Esophagus (hog)	1	0.67	0.24	c
Trachea (hog)	1	0.67	0.28	c
Skeletal muscle (rabbit)[e]	1	1.6	0.23	e

[a] As determined by densitometry of Coomassie blue-stained SDS–polyacrylamide gels of whole muscle. See Section III,C.

[b] Estimated from heavy chain peak and corrected for light chain mass by multiplying by 1.19.

[c] Excluding the peak from filamin (see Cohen and Murphy, 1978).

[d] (a)Small and Sobieszek (1977a); (b) Tregear and Squire (1973); (c) Cohen and Murphy (1978); (d) Sobieszek and Bremel (1975); (e) Potter (1974).

[e] Gels stained with Fast Green.

smooth muscle actomyosin. These correspond to filamin ($M_r \sim 240,000$; Wang, 1977), α-actinin ($M_r \sim 105,000$), and a protein with a polypeptide molecular weight of around 130,000. These proteins are discussed at more length in a later section. Suffice it to point out here that the purification of actomyosin by the use of divalent cations (Sobieszek and Small, 1976) results in the invariable loss of these additional components. From the high activity and Ca sensitivity of the resulting actomyosin, these components would appear to be involved in the structural organization of the myofilaments rather than in the regulation of the actin–myosin interaction (see also Section IV,A).

One interesting feature of smooth muscle actomyosin is the ease with which tropomyosin may be removed by ammonium sulfate fractionation (Fig. 14F; Sobieszek and Small, 1977). A similar separation cannot be readily achieved with skeletal muscle actomyosin. The ammonium sulfate fractionation not only effects the removal of tropomyosin, but also releases the myosin light chain kinase and light chain phosphatase (see Sections IV, C and IV,D), yielding an actomyosin composed of myosin and actin alone. This actomyosin corresponds in all respects to one reconstituted from purified actin and myosin and serves as an ideal system for many studies of the actin–myosin interaction. Since the satisfactory purification of the separate contractile proteins of smooth muscle has been achieved so far only for gizzard, the potential of tropomyosin-free actomyosins in investigations of other smooth muscles could be usefully employed.

B. The Thin Filaments

1. Isolation and Composition

The thin filaments of smooth muscle may be isolated directly from washed muscle homogenates in the presence of Mg and ATP at around pH 6.5 (Sobieszek and Small, 1976). Similar conditions also apply for the isolation of thin filaments from various other muscles (Lehman et al., 1973). Under these conditions, the thick filaments are not dissociated and only small amounts of myosin are released in the extract. Further purification can then be achieved by raising the pH to around 7.0 and collecting the thin filaments by high speed centrifugation.

The significant feature of smooth muscle thin filaments obtained by this mild procedure is that they contain only two primary components, actin and tropomyosin (Fig. 14G). On heavily loaded SDS gels, small amounts of impurities may also be recognized and correspond to filamin ($M_r \sim 240,000$), myosin ($M_r \sim 200,000$), and α-actinin ($M_r \sim 105,000$). In more crude preparations obtained, for example, from unwashed muscle, another band at around 130,000 MW may also be recognized. This component, which has been purified by Driska and Hartshorne, was found to produce a Ca-insensitive inhibition of the ATPase activity of actomyosin and was initially thought to be involved in the control of the actin–myosin interaction (Driska and Hartshorne, 1975). Although this view is no longer held (Hartshorne et al., 1977), the role of the 130,000 MW component, and which occurs in amounts comparable to α-actinin in whole muscle (Fig. 14A), has not been clarified.

2. Actin

The early studies of smooth muscle actin are very adequately reviewed by Murphy and Megerman (1977) and will here be considered only briefly.

Although significant differences could not be observed in the amino acid composition and subunit molecular weight of smooth muscle actin as compared to skeletal muscle actin (Gosselin-Rey et al., 1969), considerable difficulties were encountered in preparing actin by traditional procedures. Very low yields were obtained from acetone-dried residue, and the actin showed a considerable reluctance to polymerize into long filaments under standard conditions (Carsten and Mommaerts, 1963).

Various studies have shown that the successful extraction of actin from smooth muscle acetone powder critically depends on the pretreatment of the muscle prior to drying. Recently, Strzelecka-Golaszewska *et al.* (private communication), using a procedure adapted from that of Carsten and Mommaerts (1963), have found that washing of the muscle in EDTA yields a residue after drying from which actin may be extracted that shows comparable polymerization properties to skeletal muscle actin. For muscle not washed in EDTA, the extracted actin must first be passed through Sephadex G-100 before it is capable of polymerizing to the same extent; otherwise only short filaments are formed (Strzelecka-Golaszewska, private communication). This suggests that a factor affecting actin polymerization (see also Gosselin-Rey *et al.*, 1969) is removed during the EDTA washing step. Such a factor, which may be related to profilin from thymus and blood platelets (Carlsson *et al.*, 1977), could also explain the large amount of nonsedimentable actin that occurs in the washes of smooth muscle prior to actomyosin extraction (see Sobieszek and Bremel, 1975). Rather pure actin may also be obtained from smooth muscle without resorting to the use of acetone powder (Fig. 14H). This may be achieved simply by centrifuging a thin filament extract in the presence of 0.4 M KCl (Sobieszek and Small, 1976) to release tropomyosin (see Spudich and Watt, 1971). Actin, prepared either in this way or from acetone powder, activates skeletal muscle myosin only to about 50% of the level of ATPase activity reached with skeletal muscle F-actin (Próchniewicz and Strzelecka-Golaszewska; Sobieszek and Small, unpublished data).

On the basis of their different isoelectric points, three forms of actin, α, β, and γ, have been identified in muscle and nonmuscle cells (Whalen *et al.*, 1976; Rubenstein and Spudich, 1976; Garrels and Gibson, 1976; Izant and Lazarides, 1977). In skeletal muscle, only the α species is present, whereas it is depleted in smooth muscle. From these studies and recent data on the amino acid sequences of actins from different sources (Vandekerckhove and Weber, 1978), it is apparent that smooth muscle actin exhibits properties in primary structure intermediate between skeletal muscle actin and the cytoplasmic actins. Whether or not this difference is related to the noted differences in activations by the two muscle actins remains to be investigated.

3. *Tropomyosin*

Due to its marked stability in organic solvents, its high solubility, and its readiness to crystallize, tropomyosin was the first contractile protein to be obtained in purified form (Bailey, 1948). In retrospect, it seems likely that the tropomyosins obtained later from vertebrate smooth muscle (Sheng and Tsao, 1954) may in fact have been the most pure. In the case of skeletal and cardiac muscle, troponin impurities are rather tenaciously bound and can only be removed after chromatography (Eisenberg and Kielley, 1970), whereas for invertebrate muscles, extra precautions must be taken to separate tropomyosin from paramyosin (Bailey and Rüegg, 1960).

In a series of comparative studies (Sheng and Tsao, 1954; Tsao *et al.*, 1955; Jen and Tsao, 1957; Kominz *et al.*, 1957), the physical and chemical properties of smooth muscle tropomyosin were investigated and compared to those of skeletal muscle. From these and subsequent investigations (Tsao *et al.*, 1965; Carsten, 1968), it was shown that the two tropomyosins have closely similar properties except for a greater tendency of smooth muscle tropomyosin towards polymerization. This interesting difference led to the high initial estimate of the molecular weight for smooth muscle tropomyosin of around 150,000 (Tsao *et al.*, 1955).

As with tropomyosins from skeletal and cardiac muscle (Cummins and Perry, 1973; Leger *et al.*, 1976), differences have been recognized in the polypeptide molecular weights of different smooth muscle tropomyosins. Like rabbit skeletal muscle tropomyosin, gizzard tropomyosin runs as a doublet on SDS–polyacrylamide gels, whereas the tropomyosins from uterus (Cummins and Perry, 1974), vas deferens, taenia coli, carotid, and stomach (Small and Sobieszek, 1977a,b) run as a single band. This has been taken to indicate the presence of different isomers of tropomyosin made up from different combinations of the polypeptide subunits (Cummins and Perry, 1973). The subunits of smooth muscle tropomyosin differ, however, in several respects from those of skeletal muscle (Cummins and Perry, 1974; Hayashi *et al.*, 1977). In addition to differences in amino acid composition, smooth muscle tropomyosins show a lower cysteine content (about 2 moles per 68,000 gm) than skeletal muscle tropomyosin (about 3 moles per 68,000 gm; Cummins and Perry, 1974), and for gizzard, the sum of the molecular weights of the tropomyosin subunits (40,000 and 36,000) is slightly higher than for the subunits of chicken breast muscle tropomyosin (36,000 and 37,500; Hayashi *et al.*, 1977). According to the results of Cummins and Perry (1974), the subunits of gizzard and uterus tropomyosin are immunologically distinct from those of skeletal muscle tropomyosin, but some immunological cross-reaction between chicken breast and gizzard tropomyosin has been detected (Hayashi *et al.*, 1977). Further, for the whole molecule, common antigenic sites have been shown to be present in tropomyosins from a wide spectrum of vertebrate and invertebrate muscles (Hayashi and Hirabayashi, 1978) and appear also to be preserved in the smaller tropomyosins found in nonmuscle cells (Hayashi and Hirabayashi, 1978; Lazarides, 1975). These sites are presumably related to the regions involved in the specific interaction of tropomyosin with the F-actin filament.

Like skeletal muscle tropomyosin, smooth muscle tropomyosin can be induced to form paracrystals and nets with a typical repeat period of about 40 nm (Tsao *et al.*, 1965; Holtzer *et al.*, 1965; Cohen and Longley, 1966). This repeat period has been shown to correspond to the molecular length (Peng *et al.*, 1965; Caspar *et al.*, 1969). Reports (Takagi *et al.*, 1976) that the repeat period for bovine arterial tropomyosin is closer to the value of 34.5 nm characteristic of nonmuscle tropomyosins (Fine and Blitz, 1975) remains to be confirmed. For gizzard, a value of 39 nm was obtained using the 14.5 nm spacing of paramyosin filaments as an internal calibration in the electron microscope (Sobieszek and Small, 1973b).

In terms of their effect on the ATPase activity of actomyosin, some significant differences between smooth and skeletal muscle tropomyosin have been detected; discussion of these differences has been reserved for Section IV,B.

C. Myosin

The general properties of the smooth muscle myosin molecule have been well reviewed by Hamoir (1973) and Sobieszek (1977X). In its physiochemical and structural properties, the molecule compares closely with myosin from skeletal muscle. It consists of an α-helical rod-shaped section that terminates in two globular heads (Elliott *et al.*, 1976), has a molecular weight on the order of 480,000, and posesses the ability to assemble into thick filaments at low ionic strength (see Section II,D).

At the same time, some notable differences exist between skeletal and smooth muscle myosins: in solubility (see Section II,D), amino acid composition (Hamoir, 1973), molecular length (Kendrick-Jones *et al.*, 1971), susceptibility to proteolytic enzymes (Huriaux, 1965; Bailin and Bárány, 1971; Katoh and Kubo, 1977), and light chain composition. Such differences may explain the further observation that antibodies to smooth muscle myosin are unable to cross-react with myosins from cardiac or skeletal muscle (Aida *et al.*, 1968; Becker and Murphy, 1969; Gröschel-Stewart, 1971; Masaki, 1974).

The diverse methods that have been reported for the purification of smooth muscle myosin attest to the general difficulties that have been encountered in obtaining homogeneous and enzymatically active preparations. Initial extractions of actomyosin have been made in low (Huriaux *et al.*, 1967; Kotera *et al.*, 1969; Megerman and Murphy, 1975) or high (Yamaguchi *et al.*, 1970; Gröschel-Stewart, 1971; Wachsberger and Kaldor, 1971; Wang, 1977; Malik, 1978) concentrations of KCl, or, with the aim of denaturing actin, in potassium iodide (Bárány *et al.*, 1966). Subsequent purification of myosin has generally involved centrifugation, followed by precipitation with ammonium sulfate (Yamaguchi *et al.*, 1970; Gröschel-Stewart, 1971), divalent cations (Sobieszek and Bremel, 1975; Ebashi, 1976), or polyethylene glycol (Megerman and Murphy, 1975), or alternatively, by the use of gel filtration or ion-exchange chromatography (Kendrick-Jones, 1973; Wang, 1977; Malick, 1978). Since the most troublesome and tenaciously bound contaminant is actin, which is released in increased amounts in solutions of elevated salt concentration, the low ionic strength procedures would appear to be more suitable for the initial extraction of actomyosin.

From extensive studies with gizzard, we have developed a procedure for myosin purification that consistently produces myosin in high yield (about 0.2–0.5 gm/100 gm wet muscle) and purity (Sobieszek and Bremel, 1975; Sobieszek, 1977a). The method, which has been described at length (Sobieszek, 1977a), involves the prolonged centrifugation of actomyosin extracted at low ionic strength, followed by selective precipitation of the myosin in the supernatant with 40 mM $MgCl_2$ (Fig. 15A). Adaptions of this method have been reported by others (Ebashi, 1976). The only significant contaminant in such preparations is a small amount of the myosin light chain kinase (Sobieszek, 1977a; Section IV,C), which, if necessary, can be separated from myosin by gel filtration. The retention of the native properties of the molecule after this purification procedure is indicated by the high actin-activated ATPase activity (Sobieszek and Small, 1976, 1977) and ability to assemble into filamentous structures of high regularity (Section II,D,2). The application of the same method to other smooth muscles has not met with quite the same degree of success as with gizzard. For porcine stomach, the myosin obtained always shows minor actin contamination (Fig. 15B), which can only be removed by gel filtration in the presence of MgATP. In our experience this results in the complete separation of myosin from actin, rather than myosin from actomyosin (Adelstein and Conti, 1975; Chacko et al., 1977).

Estimates of the amount of myosin as compared to actin and tropomyosin have been obtained from the densitometry of Coomassie blue-stained polyacrylamide gels of whole muscle (Tregear and Squire, 1973; Small and Sobieszek, 1977a; Cohen and Murphy, 1978). These are listed in Table II and compared with the values obtained for skeletal muscle myofibrils (Potter, 1974). While the variation in these estimates is too large to allow the confident assignment of realistic values for further detailed analysis, some consistent trends in myosin content can be recognized. Relative to actin, carotid muscle shows the lowest content of myosin; intestinal muscle, intermediate values; and vas deferens, the highest. Furthermore, the amount of myosin present in smooth muscle is clearly much greater than was supposed from the early estimates based on yields of contractile proteins (Needham, 1971), and, for vas deferens, appears to be in the order of that found in skeletal muscle. More analysis, possibly using Fast Green for staining (Potter, 1974), is called for to settle the current differences between the published values.

Apart from their extractability at low ionic strength, the most consistent distinguishing feature of the smooth muscle myosins is their possession of two species of light chains that differ from those found in either skeletal or cardiac muscle myosins (see Weeds et al., 1977). The smooth muscle myosin light chains have approximate molecular weight values of 17,000 (L_{17}) and 20,000 (L_{20}), which

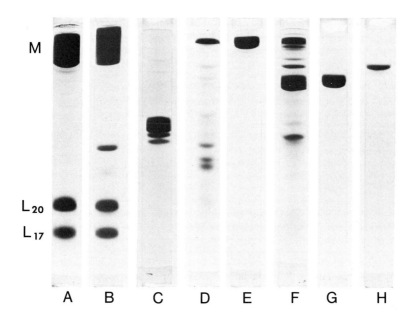

FIG. 15. SDS–polyacrylamide gels (7.5%) of the following preparations. (A) Chicken gizzard myosin (MY, Fig. 11). L_{20} and L_{17} indicate the 20,000 MW and 17,000 MW myosin light chains; M, myosin heavy chain. (B) Hog stomach myosin prepared as for (A). The small amount of contaminating actin is typical for this mammalian myosin and can only be removed by gel filtration (see text). (C) Hog stomach skeletin obtained by acetic acid extraction and further fractionation on DEAE(diethylaminoethyl)-Sepharose in the presence of 7 M urea. The minor bands correspond to proteolytic products of the major 55,000 MW component. (D) Filamin-rich fraction (Fig. 11) obtained via precipitation of actomyosin with $MgCl_2$. (E) Filamin purified from fraction shown in (D) by gel filtration on Sepharose 4B, followed by ion-exchange chromatography on DEAE Sepharose (Wang, 1977). (F) Crude α-actinin fraction obtained by fractionation of the KI extract of smooth muscle residue (Fig. 11). (G) α-Actinin purified from preparation as in (F) by one passage each over DEAE Sepharose and hydroxylapatite. (H) Purified myosin light chain kinase from chicken gizzard obtained by gel filtration and ion-exchange chromatography of the kinase rich extract (Fig. 11). Band occurs at an apparent molecular weight of around 145,000.

are essentially invariant among the various smooth muscles investigated (Kendrick-Jones, 1973; Leger and Focant 1973; Wachsberger and Pepe, 1974; Small and Sobieszek, 1977b; Katoh and Kubo, 1977). Densitometry of SDS–polyacrylamide gels of purified myosin (Sobieszek and Bremel, 1975) indicates that each molecule contains two of each of the light chains, that is, one of each type per head. As we shall see later, the L_{20} light chain is intimately involved in the process of Ca regulation of the actin–myosin interaction. By comparison, rather little has been established about the function of the L_{17} light chain. So far, conditions for the reversible association of this light chain with myosin, such as those developed for the skeletal muscle alkali light chains (Wagner and Weeds, 1977), have not been found. However, the retention of the L_{17} light chain and not the L_{20} light chain in the enzymatically active S-I subfragment from myosin (Kendrick-Jones, 1973; Sobieszek and Small, 1976; Sobieszek, 1977a) suggests that L_{17} is required for binding to actin and for ATP hydrolysis.

The important properties of smooth muscle myosin that are associated with the regulation of the actin–myosin interaction are discussed in Section IV,A.

D. ADDITIONAL PROTEINS IN ACTOMYOSIN

1. Filamin

In SDS–polyacrylamide gels of whole smooth muscle or crude actomyosin, two bands occur in the region corresponding to polypeptides of around 200,000 MW (Sobieszek and Bremel, 1975; Wang et al., 1975). One corresponds to the heavy chain of myosin and the other to a protein Wang and co-workers have chosen to call "filamin." The latter name was suggested from the staining, by antibodies to this protein, of the actin filament bundles in cultured smooth muscle cells (Wang et al., 1975). On the assumption that filamin and myosin have similar stainabilities on the polyacrylamide gels, it was estimated that filamin occurs in amounts corresponding to 30 to 50% of the myosin content.

From the studies of Wang (Wang et al., 1975; Wang, 1977) and Shizuta et al. (1976), filamin has been fairly well characterized. The native protein is an asymmetric oligomer consisting of two subunits (each about M_r 250,000) and having a frictional ratio (2.2 to 2.3) comparable to fibrinogen and spectrin. It is soluble both at high and low ionic strength and is not precipitated by high concentrations of divalent cations. In its physicochemical properties, it is closely similar to spectrin, but it differs from the latter in its polypeptide molecular weight and amino acid composition (Wang, 1977).

Filamin may be readily purified from extracts of whole muscle or from crude actomyosin. In our experience, gizzard filamin is most easily purified from the filamin-rich supernatant obtained after precipitation of actomyosin with 25 mM $MgCl_2$ (see Fig. 15D). Fractionation with ammonium sulfate between 25 and 40% saturation, followed by one fractionation each by gel filtration and ion-exchange chromatography is sufficient to obtain the protein in rather high purity (Fig. 15E; unpublished results).

The localization of filamin within the smooth muscle cell has yet to be established. Although Wang et al. (1975) described the staining of cells in primary cultures of chick gizzard, the interpretation of this result is complicated by the known phenotypic changes of smooth muscle cells in culture (Chamley et al., 1977). There seems little doubt, however, that filamin interacts with F-actin (Shizuta et al., 1976; Wang and Singer, 1977) and that it belongs to the class of actin-binding proteins including α-actinin, spectrin, and the actin-binding protein of macrophages (Stossel and Hartwig, 1976). In mixtures of skeletal muscle F-actin and filamin, a sedimentable complex is formed in which the molar ratio of filamin to actin is in the range of 1 to 8–12 (Wang and Singer, 1977). This compares with the ratio of 1 to 10 for the complex between F-actin and α-actinin (Goll *et al.*, 1972). Filamin has also been reported to inhibit the actin-activated ATPase activity of skeletal muscle heavy meromyosin (Davies *et al.*, 1977). A similar inhibition and dramatic precipitation occurs also when filamin is recombined in increasing amounts with smooth muscle actomyosin (Small and Sobieszek, unpublished results). Unlike spectrin (Pinder *et al.*, 1975), filamin has not been found to induce the polymerization of G- to F-actin (Wang and Singer, 1977).

The observation (Hiller and Weber, 1977) that spectrin is not as ubiquitous as earlier supposed suggested that the high-molecular-weight components earlier recognized in various nonmuscle cells and compared to spectrin (see Hiller and Weber, 1977) may be more closely related to filamin. Subsequent studies have tended to confirm this conclusion. Components binding antibodies to filamin have now been localized in cultured fibroblasts (Heggeness *et al.*, 1977; Wallach, *et al.*, 1978a), in blood platelets (Wallach *et al.*, 1978b), and in the terminal web of intestinal epithelial cells (Bretscher and Weber, 1978). From its similar cross-reaction with antifilamin antibodies (Wallach, Davies & Pastan, 1978b), the macrophage actin-binding protein of Stossel & Hartwig (1976) would also seem to be closely related to filamin (Wallach *et al.*, 1978a), although parallel studies (Brotschi, *et al.*, 1978) have indicated differences in the extents to which each protein can form gel complexes with F-actin *in vitro*.

While further studies will be required to establish the function of filamin, the current data indicate that it is involved in structural interactions between actin filaments, perhaps in association with other actin-binding proteins, such as α-actinin. The recent observation that filamin can also become phosphorylated (Davies *et al.*, 1977; Wallach *et al.*, 1978b) suggests that the phosphorylation may be of significance in the control of the filamin–actin interaction.

2. α-Actinin

Soon after the discovery of α-actinin in skeletal muscle (Ebashi and Ebashi, 1965), Ebashi and co-workers were able to show the presence of a similar protein both in cardiac and smooth muscle (Ebashi *et al.*, 1966). Smooth muscle α-actinin showed the same high solubility at low ionic strength and, like skeletal muscle α-actinin, potentiated the superprecipitation of skeletal muscle actomyosin. Subsequent studies (see reviews by Ebashi and Nonomura, 1973; Katz, 1970; Suzuki *et al.*, 1976) have been confined almost exclusively to skeletal muscle α-actinin, and until now a detailed investigation of the smooth muscle protein has not been undertaken. However, the preliminary data available

(Schollmeyer *et al.*, 1976; Suzuki *et al.*, 1976) would suggest that smooth muscle α-actinin shares similar properties to its skeletal muscle counterpart.

By various criteria (Suzuku *et al.*, 1976), the molecular weight of skeletal muscle α-actinin has been estimated as around 200,000. The protein has an α-helical content in the order of 74% and consists of two polypeptides of each around 100,000 MW. Electron microscopy has shown the molecule to be elongated, with dimensions from 300 × 20 Å (Podlubnaya *et al.*, 1975) to 400 × 44 Å (Suzuki *et al.*, 1976).

In skeletal muscle, α-actinin is localized exclusively in the Z-line (Masaki *et al.*, 1967) where its ability to cross-link actin filaments (Stromer and Goll, 1972; Podlubnaya *et al.*, 1975) fits well with the concept of it playing a primary role in maintaining the ordered organization of the I-band.

From investigations of smooth muscle and of certain nonmuscle cells, it is apparent that α-actinin may have a dual function (see also Tilney, 1975): to form cross-links between actin filaments and to bind actin filaments to the cell membrane. The most suggestive evidence for this has come from indirect immunofluorescence of cultured cells in which α-actinin has been found periodically arranged along the actin filament bundles, or stress fibers, and at the site of insertion of a proportion of these fibers in the plasmalemma (Lazarides and Burridge, 1975). More recently its localization in the regions of the tight junctions of intestinal epithelial cells (Craig and Pardo, 1979) and in the cleavage furrow of cells undergoing mitosis (Fujiwara *et al.*, 1978) has also been demonstrated rather clearly, as well as its association with secretory vesicles of chromaffin cells (Jockusch *et al.*, 1977). Preliminary studies on smooth muscle (Schollmeyer *et al.*, 1976) indicate that α-actinin is localized in the attachment plaques beneath the plasmalemma, as well as in the cytoplasmic dense bodies (see also Section II,F). It remains to be shown, however, whether α-actinin is the only protein involved in filament anchorage or whether other components are also required.

To date, a full procedure for the purification of smooth muscle α-actinin has remained unpublished (see Schollmeyer *et al.*, 1976; Fujiwara *et al.*, 1978; Craig and Pardo, 1979). In our experience (also unpublished results), the most suitable source for this protein is the residue remaining after the extraction of actomyosin in high and low salt (see Fig. 11). α-Actinin is then readily released in 1 *M* KI and, following dialysis to low ionic strength, may be obtained as the major component by ammonium sulfate fractionation between 20 and 40% saturation (Fig. 15F). Purification to homogeneity (Fig. 15G) then requires only a single step on an ion-exchange gel and hydroxylapatite (Suzuki *et al.*, 1976) rather than the multiple chromatographic steps that have been described (Suzuki *et al.*, 1976; Fujiwara *et al.*, 1978). Further studies with such preparations, in combination with smooth muscle actomyosin, will be required to establish more details of the α-actinin–thin filament interaction in smooth muscle. Additional studies with isolated cells and anti-α-actinin antibodies are also needed to confirm the reported localization of this protein *in vivo* (Schollmeyer *et al.*, 1976).

IV. Regulation of the Actin-Myosin Interaction via Ca^{2+}

A. Myosin Phosphorylation and Ca-Dependent Regulation

1. Evidence for Myosin-Linked Regulation in Vertebrate Smooth Muscle

In their pioneering studies of invertebrate muscle actomyosins, Kendrick-Jones, Szent-Gyorgyi, and Lehman (see reviews by Szent-Gyorgyi, 1975; and Lehmann, 1976) were able to demonstrate for the first time that the interaction between actin and myosin can be regulated via myosin. This type of regulation, found in molluscan muscles, was termed "myosin-linked" Ca-dependent regulation to distinguish it from the well characterized actin-linked (i.e., troponin-tropomyosin) Ca-dependent regulation operative in vertebrate skeletal muscle (see review by Weber and Murray, 1973). At first it was considered likely from comparative studies that myosin-linked regulation was confined only to invertebrates (Lehman *et al.*, 1973). It is now clear, however, that vertebrate smooth muscle is also regulated via myosin and that this type of regulation may be operative in various vertebrate nonmuscle systems.

The primary evidence for myosin-linked Ca-dependent regulation in molluscan muscles came from measurements of the Ca sensitivity of the actin-activated ATPase activities of actomyosins reconstituted from purified contractile proteins (Kendrick-Jones *et al.*, 1970). Thus, molluscan myosin was shown to form a Ca-sensitive actomyosin when combined with rabbit skeletal muscle F-actin. Conversely, the actomyosin formed from molluscan thin filaments and rabbit skeletal muscle myosin was Ca-insensitive, indicating the absence of actin-associated regulatory components. A further test, the "actin competition test" was also devised for probing for myosin-linked regulation in more crude native actomyosins and in myofibril preparations (Lehman *et al.*, 1973). This test was based on the assumption that the addition of pure F-actin to a system possessing myosin-linked regulation should have no effect on the measured Ca sensitivity. A converse test for actin-linked regulation was made using skeletal muscle myosin, HMM or HMM S-1, in which the loss of Ca sensitivity could be correlated with the absence of regulatory components on the thin filaments.

The observation that typical troponin-like components were absent from Ca-sensitive smooth muscle actomyosin (Sobieszek and Bremel, 1975; Driska and Hartshorne, 1975; see Sections III,A and III,B) provided the first indication that Ca-dependent regulation in this system may be myosin-linked (Sobieszek and

Bremel, 1975). Subsequent studies, involving recombination experiments and tests of the type described above, furnished direct evidence for this form of regulation (Bremel, 1974; Sobieszek and Small, 1976). Chicken gizzard myosin, in combination with rabbit striated muscle or gizzard F-actin, was shown to form an actomyosin that was Ca-sensitive, while the actomyosin formed from smooth muscle thin filaments (from chicken gizzard or hog stomach) and skeletal muscle myosin was Ca-insensitive (Sobieszek and Small, 1976; Small and Sobieszek, 1977b). The results of actin and myosin competition tests, undertaken with gizzard actomyosin and myofibrils were likewise consistent with Ca-dependent regulation being associated only with the myosin component (Bremel, 1974; Sobieszek and Small, 1976). In studies with vascular smooth muscle actomyosin, Mrwa and Rüegg (1975) also reported the results of actin and myosin competition tests that were consistent with a myosin-regulated system, although the absolute activities obtained (see Section III,A) were not given.

For molluscan muscle, the Ca-sensitive properties of myosin have been shown to be conferred by one of the two pairs of myosin light chains, the so-called "regulatory light chains" (Szent-Gyorgyi et al., 1973; Kendrick-Jones et al., 1976). In the presence of EDTA, one of these light chains is readily removed with the consequence that the myosin becomes desensitized with respect to Ca^{2+}. Furthermore, this process is reversible: the light chain may be reconstituted with desensitized myosin and the Ca sensitivity restored. The restoration of Ca sensitivity not only applies to the actin-activated ATPase activity in actomyosin, but also to the tension response of similarly treated demembranated myofibrils (Simmons and Szent-Gyorgyi, 1978).

Of particular relevance to the present discussion was the additional finding (Kendrick-Jones et al., 1976) that the isolated L_{20} light chain from smooth muscle myosin could fully restore the Ca sensitivity and Ca binding of desensitized scallop myosin; that is, it could functionally replace the scallop regulatory light chain. The same property is exhibited by the myosin regulatory light chains of other molluscan muscles that have been shown to possess myosin-linked Ca-dependent regulation. The light chains of around 20,000 MW from the myosins of vertebrate skeletal and cardiac muscle were less effective in the molluscan system in that they could recombine with scallop myosin only when it was complexed with actin and did not restore full Ca binding. Apart from their general implications in relation to the mechanism of myosin-linked regulation, these results added further weight to the suggestion (Sobieszek and Small, 1976) that the L_{20} light chain is the component involved in the regulation of smooth muscle myosin.

In the following section we show that while parallel studies have confirmed the involvement of the L_{20} light chain in Ca-dependent regulation in smooth muscle, this regulation operates via a phosphorylation of the light chain, rather than as in scallop via direct Ca^{2+} exchange.

2. L_{20} "P"-Light Chain Phosphorylation and Actin Activation

One general property of vertebrate myosins that is apparently not shared by molluscan myosin (Kendrick-Jones and Jakes, 1977) is the ability of the light chain of 19,000–20,000 MW to become phosphorylated (Perrie et al., 1973; Frearson and Perry, 1975). Frearson and Perry (1975) have proposed that this class of phosphorylatable light chains be referred to as the "P"-light chains. The P-light chain phosphorylation was first detected in vertebrate skeletal muscle myosin (Perrie et al., 1973) and was further shown to be catalyzed by a specific enzyme, "a Ca-requiring myosin light chain kinase" (Pires et al., 1974). But, although Ca dependent, no effect of the phosphorylation by the kinase could be detected on the actin-activated ATPase activity of skeletal muscle actomyosin. Subsequent studies on actomyosin from blood platelets showed, in contrast, that a similar phosphorylation of platelet myosin (Adelstein et al., 1973) could be correlated with an increased actin-activated ATPase activity (Adelstein and Conti, 1975, 1976). This phosphorylation was, however, not Ca dependent and its possible participation in the regulation of contractility was, in consequence, unclear. Nevertheless, this result was of particular interest in view of the already evident similarity between the light chain patterns and other properties of smooth muscle and nonmuscle myosins (Burridge and Bray, 1975; Adelstein and Conti, 1974; and Section III,C).

With the development of suitable procedures for the preparation of active and Ca-sensitive smooth muscle actomyosin (Sobieszek and Bremel, 1975; Sobieszek and Small, 1976), it became appropriate to investigate the possible involvement of myosin phosphorylation in the Ca regulation mechanism in this system. The results of these first studies, reported in 1975 (Sobieszek, 1977a), proved to be significant, since they showed that the phosphorylation of the "P"-light chain in smooth muscle acts as the primary switch in controlling the actin–myosin interaction.

The major evidence for such a control was as follows (Sobieszek, 1977a,b). First, the phosphorylation, which occurred exclusively in the P-light chain, was dependent on concentrations of free Ca^{2+} ion in the same range (10^{-6}–$10^{-5}M$) as required to stimulate the actin-activated ATPase activity of actomyosin; below about 10^{-7} M, no phosphorylation took place and the myosin ATPase activity was not activated by actin (Fig. 16). Second, the time course of phosphorylation was such that phosphate incorporation preceded the rise in actin-activated ATPase activity, indicating that phosphorylation was a prerequisite for actin–myosin interaction (Fig. 17A and B). The separate effects of Ca^{2+} and phosphorylation on the actin-activated ATPase activity were shown by a further experiment (Fig. 18; Sobieszek, 1977a; Small and Sobieszek 1977c). With gizzard actomyosin, first phosphorylated in the presence of MgATP and Ca^{2+}, and then treated 40 seconds later with excess EGTA (to remove Ca^{2+}), the degree of phosphorylation

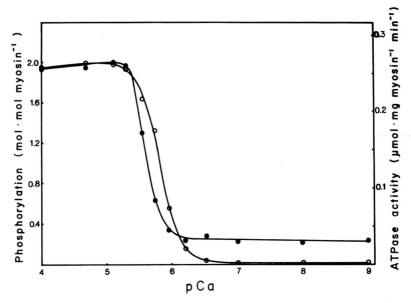

Fig. 16.   Dependence of myosin light chain phosphorylation (O–O) and actin-activated ATPase activity (●–●) of gizzard actomyosin as a function of the free Ca ion concentration. (From Sobieszek, 1977b.)

of the light chain remained essentially constant. Correspondingly, the rate of ATP hydrolysis remained unchanged after Ca^{2+} was removed. Thus, calcium was required only for myosin phosphorylation and not for the hydrolysis of ATP by myosin.

The results of these studies were rapidly confirmed by other groups, working with chicken gizzard (Aksoy et al., 1976; Górecka et al., 1976; Sherry et al., 1978; Ikebe et al., 1978), bovine aorta (DiSalvo et al., 1977), and guinea pig vas deferens (Chacko et al., 1977), as well as in parallel investigations with hog stomach (Small and Sobieszek, 1977b). The light chain phosphorylation could further be shown to be dependent on a myosin light chain kinase endogenous to smooth muscle (Frearson et al., 1976a) and enriched in a tropomyosin-containing ammonium sulfate fraction of actomyosin (Górecka et al., 1976; Sobieszek, 1977b; Sobieszek and Small, 1977; Chacko et al., 1977). The kinase was also present in varying amounts in different preparations of actomyosin and the contractile proteins (Sobieszek and Small, 1977), and could only be entirely removed from myosin by gel filtration (Sobieszek, 1977b; Adelstein et al., 1976; Sherry et al., 1978; see Section III,C). (The tight association of the P-light chain kinase to myosin purified according to Sobieszek (1977a) explains the positive results of the functional tests for myosin-linked regulation that

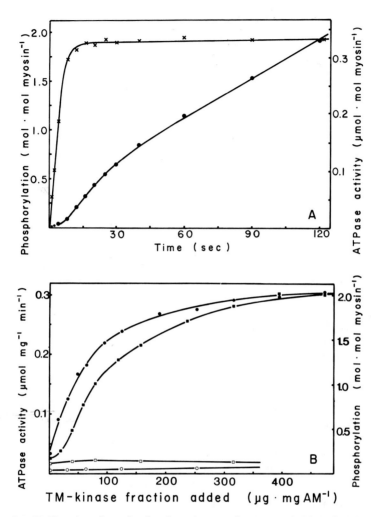

Fig 17.   (A) Time dependence for phosphorus incorporation (x–x) and actin-activated ATPase activity (●–●) in actomyosin. Note that the degree of phosphorylation of the myosin light chain is almost saturated prior to the rapid rise in ATPase activity. (B) Activating effect of a myosin light chain kinase–tropomyosin fraction on tropomyosin-free actomyosin. Circles show the phosphate incorporation into the myosin L_{20} light chain, and squares show the actin-activated ATPase activity. Closed and open symbols correspond to the presence and absence of Ca^{2+}, respectively. As in (A), the phosphorylation of the light chain precedes the rise in ATPase activity. (From Sobieszek, 1977b.)

we have already described in the preceding section). The activating effect of the myosin light chain kinase could be shown very clearly using kinase-depleted gizzard actomyosin (Fig. 19) and hog stomach actomyosin (Fig. 20A). In the

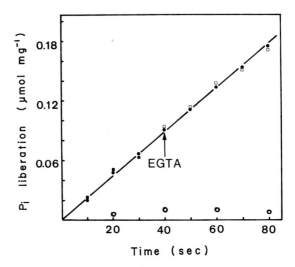

FIG. 18. Experiments showing effect of Ca^{2+} removal on the ATPase activity of phosphorylated actomyosin. Closed circles show inorganic phosphate liberation from 0 to 80 seconds in the presence of Ca^{2+} (using a Ca/EGTA buffer giving a free Ca^{2+} concentration of 2×10^{-5} M). Squares show an experiment carried out under the same conditions except that excess EGTA (2 mM) was added after 40 seconds of incubation. Open circles show a control experiment for which the same amount of EGTA was added prior to the initiation of the assays with ATP. As shown, the removal of Ca^{2+} at 40 seconds did not affect the rate of ATP hydrolysis. Other measurements showed that the extent of myosin phosphorylation was also unchanged. (From Small and Sobieszek, 1977c.)

absence of any added myosin light chain kinase, these actomyosins were completely inactive. However, full activity and Ca sensitivity was effected by the addition of purified kinase or kinase-rich extracts obtained from whole muscle or myofibrils (myofibril extract, Fig. 11).

In a series of brief communications, Ebashi and co-workers have claimed that the process underlying Ca-dependent regulation in smooth muscle is not associated with myosin phosphorylation (Ebashi *et al.*, 1975; Hirata *et al.*, 1977; Mikawa *et al.*, 1977a,b). Much of their data, however, may be explained in terms of a contamination of the different preparations with the light chain kinase and possibly also the activator protein that confers Ca sensitivity to the kinase (Section IV,C). In repeating the superprecipitation experiments described by Mikawa *et al.* (1977a) with phosphorylated gizzard actomyosin, we could not confirm the drop in absorbance concomitant with the removal of Ca^{2+} with EGTA. Since neither the methods of preparation nor the precise nature of the different fractions used by these workers are fully documented, a detailed evaluation of their data is not possible. The likelihood may be considered, however, that their "essential" 80,000 MW component corresponds to a proteolytic product of the kinase comparable to the Ca-insensitive component earlier purified from blood platelets (Daniel and Adelstein, 1976).

The role of Ca^{2+} in the activation mechanism deserves further consideration. Like scallop myosin (Kendrick-Jones et al., 1970), smooth muscle myosin binds Ca^{2+} in the same range of Ca^{2+} concentrations as required to activate the contractile process (Sobieszek, 1977a; Sobieszek and Small, 1976) with about 1.5–2.0 moles Ca^{2+} being bound per mole of myosin. But in one case, for scallop, regulation occurs via reversible Ca binding (Kendrick-Jones et al., 1970, 1976), whereas in the other, a phosphorylation is involved. Current evidence indicates that these two processes are independent. Thus, neither of the isolated light chains from scallop and smooth muscle bind Ca^{2+} (Kendrick-Jones and Jakes, 1977), but this has no effect on the ability of the isolated smooth muscle light

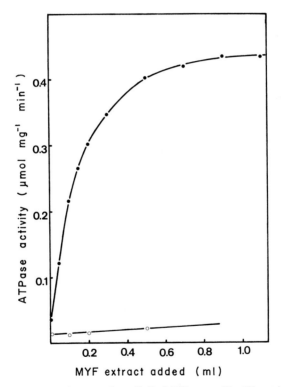

FIG. 19.   Activating effect of extract of myofibrils (MYF extract, Fig. 11) on gizzard actomyosin depleted of the myosin light chain kinase. The actomyosin was reconstituted from tropomyosin free actomyosin (see text and Fig. 14F) and purified gizzard tropomyosin. Open and closed symbols correspond to the presence and absence of Ca^{2+}, respectively. The activating effect can be attributed to the presence of the myosin light chain kinase in the myofibril extract.

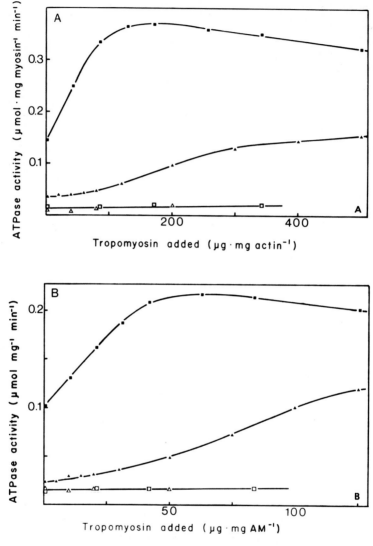

FIG. 20.   The separate effects of the light chain kinase and tropomyosin on reconstituted gizzard actomyosin (A) and tropomyosin-free actomyosin (B). The weight ratio of actin to myosin in (A) was 1:1. ($\triangle$,$\blacktriangle$) The addition of increasing amounts of the column-purified light chain kinase (arbitrary units). ($\square$,$\blacksquare$) The activating effect for the same preparation of increasing amounts of tropomyosin in the presence of an optimal and fixed amount of the myosin light chain kinase. (From Sobieszek and Small, 1977.)

chain to become phosphorylated. In fact, the isolated light chain can be phosphorylated by the light chain kinase even more readily than in its native situation in myosin. Additionally, a papain fragment of the gizzard light chain that cannot be phosphorylated is still capable of conferring Ca sensitivity to desensitized scallop myosin (Kendrick-Jones and Jakes, 1977). From more recent studies of the kinetics of Ca binding by skeletal muscle myosin (Bagshaw and Reed, 1977), the possibility is raised that the common divalent cation binding site recognized in the amino acid sequence of the different light chains (Kendrick-Jones and Jakes, 1977; Jakes *et al.*, 1976; Matsuda *et al.*, 1977) may be involved, for the P-light chains, in the structural association of the light with the heavy chain (Kendrick-Jones and Jakes, 1977). This, however, leaves open the nature of the site involved in scallop in Ca-dependent regulation, as well as the possible existence of a similar site in smooth muscle myosin. Thus, while the current data indicates that the primary role of Ca^{2+} in the regulation process is associated with its activation of the myosin light chain kinase (Section IV,C), the possibility that Ca binding on myosin may modulate the actin-activated ATPase activity, as has been suggested (Chacko *et al.*, 1977; Mrwa and Rüegg, 1977), cannot be rigorously excluded. Measurements of the affinity of binding of Ca^{2+} to phosphorylated, as compared to unphosphorylated, smooth muscle myosin may resolve this problem.

The anticipated existence of a regulatory process based on myosin phosphorylation in vertebrate nonmuscle cells has already been indicated. In addition to the first results from blood platelets (Adelstein and Conti, 1975), an increase in actin activation of myosin associated with myosin P-light chain phosphorylation has been detected in actomyosins from pulmonary macrophages (Adelstein *et al.*, 1978), cultured fibroblasts (Yerna *et al.*, 1978), and proliferative myoblasts (Scordilis and Adelstein, 1977, 1978). Blood platelet myosin phosphorylation has also been shown to be associated with the increased tension development of actomyosin threads (Lebowitz and Cooke, 1978). At issue at the present time is whether or not the lack of Ca sensitivity of the endogenous kinases in some of these systems (Adelstein *et al.*, 1977) arises from a desensitization during isolation (Bremel and Shaw, 1978; see also Section IV,C).

Although no relationship has been found between myosin phosphorylation and actin activation in skeletal and cardiac muscle actomyosins, some effects have been detected *in vivo*. For cardiac muscle, in contrast to what might be expected in smooth muscle, an increase in phosphorylation of the P-light chain has been found to be associated with decreased tension development (Frearson *et al.*, 1976b). On the other hand, the tetanic stimulation of rabbit and frog skeletal muscle has been correlated with an increase in phosphorylation of the P-light chain (Bárány and Bárány, 1977; Stull and High, 1977). The strong possibility thus exists that myosin phosphorylation in skeletal muscle, while not serving a predominant role in regulation, does perform a modulatory function that remains to be elucidated (Perrie *et al.*, 1973).

B. The Activating Effect of Tropomyosin

Tropomyosin has been found to be a constant component of muscle thin filaments, regardless of whether Ca-dependent regulation has been shown to be linked to myosin or actin (Lehman *et al.*, 1973). In skeletal muscle, tropomyosin is generally assumed to propagate the changes induced by the binding of Ca^{2+} to troponin in a cooperative manner along the thin filament (Bremel and Weber, 1972). Moreover, evidence from X-ray diffraction and electron microscopy has been taken to indicate that this process involves the azimuthal movement of tropomyosin on F-actin, which, in turn, could regulate access for the myosin head (Haselgrove, 1973; Huxley, 1973; Parry and Squire, 1973). In the relaxed configuration of the thin filament, tropomyosin, on this model, is envisaged to provide a steric block against myosin–actin interaction.

Despite the compelling nature of the arguments derived from the structural data, the noted biochemical properties of skeletal muscle actomyosin are not altogether consistent with such a steric blocking role for tropomyosin (Eaton, 1976). In particular, it has been shown that the binding of the myosin head to actin enhances the binding of tropomyosin. For the steric blocking model in which tropomyosin is presumed, under the same conditions, to move to a position in which it is less firmly bound, the opposite effect would be expected. Clearly one function of tropomyosin is to mediate the effect of Ca binding to troponin. But its presence in muscles showing myosin-linked regulation, and the similar structural changes in the thin filament that occur in these muscles on activation (Vibert *et al.*, 1972), indicate that it performs a significant and additional role in the actin–myosin interaction.

Of particular relevance for the present discussion is the pronounced stimulatory effect that tropomyosin has on the activity of smooth muscle actomyosin (Sobieszek and Small, 1977; Chacko *et al.*, 1977; Hartshorne *et al.*, 1977b). This effect is most readily demonstrated using tropomyosin-free smooth muscle actomyosin that we described in Section III (Fig. 14F). In the presence of optimal amounts of the myosin light chain kinase, tropomyosin produces a 2- to 3- fold amplification of the actin-activated ATPase activity (Sobieszek and Small, 1977; Fig. 20A and B). This stimulation holds for both smooth and skeletal muscle tropomyosin and for reconstituted actomyosins containing either skeletal or smooth muscle F-actin (Chacko *et al.*, 1977; Hartshorne *et al.*, 1977b; Sobieszek and Small, 1977). Significantly, smooth muscle tropomyosin produces a similar activation of reconstituted skeletal muscle actomyosin, composed of myosin and actin (Sobieszek and Small, 1977), while for this system, skeletal muscle tropomyosin has no effect (Bremel *et al.*, 1973; Lehman and Szent-Gyorgyi, 1972; Shigekawa and Tonomura, 1972). [At low MgATP concentrations, skeletal muscle tropomyosin does potentiate the ATPase activity (Shigekawa and Tonomura, 1972; Bremel *et al.*, 1973), but, under the same conditions, smooth muscle tropomyosin has an even greater stimulatory effect

(Sobieszek and Small, in preparation).] It can also be shown in a competitive assay that skeletal and smooth muscle myosins have a greater affinity for the thin filament complex containing smooth muscle tropomyosin than for actin alone (Sobieszek and Small, in preparation). Similarly, skeletal muscle myosin has a higher affinity for the F-actin smooth muscle tropomyosin thin filament than for its own homologous complex.

An activating effect produced by tropomyosin has previously been reported by Lehman and Szent-Gyorgyi (1972) to occur in *Limulus* striated muscle, which possesses myosin-linked Ca-dependent regulation. Here the activation was shown to be associated with an increased affinity of *Limulus* myosin for the tropomyosin–actin complex. As for smooth muscle myosin, *Limulus* myosin was also activated by thin filaments composed of rabbit skeletal muscle tropomyosin and F-actin. The effect of *Limulus* tropomyosin on rabbit skeletal muscle myosin plus F-actin was not tested. But it was found that myosin from scallop, which, like Limulus myosin, is regulated by Ca^{2+} (see Section IV,A), was not activated by tropomyosin.

These pronounced and clearly significant effects of tropomyosin on the activation of the ATPase activity of actomyosins in different situations remain to be explained. They are not simply restricted to a specific type of tropomyosin nor to myosins that are, by functional criteria, Ca regulated. It is also difficult to accept that fast rabbit skeletal muscle does not utilize the potential activating effect that its own tropomyosin shows in the smooth muscle and *Limulus* systems. As pointed out by Lehman and Szent-Gyorgyi (1972), the "regulation of cross-bridge formation may not be restricted in inhibition." The data available would rather favor a role for tropomyosin in the positive control of the actin–myosin interaction, the details of which have yet to be elucidated.

C. The Myosin Light Chain Kinase

Following the first identification of a myosin light chain kinase in skeletal muscle (ATP:myosin light chain phosphotransferase; Pires *et al.*, 1974), corresponding enzymes have been found in cardiac muscle (Frearson and Perry, 1975), smooth muscle (see following), and nonmuscle tissues (Adelstein *et al.*, 1977). In contrast to the protein kinase involved in glycogen metabolism (see review by Krebs, 1972), the myosin light chain kinases are very specific, not only for the P-light chains, but also for the myosins to which they are homologous (Frearson *et al.*, 1976a; Scordilis and Adelstein, 1978). Thus, the rate of incorporation of phosphate into the smooth muscle light chains by the skeletal muscle kinase is about one-tenth of that for the skeletal muscle light chains (Frearson *et al.*, 1976a); whereas the gizzard (Sobieszek and Small, unpublished observations) and platelet kinases (Scordilis and Adelstein, 1978) are equally ineffective in phosphorylating skeletal muscle myosin. Recently, claims have been made that a Ca^{2+}-dependent kinase isolated from skeletal muscle is capable of phosphorylating myosin light chains as well as other standard substrates, such as histone (Waisman *et al.*, 1978), but on the basis of the other data available, it would seem likely that this particular preparation contains a mixture of kinases.

In view of the primary role played by the smooth muscle light chain kinase in the activation of the contractile process, concerted efforts are being made to characterize this enzyme. As already indicated, the kinase is present in varying amounts in preparations of the contractile proteins (Sobieszek and Small, 1977) and binds with different affinities to gizzard and mammalian smooth muscle actomyosins (Small and Sobieszek, 1977b). Extracts showing high kinase activity may be obtained from washed smooth muscle using solutions of moderate ionic strength ($\mu \sim 0.2$) and slightly alkaline pH (Sobieszek and Small, 1977; Dabrowska *et al.*, 1978). As for other P-light chain kinases (Pires and Perry, 1977; Adelstein *et al.*, 1977; Scordilis and Adelstein, 1978), the smooth muscle enzyme precipitates with tropomyosin in the range of ammonium sulfate saturation between about 40 to 60% (Górecka *et al.*, 1976; Sobieszek and Small, 1977; Chacko *et al.*, 1977). Further fractionation of the enzyme has been achieved by gel filtration and ion-exchange chromatography and is now yielding the enzyme in sufficient quantities for identification on SDS–polyacrylamide gels. But the corresponding values for the molecular weight of the main kinase subunit (see following) are rather variable and range from 105,000 (Dabrowska *et al.*, 1977; 1978) and 125,000 (Adelstein *et al.*, 1978) to 145,000 (Small and Sobieszek, 1977b; Sobieszek, unpublished). While some of the variation may be attributed to the different gel systems employed the major part of the variation probably arises from some proteolytic degradation of the enzyme that occurs during preparation. Estimates of the molecular weights of other light chain kinases are even lower; for the platelet kinase, about 80,000 (Daniel and Adelstein, 1976), and for the skeletal muscle kinase, 80,000 (Pires *et al.*, 1977) and 100,000 (Yazawa and Yagi, 1977). Since the purified smooth muscle kinase consistently shows a single band of around 145,000 MW (Fig. 15H) and elutes in a position, after gel filtration, corresponding to a protein of about the same size, this kinase appears to possess one high molecular weight subunit rather than two smaller subunits of 70,000–80,000 MW.

Recent and significant investigations have shown that the skeletal and smooth muscle myosin light chain kinases possess a low-molecular-weight subunit that confers Ca sensitivity to the enzyme and that is essential for its activity (Yazawa and Yagi, 1977; Barylko *et al.*, 1978; Dabrowska *et al.*, 1978; Yagi and Yazawa, 1978; Adelstein *et al.*, 1978). This subunit is a heat-stable, calcium-binding protein similar or identical to the phosphodiesterase activator protein discovered by Cheung (1970) and Kakiuchi (Kakiuchi *et al.*, 1970). The latter protein, purified from brain, can replace the function of the native subunit in the smooth muscle (Dabrowska *et al.*, 1978) and skeletal muscle light chain kinases (Yagi and Yazawa, 1978).

The close similarity, but not identity, of the activator protein to rabbit skeletal muscle troponin C (Drabikowski *et al.*, 1978; Watterson *et al.*, 1976) provides an explanation of the earlier claims of the presence of troponin in smooth muscle (Head *et al.*, 1977; Ebashi *et al.*, 1975). According to Drabikowski *et al.* (1978) only the activator protein, which can be distinguished from troponin C on the basis of its smaller molecular weight and its stimulation of cAMP phosphodiesterase, is found in smooth muscle tissue.

From the data currently available, the stoichiometry of the large and small (activator) subunits of the kinase is unclear. In the case of phosphodiesterase a 1:1 complex is formed and is stable in the presence of calcium. In the absence of calcium, the enzyme and activator elute separately after gel filtration (Ho *et al.*, 1976). From current studies being carried out in this laboratory, it has been established that the light chain kinase from gizzard and hog stomach elutes as a complex with the activator after gel filtration (Sobieszek, unpublished). However, the complex is not stable on ion-exchange media and the two components elute separately (see also Adelstein *et al.*, 1978). Assuming a stoichiometry of 1:1 between the major kinase subunit and the activator, we arrive at a total estimated molecular weight for the complex of about 160,000. This value corresponds closely to our earlier estimates (Small and Sobieszek, 1977) and is a little higher than that obtained by Adelstein *et al.* (1978) using procedures adapted from our own (Sobieszek and Small, 1977; Sobieszek, unpublished).

So far, kinetic analysis of the phosphorylation reaction are rather limited (Scordilis and Adelstein, 1978). Most laboratories, however, are using approximately the same assay conditions with respect to the concentrations of ATP (0.5–1 mM MgATP), light chains (0.1–0.5 mg/ml), ionic composition, and pH. Therefore, the times used for the kinase assays reflect the differences in the specific activities of the enzymes and their concentrations. These times cover an extremely wide range, being in the order of seconds for the gizzard kinase (Sobieszek and Small, 1977b), and, at best, in the order of several minutes for the skeletal muscle light chain kinase (Frearson and Perry, 1975; Yagi and Yazawa, 1978). It is not clear at present whether these differences are due to different kinetic parameters of the intact light chain kinases or arise from the use of preparative methods that yield the enzymes in a less active form. With regard to the smooth muscle myosin light chain kinase, the rate of incorporation of phosphate into the P-light chain could easily account for the latency period of about 200 milliseconds defined from electrical stimulation of isolated smooth muscle cells (Fay, 1977).

D. Myosin Dephosphorylation and Relaxation

We have already discussed the evidence for the triggering of the actin–myosin interaction occurring in smooth muscle via a phosphorylation of the P-light chain of myosin by a Ca-requiring myosin light chain kinase (Section IV,C). Under all conditions, the degree of actin activation could be correlated with the extent of phosphorylation of the myosin P-light chain. Since the dephosphorylation of myosin and the deactivation of ATP hydrolysis could not be effected by Ca^{2+} removal (Sobieszek, 1977a; Small and Sobieszek, 1977c; Sherry *et al.*, 1978), it was presumed that a phosphatase must be required for relaxation (see also Górecka *et al.*, 1976; Chacko *et al.*, 1977). Direct evidence for the involvement of a phosphatase in the deactivation process came from studies of hog stomach actomyosin (Small and Sobieszek, 1977b).

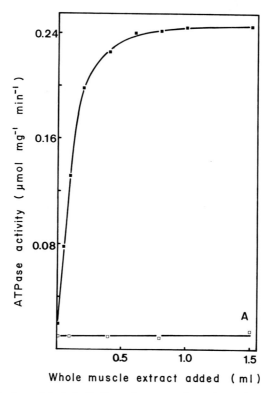

FIG. 21.   Activating and ''relaxing'' effects of an extract of smooth muscle mince on hog stomach actomyosins. The extract was obtained directly from freshly minced muscle under the conditions given in Fig. 11 for the myofibril (MYF) extract. Closed and open symbols correspond to the presence and absence of Ca^{2+}, respectively. (A) Activating effect of extract on dephosphorylated actomyosin purified via precipitation with 20 mM $CaCl_2$. (B) ''Relaxing'' effect of extract on phosphorylated actomyosin purified via precipitation with 2 mM $CaCl_2$. In the absence of Ca^{2-} (□-□), the extract causes an inhibition of the ATPase activity. The two effects can be correlated, in (A), with a phosphorylation of the myosin L_{20} light chain via the myosin light chain kinase and, in (B), with a dephosphorylation of the light chain via the myosin light chain phosphatase. (Adapted from Small and Sobieszek, 1977b.)

In experiments to define suitable procedures for the purification of stomach actomyosin, it was found that, depending on the concentration of $CaCl_2$ used for precipitation (from crude actomyosin (CAM); see Fig. 14C), actomyosin could be obtained with the myosin P-light chain in either a phosphorylated or a dephosphorylated state. The dephosphorylated actomyosin was completely inactive but could be converted to an active and Ca-sensitive actomyosin by its combination with a crude muscle extract or a partially purified preparation of the myosin light chain kinase (Fig. 21A). The phosphorylated actomyosin was, in contrast, very active both in the presence and absence of Ca^{2+} and without added kinase. On storage at 4°C, the extent of phosphorylation and exhibited ATPase activity of this actomyosin remained constant for several days. When combined, however, with the muscle extract (see legend to Fig. 21) rich in kinase activity, the activity of this actomyosin in the presence of EGTA was abolished, that is, the extract rendered the actomyosin Ca-sensitive (Fig. 21B). The abolition of activity in EGTA could further be correlated with a dephosphorylation of the P-light chain, showing the

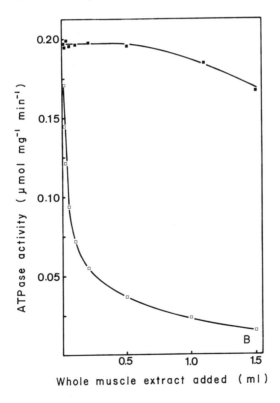

FIG. 21B.

additional presence of a phosphatase in the extract. Gel filtration of the extract from either chicken gizzard or hog stomach revealed the presence of a broad peak of light chain phosphatase activity in a position very close to the kinase and corresponding to a protein or proteins of slightly lower molecular weight. The phosphatase was shown to be active both in the presence and absence of Ca^{2+}, but in its presence it could not compete with the kinase for the P-light chain. In other studies the presence of a light chain phosphatase was deduced from the slow decay in myosin phosphorylation and actin-activated ATPase activity of reconstituted actomyosin following Ca^{2+} removal (Sherry et al., 1978) or depletion of ATP (Chacko et al., 1977).

Further characterization of the smooth muscle myosin P-light chain phosphatase has not been reported. However, Morgan et al. (1976) have described the purification and some properties of the corresponding enzyme (phosphomyosin P-light chain phosphohydrolase) from skeletal muscle. This enzyme has a reported molecular weight of about 70,000 and is rather specific for the light chain. Its activity does not depend on the presence of divalent cations but is unstable in their absence. Like the smooth muscle phosphatase, the skeletal muscle enzyme elutes in approximately the same position as the kinase. However, from the recent estimations of the molecular weight of the kinase from smooth muscle, in the order of 150,000–160,000 the smooth muscle phosphatase, like the kinase, would appear to be twice as large as its skeletal muscle counterpart. The reasons for such large discrepancies in molecular weight between the enzymes in the two muscles remain to be resolved.

E. KINASE–PHOSPHATASE REGULATION

It is not unreasonable to suppose that the light chain kinase activity can be regulated (via the activator protein) by the influx of Ca^{2+} during excitation. At the same time, Ca^{2+} could also influence the activity of other enzymes (e.g.,

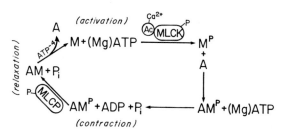

FIG. 22.   Schematic representation of the regulation of the contraction–relaxation cycle in vertebrate smooth muscle. A, Actin; M, myosin; AM, actomyosin; MLCK, myosin light chain kinase; MLCP, myosin light chain phosphatase; Ac, activator. The phosphate groups attached to the kinase and phosphatase have been included to indicate that these enzymes may be further regulated by a phosphorylation process (see text and Adelstein et al., 1978).

adenylate cyclase and phosphodiesterase) that regulate the levels of primary messengers, such as cAMP and cGMP. From physiological studies, the latter messengers have already been implicated in the modulation of smooth muscle activity (Andersson and Nilsson, 1977).

Figure 22 illustrates the primary events of the actin–myosin interaction in smooth muscle as established at the present time. Also included in this scheme is the possibility that the kinase and phosphatase themselves may be regulated through their own phosphorylation. This is suggested by the recent demonstration that the phosphorylation of the gizzard myosin light chain kinase by protein kinase produced a moderate inhibition of light chain kinase activity (Adelstein *et al.*, 1978). From the data of Morgan *et al.* (1976), the active light chain phosphatase would appear to be in a phosphorylated form. The possibility therefore exists that the cAMP and cGMP levels could influence the activity of both of these enzymes.

ACKNOWLEDGMENTS

The authors wish to acknowledge the financial support of the Muscular Dystrophy Association, the Volkswagen Foundation, the Austrian "Fonds zur Förderung der wissenschaftlichen Forschung", and the Austrian National Bank. We also thank Mrs. M. R. Small for typing and Miss G. Langanger and Mr. P. Jertschin for assistance in preparation of the manuscript.

REFERENCES

Adelstein, R. S., and Conti, M. A. (1974). *In* "Exploratory Concepts in Muscular Dystrophy" (A. T. Milhorat, ed.), Vol. 2, pp. 70–78. Excerpta Medica, Amsterdam.

Adelstein, R. S., and Conti, M. A. (1975). *Nature (London)* **256,** 597.

Adelstein, R. S., and Conti, M. A. (1976). *In* "Cell Motility" (R. Goldman, T. Pollard, and J. Rosenbaum, eds.), Vol. 3, pp. 725–737. Cold Spring Harbor Laboratory, Cold Spring Harbor, New York.

Adelstein, R. S., Conti, M. A., and Anderson, W. (1973). *Proc. Natl. Acad. Sci. U.S.A.* **70,** 3115.

Adelstein, R. S., Chacko, S., Barylko, B., Scordilis, S., and Conti, M. A. (1976). *In* "Contractile Systems in Non-Muscle Tissues" (S. V. Perry, A. Margreth, and R. S. Adelstein, eds.), pp. 153–163. North-Holland Publ., Amsterdam.

Adelstein, R. S., Conti, M. A., Scordilis, S. P., Chacko, S., Barylko, B., and Trotter, J. A. (1977). *In* "Excitation-Contraction Coupling in Smooth Muscle" (R. Casteels, T. Godfraind, and J. C. Rüegg, eds.), pp. 359–366. Elsevier/North Holland, Amsterdam.

Adelstein, R. S., Conti, M. A., and Hathaway, D. R. (1978). *J. Biol. Chem.* **253,** 8347.

Aida, M., Conti, G., Laszt, L., and Mandi, B. (1968). *Angiologica* **5,** 322.

Aksoy, M. O., Williams, D., Sharkey, E. M., and Hartshorne, D. J. (1976). *Biochem. Biophys. Res. Commun.* **69,** 35.

Anderson, R. G. G., and Nilsson, K. B. (1977). *In* "The Biochemistry of Smooth Muscle" (N. L. Stephens, ed.), pp. 263–291. Univ. Park Press, Baltimore, Maryland.

Ashton, F. T., Somlyo, A. V., and Somlyo, A. P. (1975). *J. Mol. Biol.* **98,** 17.

Bagby, R. M., Young, A. M., Dotson, R. S., Fisher, B. A., and McKinnon, K. (1971). *Nature (London)* **234**, 351.

Bagshaw, C. R., and Reed, G. H. (1977). *FEBS Lett.* **81**, 386.

Bailin, G., and Bárány, l. (1971). *Biochim. Biophys. Acta* **236**, 292.

Bailey, K. (1948). *Biochem. J.* **43**, 271.

Bailey, K., and Rüegg, J. C. (1960). *Biochim. Biophys. Acta* **38**, 239.

Bárány, K., and Bárány, M. (1977). *J. Biol. Chem.* **252**, 4752.

Bárány, M., Bárány, K., Gaetjens, E., and Bailin, G. (1966). *Arch. Biochem. Biophys.* **113**, 205.

Barylko, B., Kúznicki, J., and Drabikowski, W. (1978). *FEBS Lett.* **90**, 301.

Becker, C. G., and Murphy, G. E. (1969). *Am. J. Pathol.* **55**, 1.

Benitz, W. E., Dahl, D., Williams, K. W., and Bignami, A. (1976). *FEBS Lett.* **66**, 285.

Blose, S. H., Shelanski, M. L. and Chacko, S. (1977). *Proc. Natl. Acad. Sci. USA* **74**, 662.

Bose, R., and Stephens, N. L. (1977). *In* "The Biochemistry of Smooth Muscle" (N. L. Stephens, ed.), pp. 499–512. Univ. Park Press, Baltimore, Maryland.

Bozler, E., and Cottrell, C. L. (1937). *J. Cell. Comp. Physiol.* **10**, 165.

Brecher, S. (1975). *Exp. Cell Res.* **96**, 303.

Bremel, R. D. (1974). *Nature (London)* **252**, 405.

Bremel, R. D., and Shaw, M. E. (1978). *FEBS Lett.* **88**, 242.

Bremel, R. D., and Weber, A. (1972). *Nature (London)* **238**, 97.

Bremel, R. D., Murray, J. M., and Weber, A. (1973). *Cold Spring Harbor Symp. Quant. Biol.* **37**, 267.

Bretscher, A., and Weber, K. (1978). *J. Cell Biol.* **79**, 839.

Brotschi, E. A., Hartwig, J. H., and Stossel, T. P. (1978). *J. Biol. Chem.* **253**, 8988.

Büllbring, E., and Shuba, M. F. (1976). "Physiology of Smooth Muscle." Raven, New York.

Büllbring, E., Brading, A. F., Jones, A. W., and Tomita, T. (1970). "Smooth Muscle." Arnold, London.

Burnstock, G. (1970). *In* "Smooth Muscle" (E. Büllbring, A. F. Brading, A. W. Jones, and T. Tomita, eds.), pp. 1–69. Arnold, London.

Burridge, K., and Bray, D. (1975). *J. Mol. Biol.* **99**, 1.

Campbell, G. R., and Chamley, J. H. (1975). *Cell Tissue Res.* **156**, 2.

Campbell, G. R., Chamley-Campbell, J., Gröschel-Stewart, U., Small, J. V., and Anderson, P. (1979). *J. Cell Sci.* **37**, 303.

Carlsson, L., Nyström, L.-E., Sundkvist, I., Markey, F., and Lindberg, U. (1977). *J. Mol. Biol.* **115**, 465.

Carsten, M. E. (1968). *Biochemistry* **7**, 960.

Carsten, M. E., and Mommaerts, W. F. H. M. (1963). *Biochemistry* **2**, 28.

Caspar, D. L. D., Cohen, C., and Longley, W. (1969). *J. Mol. Biol.* **41**, 87.

Chacko, S., Conti, M. A., and Adelstein, R. S. (1977). *Proc. Natl. Acad. Sci. U.S.A.* **74**, 129.

Chamley, J. H., Campbell, G. R., and McConnell, J. D. (1977). *Cell Tissue Res.* **177**, 503.

Cheung, W. Y. (1970). *Biochem. Biophys. Res. Commun.* **38**, 533.

Choi, J. K. (1962). *Int. Congr. Electron Microsc., 5th, Philadelphia* M9. Academic Press, New York.

Cohen, C., and Longley, W. (1966). *Science* **152**, 794.

Cohen, D. M., and Murphy, R. A. (1978). *J. Gen. Physiol.* **72**, 369.

Cooke, P. H. (1975). *Cytobiologie* **11**, 346.

Cooke, P. H. (1976). *J. Cell Biol.* **68**, 539.

Cooke, P. H., and Chase, R. H. (1971). *Exp. Cell Res.* **66**, 417.

Cooke, P. H., and Fay, F. S. (1972a). *Exp. Cell Res.* **71**, 265.

Cooke, P. H., and Fay, F. S. (1972b). *J. Cell Biol.* **52**, 105.

Craig, R., and Megerman, J. (1977). *J. Cell Biol.* **75**, 990.

Craig, S. W., and Pardo, J. V. (1979). *J. Cell Biol.* **80**, 203.

Croop, J., and Holtzer, H. (1975). *J. Cell Biol.* **65**, 271.

Csapó, A. (1948). *Nature (London)* **962**, 218.

Csapó, A. (1950a). *Acta Physiol. Scand.* **19**, 100.

Csapó, A. (1950b). *Am. J. Physiol.* **160**, 46.

Csapó, A., Erdös, T., Naeslund, J., and Snellman, O. (1950). *Biochim. Biophys. Acta* **5**, 53.

Cummins, P., and Perry, S. V. (1973). *Biochem. J.* **133**, 765.

Cummins, P., and Perry, S. V. (1974). *Biochem. J.* **141**, 43.

Dabrowska, R., Aromatorio, D., Sherry, J. M. F., and Hartshorne, D. J. (1977). *Biochem. Biophys. Res. Commun.* **78**, 1263.

Dabrowska, R., Sherry, J. M. F., Aramatorio, D. K., and Hartshorne, D. J. (1978). *Biochemistry* **17**, 253.

Dahl, D., and Bignami, A. (1976). *FEBS Lett.* **66**, 281.

Daniel, J. L., and Adelstein, R. S. (1976). *Biochemistry* **15**, 2370.

Daniel, E. E., and Paton, D. M. (1975). "Methods in Pharmacology," Vol. 3. Plenum, New York.

Davies, P., Bechtel, P., and Pastan, J. (1977). *FEBS Lett.* **77**, 228.

Devine, C. E., and Somlyo, A. P. (1971). *J. Cell Biol.* **49**, 636.

Dewey, M. M., and Barr, L. (1968). *In* "Handbook of Physiology" Section 6. "Alimentary Canal" (C. F. Code, ed.), pp. 1629–1654. American Physiological Society, Washington, D.C.

DiSalvo, J., Gruenstein, E., and Silver, P. (1978). *Proc. Soc. Exp. Biol. Med.* **158**, 410.

Drabikowski, W., Kźuznicki, J., and Garbarek, Z. (1978). *Comp. Biochem. Physiol.* **60C**, 1.

Driska, S., and Hartshorne, D. J. (1975). *Arch. Biochem. Biophys.* **167**, 203.

Eaton, B. L. (1976). *Science* **192**, 1337.

Ebashi, S. (1976). *J. Biochem. (Tokyo)* **79**, 229.

Ebashi, S., and Ebashi, F. (1965). *J. Biochem. (Tokyo)* **58**, 7.

Ebashi, S., and Nonomura, Y. (1973). *In* "The Structure and Function of muscle" (G. H. Bourne, ed.), pp. 286–352. Academic Press, New York.

Ebashi, S., Iwakura, H., Nakajima, H., Nakamura, R., and Ooi, Y. (1966). *Biochem. Z.* **345**, 201.

Ebashi, S., Toyo-oka, T., and Nonomura, Y. (1975). *J. Biochem. (Tokyo)* **78**, 859.

Eisenberg, E., and Kielley, W. W. (1970). *Biochem. Biophys. Res. Commun.* **40**, 50.

Elliott, A., Offer, G., and Burridge, K. (1976). *Proc. R. Soc. B.* **193**, 43.

Elliott, G. F., and Lowy, J. (1968). *Nature (London)* **219**, 156.

Eng, L. F., Vanderhaeghen, J. J., Bignami, A., and Gerstl, B. (1971). *Brain Res.* **28**, 351.

Eriksson, A., and Thornell, L.-E. (1979). *J. Cell Biol.* **80**, 231.

Fawcett, D. W. (1966). *In* "An Atlas of Fine Structure; The Cell: Its Organelles and Inclusions," p. 243. Saunders, Philadelphia, Pennsylvania.

Fay, F. S. (1977). *In* "Excitation-Contraction Coupling in Smooth Muscle" (R. Casteel, T. Godfraind, and J. C. Rüegg, eds.), pp. 433–439. Elsevier/North-Holland, Amsterdam.

Fay, F. S., and Delise, C. M. (1973). *Proc. Natl. Acad. Sci. U.S.A.* **70**, 641.

Fay, F. S., Cooke, P. H., and Canaday, P. G. (1976). *In* "Physiology of Smooth Muscle" (E. Bullbring and M. F. Shuba, eds.), pp. 249–264. Raven, New York.

Filo, R. S., Rüegg, J. C., and Bohr, D. F. (1963). *Am. J. Physiol.* **205**, 1247.

Fine, R. E., and Blitz, A. L. (1975). *J. Mol. Biol.* **95**, 447.

Fischer, E. (1944). *Physiol. Rev.* **24**, 467.

Fischer, E. (1946). *Ann. N.Y. Acad. Sci.* **47**, 783.

Fisher, B. A., and Bagby, R. M. (1977). *Am. J. Physiol.* **232**, C5.

Franke, W. W., Schmid, E., Osboin, M., and Weber, к. (1978a). *Proc. Natl. Acad. Sci. U.S.A.* **75**, 5034.

Franke, W. W., Schmid, E., Osborn, M., and Weber, K. (1978b). *Cytobiologie* **17**, 392.

Frearson, N., and Perry, S. V. (1975). *Biochem. J.* **151**, 99.

Frearson, N., Focant, B. W. W., and Perry, S. V. (1976a). *FEBS Lett.* **63**, 27.

Frearson, N., Solaro, R. J., and Perry, S. V. (1976b). *Nature (London)* **264,** 801.
Frederiksen, D. W. (1976). *Proc. Natl. Acad. Sci. U.S.A.* **73,** 2706.
Fujiwara, K., Porter, M. E., and Pollard, T. D. (1978). *J. Cell. Biol.* **79,** 268.
Garamvölgyi, N., Vizi, E. S., and Knoll, J. (1971). *J. Ultrastruct. Res.* **34,** 135.
Garrels, J. I., and Gibson, W. (1976). *Cell* **9,** 793.
Gilbert, D. S., Newby, B. J., and Anderton, B. H. (1975). *Nature (London)* **265,** 586.
Goldman, R., and Follett, E. (1970). *Science* **169,** 286.
Goll, D. E., Suzuki, A., Temple, J., and Holmes, G. R. (1972). *J. Mol. Biol.* **67,** 469.
Gordon, W. E., Bushnell, A., and Burridge, K. (1978). *Cell* **13,** 249.
Górecka, A., Aksoy, M. O., and Hartshorne, D. J. (1976). *Biochem. Biophys. Res. Commun.* **71,** 325.
Gosselin-Rey, C., Gerday, C., Gaspar-Godfroid, A., and Carsten, M. E. (1969). *Biochim. Biophys. Acta* **175,** 165.
Gröschel-Stewart, U. (1971). *Biochim. Biophys. Acta* **229,** 322.
Hamoir, G. (1973). *Phil. Trans. R. Soc. London B.* **265,** 169.
Hanson, J., and Huxley, H. E. (1955). *Symp. Soc. Exp. Biol.* **9,** 228.
Hanson, J., and Lowy, J. (1960). *In* "Structure and Function of Muscle" (G. H. Bourne, ed.), Vol. I, pp. 265–335. Academic Press, New York.
Hanson, J., and Lowy, J. (1961). *Proc. Soc. London B.* **154,** 173.
Hanson, J., and Lowy, J. (1964). *Proc. R. Soc. London B.* **160,** 449.
Hanson, J., O'Brien, E. J., and Bennett, P. N. (1971). *J. Mol. Biol.* **58,** 865.
Hartshorne, D. J., and Aksoy, M. (1977). *In* "Biochemistry of Smooth Muscle" (N. L. Stephens, ed.), pp. 363–378. Univ. Park Press, Baltimore, Maryland.
Hartshorne, D. J., Abrams, L., Aksoy, M., Dabrowska, R., Driska, S., and Sharkey, E. (1977a). *In* "Biochemistry of Smooth Muscle" (N. L. Stephens, ed.), pp. 513–532. Univ. Park Press, Baltimore, Maryland.
Hartshorne, D. J., Górecka, A., and Aksoy, M. O. (1977b). *In* "Excitation-Contraction Coupling in Smooth Muscle" (R. Casteels, T. Godfraind, and J. C. Rüegg, eds.), pp. 377–384. Elsevier/North Holland, Amsterdam.
Haselgrove, J. C. (1973). *Cold Spring Harbor Symp. Quant. Biol.* **37,** 341.
Hayashi, J. I., and Hirabayashi, T. (1978). *Biochim. Biophys. Acta* **533,** 362.
Hayashi, J.-I., Ishimoda, T., and Hirabayashi, T. (1977). *J. Biochem. (Tokyo)* **81,** 1487.
Head, J. F., Weeks, R. A., and Perry, S. V. (1977). *Biochem. J.* **161,** 465.
Heggeness, M. H., Wang, K., and Singer, S. J. (1977). *Proc. Natl. Acad. Sci. U.S.A.* **74,** 3883.
Heumann, H.-G. (1971). *Cytobiologie* **3,** 259.
Heumann, H.-G. (1973). *Phil. Trans. R. Soc. London B.* **265,** 213.
Hiller, G., and Weber, K. (1977). *Nature (London)* **266,** 181.
Hinssen, H., D'Haese, J., Small, J. V., and Sobieszek, A. (1978). *J. Ultrastruct. Res.* **64,** 282.
Hirata, M., Mikawa, T., Nonomura, Y., and Ebashi, S. (1977). *J. Biochem. (Tokyo)* **82,** 1793.
Ho, H. C., Wirch, E., Stevens, F. C., and Wang, J. H. (1976). *J. Biol. Chem.* **252,** 43.
Holtzer, A., Clark, R., and Lowey, S. (1965). *Biochemistry* **4,** 2401.
Hubbard, B. D., and Lazarides, E. (1979). *J. Cell Biol.* **80,** 166.
Huriaux, F. (1965). *Angiologica* **2,** 153.
Huriaux, F. (1972). *Arch. Int. Physiol. Chim.* **80,** 541.
Huriaux, F., Hamoir, G., and Oppenheimer, G. (1967). *Arch. Biochem. Biophys.* **120,** 274.
Huxley, H. E. (1963). *J. Mol. Biol.* **7,** 281.
Huxley, H. E. (1973). *Cold Spring Harbor Symp. Quant. Biol.* **37,** 361.
Huxley, H. E., and Brown, W. (1967). *J. Mol. Biol.* **30,** 383.
Hynes, R. O., and Destree, A. T. (1978). *Cell* **13,** 151.
Ikebe, M., Onishi, H., and Watanabe, S. (1977). *J. Biochem. (Tokyo)* **82,** 299.

Ikebe, M., Aiba, T., Onishi, H., and Watanabe, S. (1978). *J. Biochem. (Tokyo)* **83**, 1643.

Ishikawa, H. (1974). *In* "Explanatory Concepts in Muscular Dystrophy"(A.T. Milhorat ,ed.),Vol. II, pp. 37–50. Excerpta Medica, Amsterdam.

Ishikawa, H., Bischoff, R., and Holtzer, H. (1968). *J. Cell Biol.* **38**, 538.

Ishikawa, H., Bischoff, R., and Holtzer, H. (1969). *J. Cell Biol.* **43**, 312.

Izant, J. G., and Lazarides, E. (1977). *Proc. Natl. Acad. Sci. U.S.A.* **74**, 1450.

Jakes, R., Northrop, F., and Kendrick-Jones, J. (1976). *FEBS Lett.* **70**, 229.

Jen, M. H., and Tsao, T. C. (1957). *Scientia Sinica* **6**, 317.

Jockusch, B. M., Burger, M. M., DaPrada, M., Richards, J. G., Chaponnier, C., and Gabbiani, G. (1977). *Nature (London)* **270**, 628.

Jones, A. W., Somlyo, A. P., and Somlyo, A. V. (1973). *J. Physiol. (London)* **232**, 247.

Kakiuchi, S., Yamazaki, R., and Nakajima, H. (1970). *Proc. Jpn. Acad.* **46**, 587.

Kaminer, B. (1969). *J. Mol. Biol.* **39**, 257.

Kaminer, B., Szonyi, E., and Belcher, C. D. (1976). *J. Mol. Biol.* **100**, 379.

Katoh, N., and Kubo, S. (1977). *J. Biochem. (Tokyo)* **81**, 1497.

Katz, A. M. (1970). *Physiol. Rev.* **50**, 63.

Kelly, R. E., and Rice, R. V. (1968). *J. Cell Biol.* **37**, 105.

Kelly, R. E., and Rice, R. V. (1969). *J. Cell Biol.* **42**, 683.

Kendrick-Jones, J. (1973). *Phil. Trans. R. Soc. London B.* **263**, 183.

Kendrick-Jones, J., and Jakes, R. (1977). *In* "Excitation-Contraction Coupling in Smooth Muscle" (R. Casteels, T. Godfraind, and J. C. Rüegg, eds.), pp. 343–352. Elsevier/North Holland, Amsterdam.

Kendrick-Jones, J., Lehman, W., and Szent-Györgyi, A. G. (1970). *J. Mol. Biol.* **54**, 313.

Kendrick-Jones, J., Szent-Györgyi, A. G., and Cohen, C. (1971). *J. Mol. Biol.* **59**, 527.

Kendrick-Jones, J., Szent-Kiralyi, E. M., and Szent-Györgyi, A. G. (1976). *J. Mol. Biol.* **104**, 747.

Kominz, D. R., Saad, F., Gladner, J. A., and Laki, K. (1957). *Arch. Biochem. Biophys.* **70**, 16.

Kotera, A., Yokoyama, M., Yamaguchi, M., Miyazawa, Y. (1969). *Biopolymers* **7**, 99.

Krebs, E. G. (1972). *Curr. Top. Cell. Reg.* **5**, 99.

Kurki, P., Lindner, E., Virtanen, I., and Stenman, (1977). *Nature (London)* **268**, 240.

Lane, B. P. (1965). *J. Cell Biol.* **27**, 199.

Lasek, R. J., and Hoffmann, P. N. (1976). *In* "Cell Motility" (R. Goldman, T. Pollard, and J. Rosenbaum, eds.), Vol. 3, pp. 1021–1049. Cold Spring Harbor Laboratory, Cold Spring Harbor, New York.

Laszt, L., and Hamoir, G. (1961). *Biochim. Biophys. Acta* **50**, 430.

Lazarides, E. (1975). *J. Cell Biol.* **65**, 549.

Lazarides, E., and Burridge, K. (1975). *Cell* **6**, 289.

Lazarides, E., and Granger, B. L. (1978). *Proc. Natl. Acad. Sci. U.S.A.* **75**, 3683.

Lazarides, E., and Hubbard, B. D. (1976). *Proc. Natl. Acad. Sci. U.S.A.* **73**, 4344.

Lebowitz, E. A., and Cooke, R. (1978). *J. Biol. Chem.* **253**, 5443.

Leger, J. J., and Focant, B. (1973). *Biochim. Biophys. Acta* **328**, 166.

Leger, J., Bouvert, P., Schwartz, K., and Swynghedauw, B. (1976). *Pflügers Arch.* **362**, 271.

Lehman, W. (1976). *Int. Rev. Cytol.* **44**, 55.

Lehman, W., and Szent-Gyorgyi, A. G. (1972). *J. Gen. Physiol.* **59**, 375.

Lehman, W., Kendrick-Jones, J., and Szent-Györgyi, A. G. (1973). *Cold Spring Harb. Symp. Quant. Biol.* **37**, 319.

Lehto, V.-P., Virtanen, I., and Kurki, P. (1978). *Nature (London)* **272**, 175.

Liem, R. K., and Shelanski, M. L. (1978). *Brain Res.* **145**, 196.

Litten, R. Z., Salaro, R. J., and Ford, G. D. (1977). *Arch. Biochem. Biophys.* **182**, 24.

Lowey, S. (1971). *In* "Biological Macromolecules" (S. N. Timasheff and G. D. Fasman, eds.), Vol. V, Part A, pp. 201–259. Dekker, New York.

Lowy, J., and Small, J. V. (1970). *Nature (London)* **227**, 46.
Lowy, J., Poulsen, F. R., and Vibert, P. J. (1970). *Nature (London)* **225**, 1053.
Lowy, J., Vibert, P. J., Haselgrove, J. C., and Poulsen, F. R. (1973). *Phil. Trans. R. Soc. London B.* **265**, 191.
McGill, C. (1909). *Am. J. Anat.* **9**, 493.
Malik, M. N. (1978). *Biochemistry* **17**, 27.
Masaki, T. (1974). *J. Biochem. (Tokyo)* **76**, 441.
Masaki, T., Endo, M., and Ebashi, S. (1967). *J. Biochem. (Tokyo)* **62**, 630.
Matsuda, G., Suzuyama, Y., Maita, T., and Umegane, T. (1977). *FEBS Lett.* **84**, 53.
Maxwell, L. C., Bohr, D. F., and Murphy, R. A. (1971). *Am. J. Physiol.* **220**, 1871.
Megerman, J., and Murphy, R. A. (1975). *Biochim. Biophys. Acta* **412**, 241.
Mehl, J. W. (1938). *J. Biol. Chem.* **123**, Lxxxiii.
Mikawa, T., Naramura, Y., and Ebashi, S. (1977a). *J. Biochem. (Tokyo)* **82**, 1789.
Mikawa, T., Toyo-oka, T., Nonomura, Y., and Ebashi, S. (1977b). *J. Biochem. (Tokyo)* **81**, 273.
Moos, C. (1973) *Cold Spring Harb. Symp. Quant. Biol.* **37**, 93.
Moos, C., Offer, G., Starr, R., and Bennett, P. (1975). *J. Mol. Biol.* **97**, 1.
Morgan, M., Perry, S. V., and Ottoway, J. (1976). *Biochem. J.* **157**, 687.
Mrwa, U., and Rüegg, J. C. (1975). *FEBS Lett.* **60**, 81.
Mrwa, U., and Rüegg, J. C. (1977). *In* "Excitation-Contraction Coupling in Smooth Muscle" (R. Casteels, T. Godfraind, and J. C. Rüegg, eds.), pp. 353–357. Elsevier/North Holland, Amsterdam.
Murphy, R. A., and Megerman, J. (1977). *In* "Biochemistry of Smooth Muscle" (N. L. Stephens, ed.), pp. 473–498. Univ. Park Press, Baltimore, Maryland.
Murphy, R. A., Herlihy, J., and Megerman, J. (1974). *J. Gen. Physiol.* **64**, 691.
Needham, D. M. (1971). "Machina Carnis." Cambridge Univ. Press, London.
Needham, D. M., and Williams, J. M. (1963). *Biochem. J.* **89**, 534.
Nonomura, Y. (1968). *J. Cell Biol.* **39**, 741.
Osborn, M., Franke, W. W., and Weber, K. (1977). *Proc. Natl. Acad. Sci. U.S.A.* **74**, 490.
Panner, B. J., and Honig, C. R. (1967). *J. Cell Biol.* **35**, 303.
Parry, D. A. D., and Squire, J. M. (1973). *J. Mol. Biol.* **75**, 33.
Pavlova, I. B., Shestopavlova, N. M., and Raingold, V. N. (1961). *Arch. Anat. Hist. Embriol. (Leningrad)* **15**, 64.
Pease, D. C. (1968). *J. Ultrastruct. Res.* **23**, 280.
Pease, D. C., and Molinari, S. (1960). *J. Ultrastruct. Res.* **3**, 447.
Peng, C. M., Kung, T. H., Hsiung, L. M., and Tsao, T. C. (1965). *Scientia Sinica* **14**, 219.
Perrie, W. T., Smillie, L. B., and Perry, S. V. (1973). *Biochem. J.* **135**, 151.
Pinder, J. C., Bray, D., and Gratzer, W. B. (1975). *Nature (London)* **258**, 765.
Pires, E. M. V., and Perry, S. V. (1977). *Biochem. J.* **167**, 137.
Pires, E., Perry, S. V., and Thomas, M .A. W. (1974). *FEBS Lett.* **41**, 292.
Podlubnaya, Z. A., Tskhovrebova, L. A., Zaalishvili, M. M., and Stefanenko, G. A. (1975). *J. Mol. Biol.* **92**, 357.
Pollard, T. D. (1975). *J. Cell Biol.* **67**, 93.
Potter, J. D. (1974). *Arch. Biochem. Biophys* **162**, 436.
Prosser, C. L., Burnstock, G., and Kahn, J. (1960). *Am. J. Physiol.* **199**, 545.
Rice, R. V., Moses, J. A., McManus, G. M., Brady, A. C., and Blasik, L. M. (1970). *J. Cell Biol.* **47**, 183.
Rice, R. V., McManus, G. M., Devine, C. E., and Somlyo, A. P. (1971). *Nature (London)* **231**, 242.
Rosenbluth, J. (1965). *Science* **148**, 1337.
Rosenbluth, J. (1973). *In* "The Structure and Function of Muscle" (G. H. Bourne, ed.), pp. 389–420. Academic Press, New York.
Rubenstein, P. A., and Spudich, J. A. (1977). *Proc. Natl. Acad. Sci. U.S.A.* **74**, 120.

Russell, W. E. (1973). *Eur. J. Biochem.* **33**, 459.

Schirmer, R. H. (1965). *Biochem. Z.* **343**, 269.

Schollmeyer, J. E., Furcht, L. T., Goll, D. E., Robson, R. M. and Stromer, M. H. (1976). *In* "Cell Motility" (R. Goldman, T. Pollard, and J. Rosenbaum, eds), Vol. A, pp. 361–388. Cold Spring Harbor Laboratory, Cold Spring Harbor, New York.

Scordilis, S. P., and Adelstein, R. S. (1977). *Nature (London)* **268**, 558.

Scordilis, S. P., and Adelstein, R. S. (1978). *J. Biol. Chem.* **253**, 9041.

Sheng, P. K., and Tsao, T. C. (1954). *Acta Physiol. Sinica* **19**, 203.

Sherry, J. M. F., Górecka, A., Aksoy, M. O., Dabrowska, R., and Hartshorne, D. J. (1978). *Biochemistry* **17**, 4411.

Shigekawa, M., and Tonomura, Y. (1972). *J. Biochem. (Tokyo)* **71**, 147.

Shizuta, Y., Shizuta, H., Gallo, M., Davies, P., and Pastan, J. (1976). *J. Biol. Chem.* **251**, 6562.

Shoenberg, C. F. (1969). *Tissue Cell* **1**, 83.

Shoenberg, C. F. (1973). *Phil. Trans. R. Soc. London B.* **265**, 197.

Shoenberg, C. F., and Haselgrove, J. C. (1974). *Nature (London)* **249**, 152.

Shoenberg, C. F., and Needham, D. M. (1976). *Biol. Rev.* **51**, 33.

Simmons, R. M., and Szent-Györgyi, A. G. (1978). *Nature (London)* **273**, 62.

Small, J. V. (1974). *Nature (London)* **249**, 324.

Small, J. V. (1977a). *J. Cell Sci.* **24**, 327.

Small, J. V. (1977b). *In* "Biochemistry of Smooth Muscle" (N. L. Stephens, ed.), pp. 379–411. Univ. Park Press, Baltimore, Maryland.

Small, J. V. (1977c). *In* "Excitation-Contraction Coupling in Smooth Muscle" (R. Casteels, T. Godfraind, and J. C. Rüegg, eds.), pp. 305–315. Elsevier/North Holland, Amsterdam.

Small, J. V., and Celis, J. E. (1978). *J. Cell. Sci.* **31**, 393.

Small, J. V., and Sobieszek, A. (1977a). *J. Cell. Sci.* **23**, 243.

Small, J. V., and Sobieszek, A. (1977b). *Eur. J. Biochem.* **76**, 521.

Small, J. V., and Sobieszek, A. (1977c). *In* "Excitation-Contraction Coupling in Smooth Muscle" (R. Casteels, T. Godfraind, and J. C. Rüegg, eds.), pp. 385–393. Elsevier/North Holland, Amsterdam.

Small, J. V., and Squire, J. M. (1972). *J. Mol. Biol.* **67**, 117.

Sobieszek, A. (1972). *J. Mol. Biol.* **70**, 741.

Sobieszek, A. (1973). *J. Ultrastruct. Res.* **43**, 313.

Sobieszek, A. (1977a). *In* "Biochemistry of Smooth Muscle" (N. L. Stephens, ed.), pp. 413–443. Univ. Park Press, Baltimore, Maryland.

Sobieszek, A. (1977b). *Eur. J. Biochem.* **73**, 477.

Sobieszek, A., and Bremel, R. D. (1975). *Eur. J. Biochem.* **55**, 49.

Sobieszek, A., and Small, J. V. (1973a). *Cold Spring Harbor Symp. Quant. Biol.* **37**, 109.

Sobieszek, A., and Small, J. V. (1973b). *Phil. Trans. Soc. London B.* **265**, 203.

Sobieszek, A., and Small, J. V. (1976). *J. Mol. Biol.* **102**, 75.

Sobieszek, A., and Small, J. V. (1977). *J. Mol. Biol.* **112**, 559.

Somlyo, A. P., Somlyo, A. V., Devine, C. E., and Rice, R. V. (1971). *Nature (London) New Biol.* **231**, 243.

Somlyo, A. P., Devine, C. E., Somlyo, A. V., and Rice, R. V. (1973). *Phil. Trans. R. Soc. London B.* **265**, 223.

Sparrow, M. P., Maxwell, L. C., Rüegg, J. C., and Bohr, D. F. (1970). *Am. J. Physiol.* **219**, 1366.

Sparrow, M. P., and van Bockxmeer, F. M. (1972). *J. Biochem. (Tokyo)* **72**, 1075.

Spudich, J. A., and Watt, S. (1971). *J. Biol. Chem.* **246**, 4866.

Starger, J. M., and Goldman, R. D. (1977). *Proc. Natl. Acad. Sci. U.S.A.* **74**, 2422.

Starger, J. M., Brown, W. E., Goldman, A. E., and Goldman, R. D. (1978). *J. Cell Biol.* **78**, 93.

Steinert, P. M., and Gullino, M. I. (1976). *Biochem. Biophys. Res. Commun.* **70**, 221.

Stigbrand, T., Eriksson, A., and Thornell, L.-E. (1979). *Biochim. Biophys. Acta* **577**, 52.

Stossel, T. P., and Hartwig, J. H. (1975). *J. Biol. Chem.* **250**, 5706.

Stossel, T. P., and Hartwig, J. H. (1976). *J. Biol. Chem.* **68**, 602.

Stromer, M. H., and Goll, D. E. (1972). *J. Mol. Biol.* **67**, 489.

Stull, J. T., and High, C. W. (1977). *Biochem. Biophys. Res. Commun.* **77**, 1078.

Suzuki, A., Goll, D. E., Singh, J., Allen, R. E., Robson, R. M., and Stromer, M. H. (1976). *J. Biol. Chem.* **251**, 6860.

Szent-Györgyi, A. (1975). *Biophys. J.* **15**, 707.

Szent-Györgyi, A. G., Cohen, C., and Kendrick-Jones, J. (1971). *J. Mol. Biol.* **56**, 239.

Szent-Györgyi, A. G., Szent-Kiralyi, E. M., and Kendrick-Jones, J. (1973). *J. Mol. Biol.* **74**, 179.

Takagi, T., Nagai, R., Hotta, K., and Itoh, N. (1976). *J. Electron Microsc.* **25**, 91.

Taxi, J. (1961). *CR. Acad. Sci.* **252**, 331.

Tilney, L. G. (1975). *In* "Molecules and Cell Movement" (S. Inoue and R. E. Stephens, eds.), pp. 339–388. Raven, New York.

Toh, B.-H., Yildiz, A., Sotelo, J., Osung, O., Holborow, E. J., and Kanakoudi, F. (1979). *Clin. Exp. Immunol.* **37**, 76.

Tregear, R. T., and Squire, J. M. (1973). *J. Mol. Biol.* **77**, 279.

Tsao, T. C., Tan, P. H., and Peng, C. M. (1955). *Acta Physiol. Sinica* **19**, 389.

Tsao, T. C., Kung, T. H., Peng, C. M., Chang, Y. S., and Tsou, Y. S. (1965). *Scientia Sinica* **14**, 91.

Uehara, Y., Campbell, G. R., and Burnstock, G. (1971). *J. Cell. Biol.* **50**, 484.

Vandekerckhove, J., and Weber, K. (1978). *Proc. Natl. Acad. Sci. U.S.A.* **75**, 1106.

Vibert, P. J., Haselgrove, J. C., Lowy, J., and Poulsen, F. R. (1972). *J. Mol. Biol.* **71**, 489.

Wachsberger, P., and Kaldor, G. (1971). *Arch. Biochem. Biophys.* **143**, 127.

Wachsberger, P. R., and Pepe, F. A. (1974). *J. Mol. Biol.* **88**, 385.

Wagner, P. D., and Weeds, A. G. (1977). *J. Mol. Biol.* **109**, 455.

Waisman, D. M., Singh, T. J., and Wang, J. H. (1978). *J. Biol. Chem.* **253**, 3387.

Wallach, D., Davies, P. J. A., and Pastan, I. (1978a). *J. Biol. Chem.* **253**, 3328.

Wallach, D., Davies, P. J. A., and Pastan, J. (1978b). *J. Biol. Chem.* **253**, 4739.

Wang, K. (1977). *Biochemistry* **16**, 1857.

Wang, K., and Singer, S. J. (1977). *Proc. Natl. Acad. Sci. U.S.A.* **74**, 2021.

Wang, K., Ash, J. F., and Singer, S. J. (1975). *Proc. Natl. Acad. Sci. U.S.A.* **72**, 4483.

Watterson, D. M., Harrelson, W. G., Keller, P. M., Sharief, F., and Vanaman, T. C. (1976). *J. Biol. Chem.* **25**, 4501.

Weber, A., and Murray, J. (1973). *Physiol. Rev.* **53**, 612.

Weeds, A., Wagner, P., Jakes ,R., and Kendrick-Jones, J. (1977). *In* "Calcium Binding Proteins and Calcium Function" (R. H. Wasserman, R. A. Corradino, E. Carafoli, and R. H. Kretsinger, eds.). Elsevier/North-Holland, Amsterdam.

Whalen, R. G., Butler-Browne, G. S. and Gros, F. (1976). *Proc. Natl. Acad. Sci. U.S.A.* **73**, 2018.

Yagi, K., and Yazawa, M. (1978). *J. Biol. Chem.* **253**, 1338.

Yamaguchi, M., Miyazawa, Y., and Sekine, T. (1970). *Biochim. Biophys. Acta* **216**, 411.

Yazawa, M., and Yagi, K. (1977). *J. Biochem. (Tokyo)* **82**, 287.

Yerna, M. J., Aksoy, M. O., Hartshorne, D. J., and Goldman, R. D. (1978). *J. Cell Sci.* **31**, 411.

INTERNATIONAL REVIEW OF CYTOLOGY, VOL. 64

Cytophysiology of the Adrenal Zona Glomerulosa

GASTONE G. NUSSDORFER

Department of Anatomy, University of Padua, Padua, Italy

I.	Introduction	307
II.	Fine Structure of the Normally Functioning Zona Glomerulosa	308
	A. The Mammalian Zona Glomerulosa	309
	B. The Mammalian Zona Intermedia	321
	C. Zona Glomerulosa or Zona Glomerulosa-like Cells in Lower Vertebrates	323
	D. General Remarks	325
III.	Fine Structure of the Hyperfunctioning and Hypofunctioning Zona Glomerulosa	328
	A. Multifactorial Regulation of the Zona Glomerulosa Functions	328
	B. Stimulation and Inhibition of the Renin–Angiotensin System	329
	C. Alteration of the Na/K Balance	332
	D. Alteration of the Hypophyseal–Adrenal Axis	336
	E. Other Experimental Procedures	339
	F. Aldosterone-Secreting Tumors	341
IV.	Morphological–Functional Correlations in Zona Glomerulosa Cells	342
	A. Subcellular Localization of the Enzymes of Aldosterone Synthesis	343
	B. Mechanisms of Action of the Adrenoglomerulotropic Factors	344
	C. The Functional Significance of the Ultrastructural Changes in Hyperfunctioning and Hypofunctioning Zona Glomerulosa Cells	345
	D. The Mechanism of the Hormone Release by Zona Glomerulosa Cells	350
V.	Zona Glomerulosa and the Maintenance of the Adrenal Cortex	353
	A. Theories on the Cytogenesis in the Adrenal Cortex	353
	B. The Normal Maintenance of the Adrenal Cortex	355
	C. Supporting Evidence for the Cell Migration Theory	355
VI.	Concluding Remarks	359
	References	360
	Note Added in Proof	368

I. Introduction

Many review articles have appeared concerning the cytophysiology of the mammalian adrenal cortex (Symington, 1969; Idelman, 1970, 1978; Neville and McKay, 1972; Malamed, 1975; Nussdorfer *et al.*, 1978a), but none has so far dealt specifically with the zona glomerulosa.

This may be because zona glomerulosa cytophysiology raises some intriguing problems concerning its physiological regulation (e.g., its independence of or dependence on the hypothalamo-hypophyseal axis) as well as its functional significance (e.g., is the zona glomerulosa only a mineralocorticoid hormone-producing layer, or also a site of formation of new adrenocortical cells, thus participating in the normal maintenance of the adrenal gland?). Moreover, our present knowledge of the biochemical mechanism(s) underlying the multifactorial regulation of both the secretory activity and the growth maintenance of the zona glomerulosa is less extended than that of the mechanism(s) controlling the zonae fasciculata–reticularis, which are almost exclusively ACTH-dependent.

However, there are abundant morphological data on the zona glomerulosa in normal and experimental conditions, and in recent years many contributions have thrown light on the biochemical mechanisms regulating zona glomerulosa functions (see Müller, 1971; Davis, 1975; and Vinson and Kenyon, 1978, for comprehensive references). In addition some investigations, mainly employing stereological techniques, have attempted to correlate morphological and biochemical findings in the zona glomerulosa.

The aim of this article is to review the morphological data on the zona glomerulosa and to try to correlate them with the more recent biochemical findings, in order to provide a basis for further research on the cytophysiology of the zona glomerulosa.

II. Fine Structure of the Normally Functioning Zona Glomerulosa

The mammalian adrenal cortex shows a clear histologic zonation: the outer subcapsular layer, the zona glomerulosa, consists of many cords of irregular-shaped cells, which are arranged in various patterns according to the species examined (Deane, 1962). Zona glomerulosa cells possess the distinctive ultrastructural features of the steroid-producing cells: (1) mitochondria with tubular cristae, although in many species only lamellar or plate-like cristae are observed; (2) a well developed smooth endoplasmic reticulum (SER); and (3) some lipid droplets (Christensen and Gillim, 1969; Fawcett et al., 1969; Idelman, 1970, 1978; Malamed, 1975; Nussdorfer et al., 1978a). However, there are some differences in morphology among the various species. In some mammalian species a lipid-free layer between zona glomerulosa and zona fasciculata has been described and named zona intermedia (see Deane, 1962, for review).

In the following subsections, I shall review the fine structure of the normally functioning zona glomerulosa of various mammalian species as a basis for a clearer understanding of its cytophysiology and discuss the morphology of the zona intermedia cells. Finally, I shall provide a brief account of the evidence suggesting that there is histological zonation or, at least, zona glomerulosa-like cells, in the interrenal gland of some lower vertebrate species.

A. The Mammalian Zona Glomerulosa

As nearly half of the research on the ultrastructure of the mammalian adrenal cortex concerns the rat (Nussdorfer *et al.*, 1978a), I shall first give a full description of the zona glomerulosa in this animal, and then a brief comparative description of this zone in other species.

1. *Rat*

A comprehensive review of the pioneering works on the ultrastructure of the rat adrenal zona glomerulosa was given by Idelman (1970). Some ultrastructural stereological studies providing the baseline data for the zona glomerulosa cells of some rat strains are also available (Frühling *et al.*, 1973; Nussdorfer *et al.*, 1973, 1974b, 1977c; Lustyik and Szabò, 1975; Rohr *et al.*, 1975, 1976; Kasemsri and Nickerson, 1976; Nickerson, 1976, 1977; Mazzocchi *et al.*, 1977).

The zona glomerulosa consists of layers of irregularly arranged cells and occupies about 10–15% of the gland volume. The cell volume averages 600–750 μm^3, according to the strain. The nuclei are round or oval with an evident nucleolus (Rhodin, 1971), and some mitoses can occasionally be observed, especially in the inner portion of the zone (see Section V).

The mitochondria are numerous and occupy about 25–30% of the cytoplasmic volume. They are rather elongated with mainly tubular cristae whose surface per cell averages 2000–3800 μm^2 (Kasemsri and Nickerson, 1976; Nickerson, 1976; Rohr *et al.*, 1976; Mazzocchi *et al.*, 1977). According to many investigators, zona glomerulosa mitochondria show noticeable pleomorphism, especially as far as their cristal pattern is concerned. In the subcapsular cells mitochondria are ovoid or elongated and contain regular arrays of tubular cristae that frequently run parallel to the long axis of the organelle (Wassermann and Wassermann, 1974; Mazzocchi *et al.*, 1977). In cross section they appear as a hexagonal array of circles of about 225 Å in diameter that have a center-to-center spacing of about 300 Å (Fig. 1) (Giacomelli *et al.*, 1965; Propst and Müller, 1966; Wheatley, 1968; Frühling, 1977; Mazzocchi *et al.*, 1977). In the outer zona glomerulosa cells, the mitochondria are elongated and contain short plate-like cristae whose length rarely fills the central intramatrical portion of the organelle (Fig. 2) (Wassermann and Wassermann, 1974; Frühling, 1977; Mazzocchi *et al.*, 1977). In the cells of the inner zona glomerulosa, the mitochondria are more regular in shape (ovoid or cylindric) and display tubular cristae and a few dense intramatrical granules (Fig. 3) (Frühling, 1977; Mazzocchi *et al.*, 1977). Some organelles, except those in the subcapsular cells, contain amorphous lipid-like intramatrical bodies (Friend and Brassil, 1970; Wassermann and Wassermann, 1974; Frühling, 1977), while intramatrical paracrystalline inclusions are exceptional (Frühling *et al.*, 1968); on the contrary, they are frequently seen in the zona fasciculata mitochondria (see Nussdorfer *et al.*, 1978a, for review).

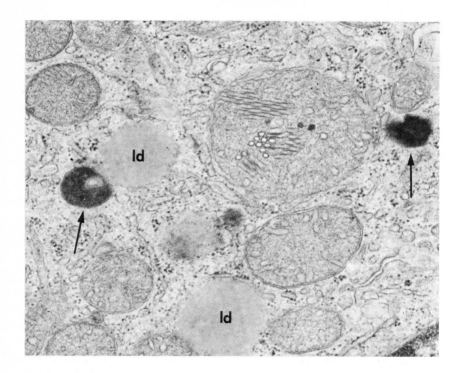

FIG. 1. Cell of the outer portion of the rat zona glomerulosa, showing a mitochondrion that contains clusters of parallel cristal tubules. ld, Lipid droplets; the arrows indicate two lyosomes. ×34,000. (From Mazzocchi *et al.*, 1977.)

The SER is abundant, occupying about 35–40% of the cytoplasmic volume; it is in the form of a network of anastomosing branching tubules that frequently surround the mitochondria and the lipid droplets (Fig. 3). It seems to be more developed in the cells of the middle-inner portions of the zona glomerulosa than in the subcapsular ones. Sparse rough endoplasmic reticulum (RER) profiles and numerous free ribosomes have been observed (Nussdorfer *et al.*, 1973, 1974b; Frühling, 1977), although Rhodin (1971) reports RER to be absent.

Lipid droplets occupy about 6–10% of the cytoplasmic volume; again, a noticeable inhomogeneity in the lipid droplet content has been described in the zona glomerulosa cell population (Lustyik and Szabò, 1975; Frühling, 1977): lipid droplets seem to be more abundant in the inner zona glomerulosa cells.

The Golgi apparatus is always present: small and sparse according to Rhodin (1971); well-developed according to Nussdorfer *et al.* (1973), Wassermann and Wassermann (1974), and Frühling (1977). It consists of many stacks of cisternae, usually in a juxta-nuclear location, and of numerous vesicles, some of which, having a "coated" appearance, seem to arise from the dilated endings of

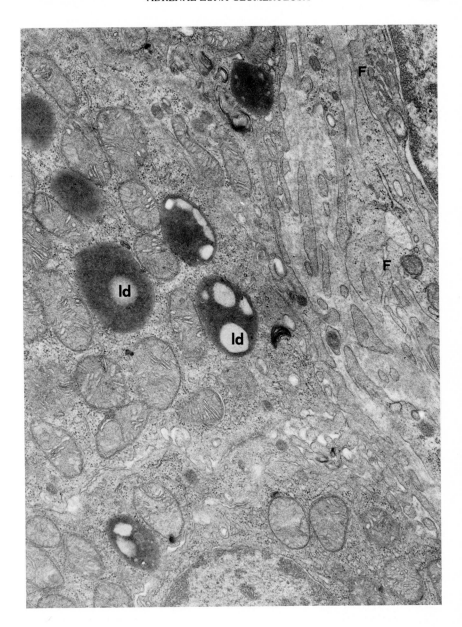

FIG. 2. Subcapsular cells of the rat zona glomerulosa. Mitochondria contain laminar cristae. SER profiles are very scarce, and free ribosomes are abundant. ld, Lipid droplets; F, capsular fibroblast. ×21,000. (From Nussdorfer *et al.*, 1973.)

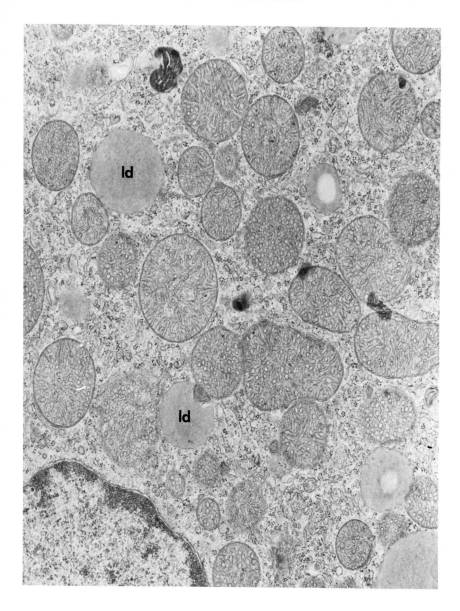

FIG. 3.   Cell of the inner portion of the rat zona glomerulosa. Round or elongated mitochondria display tubular or tubulo-vesicular cristae. ld, Lipid droplets. ×20,000. (From Mazzocchi *et al.*, 1977.)

the golgian saccules (Fig. 4). A few β-glycogen particles are scattered in the cytoplasm (Wassermann and Wassermann, 1974; Frühling, 1977). Electron-dense bodies of a lysosomal and peroxisomal nature have been seen (Frühling, 1977). Gabbiani *et al.* (1975) report that zona glomerulosa cells contain, in addition to microtubules, a network of 40–80 Å thick microfilaments just beneath the plasma membrane, and by the use of immunofluorescent techniques they demonstrate that this network is made up of actin filaments.

The cells display a few short microvilli and some coated pits (caveolae) at their plasma membrane (Fig. 4) (Rhodin, 1971; Nussdorfer *et al.*, 1973, 1974b; Palacios and Lafarga, 1976), as well as cell-to-cell attachments that are completely analogous to those found in the zona fasciculata cells (see Nussdorfer *et al.*, 1978a, for review). According to Friend and Gilula (1972), they include tight junctions (*zonulae occludentes*), small desmosomes (*maculae adhaerentes*), intermediate junctions (*zonulae adhaerentes*), gap junctions (*nexus*), and septate-like zonulae adhaerentes. Some researchers (Propst and Müller, 1966; Nussdorfer, 1970) also describe occasional rudimentary cilia of the 9+0 fiber arrangement associated with paired centrioli (Wheatley, 1967).

2. Man

The present account is based on the works of Long and Jones (1967b), Kawaoi (1969), and on Symington's review (1969), as well as on personal observations on the adult gland. Hashida and Yunis (1972) described zona glomerulosa of two children, but their description does not significantly differ from that of the previous investigators.

The cells (Fig. 5) contain round or elongated mitochondria with tubular or lamellar cristae. SER tubules are well represented, and RER is better developed than in the zona glomerulosa cells of the other species examined. Free ribosomes are plentiful and lipid droplets are quite numerous. In the cytoplasm there are many electron-dense bodies that according to Magalhães (1972), are lysosomes since they display acid phosphatase activity. A well developed Golgi apparatus and a few scattered β-glycogen particles are also present. Microvilli seem to be an exceptional feature of these cells.

3. Monkey

No significant differences were observed between zona glomerulosa cells of the two species examined: rhesus monkey (Brenner, 1966) and squirrel monkey (Penney and Brown, 1971).

The cells show an oval or scalloped nucleus. The mitochondria are elongated, with the cristae generally displaying a laminar pattern. In the squirrel monkey, SER is well developed and RER is scarce, whereas in the rhesus monkey the contrary is observed. There are abundant free ribosomes and many lipid droplets. The Golgi apparatus is prominent and numerous dense bodies of probable

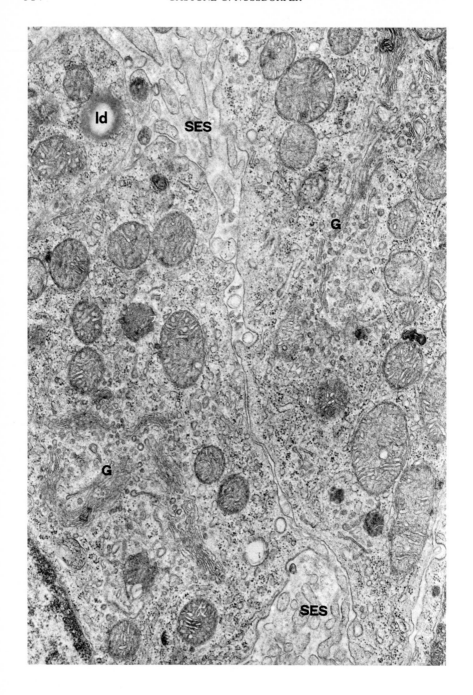

lysosomal nature can be seen in the Golgi area. The microvilli project into the subendothelial space.

4. Dog

The ultrastructure of the dog adrenals was described by Luse (1967) and Bloodworth and Powers (1968), who unfortunately used only osmic acid fixation in their investigations, so that some typical preservation artifacts (e.g., SER vesiculation, dark cells) can be observed in their electron micrographs.

The cells have a small oval nucleus with dense clumped chromatin. The mitochondria are numerous, frequently elongated, and their inner membrane is arranged in tubular or laminar cristae. SER is well developed, whereas RER is absent. There is a moderate amount of lipid droplets. A juxta-nuclear Golgi apparatus and numerous free ribosomes can be seen. Membrane-bound electron-dense bodies (1000–4000 Å in diameter) were found and interpreted by Bloodworth and Powers (1968) as microbodies. Cell-to-cell attachments and some rare finger-like microvilli have also been observed.

5. Ox

There are very few observations on the ox adrenal cortex (Frühling et al., 1971, 1973). The cells show spherical or irregular mitochondria with lamellar cristae. SER is well developed, and rare, sparse RER cisternae along with free ribosomes can be observed. Lipid droplets are very scarce, occupying no more than 0.6% of the cell volume (Frühling et al., 1973). The Golgi apparatus is well represented.

6. Rabbit

Only pioneering works suffering from inadequate fixation and embedding procedures are available (see Nussdorfer et al., 1978a, for references); therefore, this description is based on personal unpublished observations.

The cells (Fig. 6) contain an oval or scalloped nucleus. The mitochondria are ovoid or rather elongated and their cristal arrangement shows noticeable pleomorphism: in the subcapsular cells the mitochondria display poor plate-like cristae, whereas in the inner zona glomerulosa cells the mitochondrial cristae are usually of the tubular type. SER is very well developed, there are numerous free ribosomes, but RER profiles are completely absent. Lipid droplets are abundant, especially in the cells of the inner portion of the zone. The plasma membranes of adjacent cells show numerous desmosome-like attachment devices; frequently

---

FIG. 4. Cells of the middle portion of the rat zona glomerulosa. The mitochondrial cristae are of the tubulo-vesicular type. SER tubules and the Golgi apparatus (G) are quite well developed. Numerous microvilli protrude into the subendothelial space (SES). ld, Lipid droplet. ×21,000. (From Nussdorfer et al., 1974b.)

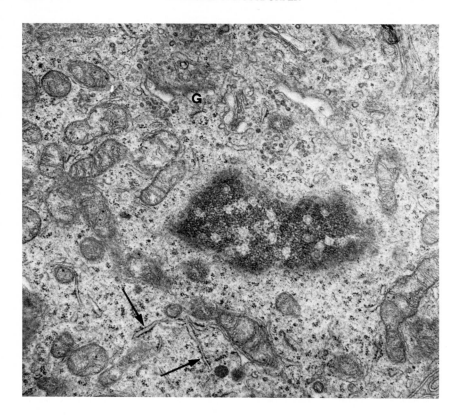

Fig. 5. Portion of a cell from the human zona glomerulosa. Elongated mitochondria contain plate-like cristae. SER is scarce, but free ribosomes are abundant. The arrows indicate RER profiles. G, Golgi apparatus. ×20,000.

numerous intercellular canaliculi, into which a few short microvillous specializations project, can be observed; intercellular canaliculi seem to open in the subendothelial spaces (SES). Another typical feature are clumps of electron-dense granules, which cytochemically do not display positive reaction to acid phosphatase, located at the juxta-sinusoidal poles of the cells and near the intercellular canaliculi. These granules seem to arise in the Golgi area.

7. Mouse

Although the fine structure of the mouse adrenal cortex has been described in several contributions (see Nussdorfer et al., 1978a, for references), we shall follow here the work by Shelton and Jones (1971) and the stereological description by Nickerson (1975).

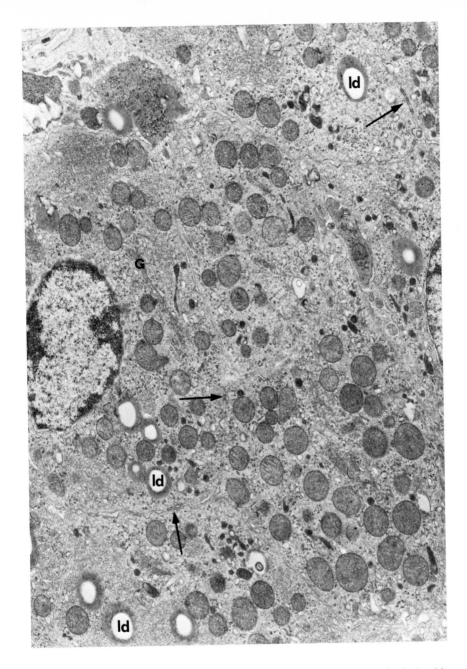

FIG. 6.   Low-power electron micrograph of the rabbit zona glomerulosa. Round mitochondria show tubular and lamellar cristae. SER is quite well represented. The arrows indicate cell-to-cell attachments. Note several clumps of electron-dense bodies at the cell periphery. G, Golgi apparatus; ld, lipid droplets. ×12,000.

The zona glomerulosa consists of many clusters of cells with scarce cytoplasm, whose volume averages 1200 μm³. The mitochondria are round or elongated and contain laminar or tubulo-vesicular cristae (Fig. 7); they occupy about 26% of the cell volume. SER tubules are present in a moderate amount and occupy no more than 35% of the cell volume; occasional profiles of RER and numerous free ribosomes can be observed. Lipid droplets are scarce (about 8% of the cell volume). The juxta-nuclear Golgi apparatus is usually small. Other features are microvilli projecting into the SES, coated pits, and cell-to-cell attachments between contiguous parenchymal cells.

8. *Guinea Pig*

The zona glomerulosa of this species was described by Sheridan and Belt (1964) and by Frühling's group (Frühling *et al.*, 1973; Sand *et al.*, 1973), who also employed stereological techniques.

The cells have spherical or ovoid mitochondria with typical lamellar cristae. SER is exceedingly well represented, occupying nearly 42% of the cell volume, and RER stacks are quite numerous. Free ribosomes and polysomes are abundant. Lipid droplets are very scarce (about 3.7% of the cell volume). The Golgi apparatus is well developed.

9. *Hamster*

No recent contributions are available: this description is based on works by DeRobertis and Sabatini (1958), Belt (1960), Cotte *et al.* (1963), and Yonetsu (1966).

The cells display elongated mitochondria with laminar cristae, abundant SER, some stacks of RER, and few lipid droplets. The Golgi apparatus is obvious. Some electron-dense bodies of lysosomal nature and rare microvilli can be observed.

10. *Mongolian Gerbil*

The fine structure of the adrenal cortex in this species was described by Nickerson (1972 ,1972) and by Kadioglu and Harrison (1975).

The cells are small and contain elongated mitochondria with a few lamellar or plate-like cristae (Fig. 8). SER is quite well developed and there are some scattered RER profiles along with numerous free ribosomes. The juxta-nuclear Golgi apparatus is prominent, with abundant vesicles arising from its cisternae. Few lipid droplets and lysosomes can be seen. Microtubules are frequently encountered.

11. *Opossum*

According to Long and Jones (1967a, 1970), zona glomerulosa cells of the opossum show short rod-like mitochondria with lamelliform cristae and dense granules in the matrix. SER tubules are plentiful and RER profiles are sparse in

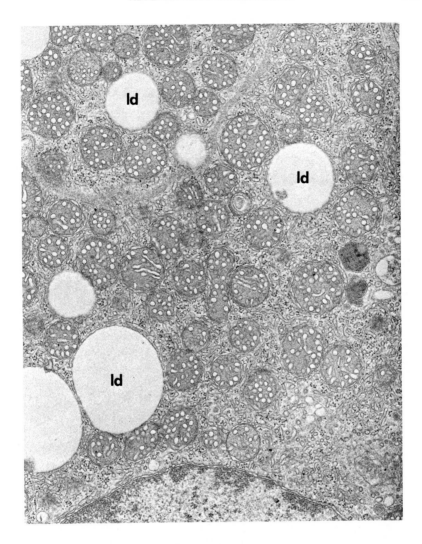

F ɪ ɢ. 7.   Cells of the mouse zona glomerulosa. Mitochondria are circular and contain tubulo-vesicular cristae. SER tubules are dispersed among the cell organelles. ld, Lipid droplets. ×20,000. (From Nickerson, 1975.)

the cytoplasm; in many instances SER/RER continuity is observed. Many clusters of free ribosomes and several lipid droplets are present. The juxta-nuclear Golgi apparatus is well developed and contains many vesicles, some of which are coated. Lysosomes are quite numerous. Other features are microvilli, coated pits, and poorly differentiated desmosomes.

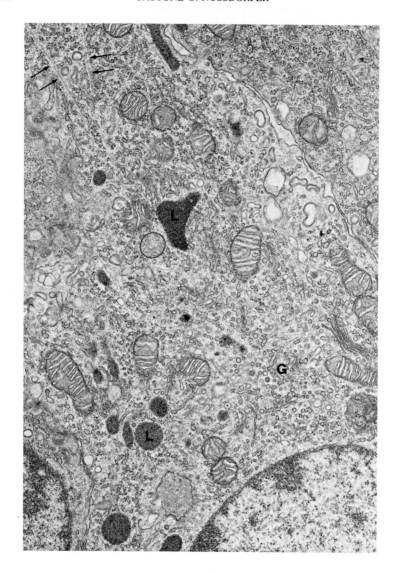

Fig. 8. Zona glomerulosa cells from a Mongolian gerbil adrenal cortex. Mitochondria contain lamelliform cristae. Numerous vesicles are present in the Golgi apparatus (G). SER profiles and lysosomes (L) are dispersed throughout the cytoplasm. The arrows indicate microtubules. ×12,000. (From Nickerson, 1971.)

12. Other Species

Only sporadic observations are available. Unsicker (1971) studied the innervation of the *pig* adrenal cortex by electron microscopy. He made unpublished observations on the zona glomerulosa of this species. The cells seem to be small and irregular in shape. The mitochondria are ovoid and show plate-like cristae. SER is not well developed, but RER profiles and free ribosomes are abundant. The Golgi apparatus is well developed and contains many coated vesicles. Lipid droplets are not seen in his electron micrographs (Unsicker, personal communication).

Luthman (1971) described intramitochondrial bodies in the zona glomerulosa cells of the *sheep*. These bodies are always in the matrix away from the cristae and show more electron-density than lipid droplets, so that they were interpreted as proteinaceous in nature. Romagnoli (1972) showed that zona glomerulosa cells of the *hedgehog* contain spherical mitochondria with long, parallel tubular cristae, SER tubules, and lipid droplets.

B. The Mammalian Zona Intermedia

In the earlier investigations on the histological zonation of the mammalian adrenal cortex, a sudanophobic layer between the zonae glomerulosa and fasciculata was recognized in many species and named zona intermedia (see Deane, 1962, for review). Electron microscopic investigations have provided a description of the zona intermedia in only a few species: the rat (Ito, 1959; Frühling and Claude, 1968; Yoshimura *et al.*, 1968; Nickerson, 1976; Frühling, 1977), the dog (Bloodworth and Powers, 1968), and the mouse (Shelton and Jones, 1971). It seems worthwhile to present a brief account of the ultrastructure of the zona intermedia as a basis for further discussion on its possible significance (see Section V,A).

1. Rat

The zona intermedia consists of three to five layers of small cells containing a regularly ovoid nucleus and scarce cytoplasm (Fig. 9). The mitochondria are small and pleomorphic (round, annular, or elongated) and display tubulo-convolute cristae occupying all the intramatrical space. Occasional intramatrical paracrystalline inclusions have also been observed (Frühling, 1977). SER is very well developed, and so are RER profiles. According to Frühling (1977), this is the only adrenocortical layer in the rat with cells showing some classic RER cisternae. Free ribosomes and polysomes are present. The most striking feature is the virtual absence of lipid droplets (1.4% of the cell volume) (Frühling, 1977).

Nickerson (1976) stereologically described in the Wistar rats a lipid-free subglomerulosa zone, containing small cells, whose volume is quite similar to that of the zona glomerulosa elements (about 900 μm^3). The volume of the mitochondrial compartment and the morphology of organelles are analogous to those of the zona glomerulosa cells, whereas the volume and surface of SER are of the same order of magnitude as those of the zona fasciculata cells. Nickerson maintains that these are transitional cells between zona glomerulosa and zona fasciculata elements.

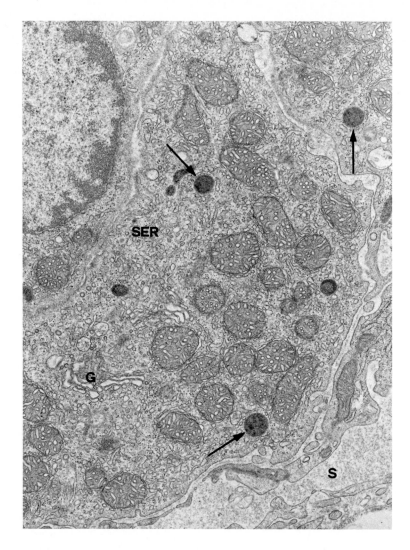

FIG. 9.   Cells from the rat zona intermedia. Mitochondria show tubulo-vesicular cristae. SER profiles are abundant. Note the complete absence of lipid droplets and the presence of numerous lysosomes (arrows). G, Golgi apparatus; S, sinusoid. ×16,000. (From Nickerson, 1976.)

2. *Dog*

The cells show an irregularly shaped nucleus and numerous pleomorphic mitochondria with plate-like and vesicular cristae. There are plenty of SER and free ribosomes. Neither the Golgi apparatus nor RER profiles are observed. Lipid droplets are present in a very moderate amount.

3. *Mouse*

The cells display spherical or elongated small mitochondria with a few short dilated lamellar cristae. SER is poorly developed. RER is lacking, whereas free ribosomes are numerous. Lipid droplets can be observed only occasionally. The Golgi apparatus is small and fragmented. Microvilli and cell-to-cell attachments are usually absent.

C. Zona Glomerulosa or Zona Glomerulosa-like Cells in Lower Vertebrates

The existence of a true zonation in the interrenal glands of the lower vertebrates is generally denied. However, I shall review here some lines of evidence that suggest that a zonal arrangement or at least two distinct cell types involved in gluco- or mineralocorticoid hormone production can be recognized in the interrenal tissue of some orders.

1. *Aves*

In birds, the glands are enclosed by a lose connective tissue capsule and are formed by cell strands arranged in radial double cords, surrounded by sinusoids, with well intermingled clusters of chromaffin elements (see Holmes and Phillips, 1976, for review). A clear histological zonation, superposable to that found in the mammalian adrenals, has been reported only in *Pelecanus occidentalis,* but conspicuous ultrastructural regional differences in the interrenal cells were not observed (Sheridan *et al.,* 1963).

In the interrenal glands of many avian species (*Columba livia, Gallus domesticus, Anas platyrhynchos*) two zones can be recognized: an outer subcapsular zone and an inner zone (Kondics, 1965; Haack *et al.,* 1972; Unsicker, 1973). On the grounds of some ultrastructural features, Kondics and Kjaerheim (1966), Kjaerheim (1968a, 1969), and Unsicker (1973) claim that the outer subcapsular cells are poorly differentiated elements that are not able to secrete steroid hormones. Haack *et al.* (1972), on the contrary, showed numerous mitoses in the inner zone of the interrenal gland of the duck and hypothesized that it contains undifferentiated stem cells that migrate toward the periphery, where they differentiate, synthesize and secrete steroid hormones, and then die. Some evidence also indicates that the outer subcapsular zone atrophies following high NaCl intake (Péczely, 1972) and hypertrophies following Na^+ depletion (Taylor *et al.,* 1970), whereas it does not display any evident change after alteration of the hypophyseal-adrenal axis (hypophysectomy, steroid-hormone, or ACTH exogenous administration) (Boissin, 1967; Péczely, 1972). These last findings suggest that the subcapsular zone may be associated with mineralocorticoid activity (see Section III,A). More recent results, however, demonstrate that noticeable ultrastructural differences between inner and outer interrenal cells do not exist, at least in *Anas platyrhynchos* (Cronshaw *et al.*, 1974; Bhattacharyya, 1975a; Pearce *et al.*, 1977), *Coturnix coturnix japonica* and *Columba livia* (Bhattacharyya, 1975b). According to Kjaerheim (1968b,c) and Bhattacharyya *et al.* (1975a,b), both cell types are able to respond to the alteration of the electrolytic balance and of the hypothalamo-hypophyseal axis. Recently, however, Pearce *et al.* (1978) stereologically described zonation in the interrenal gland of the duck after hypophysectomy and treatment with corticosterone: the cells in the inner portion atrophy and the plasma corticosterone declines, whereas no changes can be seen in the outer layer elements.

2. *Reptilia*

No evidence for histological zonation is reported in the interrenal glands of reptiles (Deane, 1962). However, Del Conte (1972a,b, 1975) observed two zones in the lizard interrenals: an inner layer made up of large "active" cells, and a peripheral one, consisting of small, "less active" cells that are greatly sensitive to ACTH (reactive zone). Quite analogous findings were reported in some snake species (Lofts and Phillips, 1965). For a comprehensive review, see Lofts (1978).

3. *Amphibia*

In neither of the two amphibian orders were structural zonation of the interrenal glands (Deane, 1962) or two distinct cell types described, at least in normal conditions (see Nussdorfer *et al.*, 1978a, for review). However, it seems quite well established that the amphibian interrenal tissue produces both corticosterone and aldosterone (see Sandor *et al.*, 1976, for review).

More recently, Varna (1977) has presented evidence that the homogeneous cell population of the *Rana catesbeiana* interrenal glands differentiates according to the experimental treatment employed. By maintaining the frogs in distilled water or by treating them with spironolactone (see Section III,A), in addition to the typical cells containing mitochondria with tubulo-vesicular cristae, some cellular elements appear displaying mitochondria with lamellar cristae. Varna concludes that this can be considered as ultrastructural proof that corticosterone and aldosterone are secreted by different cell types in the bullfrog. However, Berchtold (1973, 1975) demonstrated that, in the urodela *Triturus cristatus,* the interrenal cells exhibit ultrastructural changes, suggestive of an increased function (see Nussdorfer *et al.*, 1978a, for review), in response to both ACTH and 0.5% KCl. An excellent review on this subject is available (Hanke, 1978).

4. *Chondrichthyes and Osteichthyes*

No histological zonation is found in the interrenal tissue of fishes (Deane, 1962). The only exception is reported by Taylor *et al.* (1975), who described, in the interrenal gland of the selachiian shark *Ginglymostoma cirratum,* ultrastructural regional differences that allowed the division of the gland into three concentric zones: an external germinative zone, an intermediate mature cell zone, and an internal degenerating cell zone. The cells of the two inner layers show the typical appearance of actively steroid-producing elements, since they contain elongated or spherical mitochondria with tubulo-vesicular cristae, extensive SER, numerous free ribosomes and polysomes, and abundant lipid droplets. The external germinative zone consists of cells having a spherical nucleus with an evident nucleolus, spherical or elongated mitochondria with short lamellar cristae, an obvious Golgi apparatus, and many free ribosomes. Characteristically, SER and lipid droplets are scarce or completely absent. Whether this zone is only a reservoir of new interrenal cells or is involved in steroid hormone secretion is not settled at present. It is to be stressed, however, that the shark's interrenals do not seem to secrete aldosterone, but only deoxycorticosterone (DOC) and corticosterone (see Sandor *et al.,* 1976, for review).

In this connection, it is to be recalled that the interrenal cells of the teleostean *Carassius auratus* show signs of hyperfunction following alteration of the electrolytic environment (Ogawa, 1967).

5. *Agnatha*

In the cyclostome species examined (*Petromyzon marinus* and *Lampetra fluviatilis*), the interrenal tissue is located both in the pronephros and in the opisthonephric kidney (Deane, 1962). Hardisty (1972a,b) and Hardisty and Baines (1971) studied the fine structure of the presumptive interrenal cells located in the pronephros, while Youson (1972) described that located in the opisthonephros. Their results are quite analogous; but Youson reports mitochondria with tubulo-vesicular cristae, and Hardisty and Baines describe organelles with tubular cristae. ACTH induces hypertrophy and hyperplasia in the opisthonephric interrenal tissue of *Petromyzon marinus* (Youson, 1973) but not hyperplasia in the pronephric interrenal tissue of both *Lampetra fluviatilis* (Hardisty, 1972b) and *Petromyzon marinus* (Youson, 1973). Hardisty and Baines (1971) claim that lamprey pronephric interrenal cells closely resemble those of the mammalian zona glomerulosa.

D. GENERAL REMARKS

From the above survey it can be concluded that the mammalian zona glomerulosa cells, at variance with those of the zona fasciculata (see Nussdorfer *et al.,* 1978a for review), do not display conspicuous ultrastructural differences among the various species. This seems quite easily explicable if one takes into account that zona glomerulosa in all the mammals is involved in the aldosterone secretion, whereas zona fasciculata produces various types of glucocorticoid hormones, according to the species considered (see Sandor *et al.*, 1976, for review).

To summarize, zona glomerulosa cells contain mitochondria with predominantly plate-like cristae (though organelles showing tubular cristae are found in the rat, man, dog, and rabbit), a well developed SER, and various amounts of lipid droplets. The significance of these features in relation to steroid synthesis was discussed in a previous review article (Nussdorfer *et al.*, 1978a). However, the absence of the vesicular cristal arrangement in the zona glomerulosa mitochondria must be briefly discussed. In fact, these organelles contain many enzymes of the aldosterone synthetic pathway, a part of which is the same as that involved in corticosterone and cortisol production by zona fasciculata cells whose mitochondria display vesicular cristae (Section IV,A). This observation is in contrast with the view that the vesicular cristal configuration is the morphological counterpart of the peculiar enzymatic content of adrenocortical mitochondria that is hypothesized to be the classic chain coupling the oxidative electron transfer chain to the ATP synthetic apparatus contained on the tubular or lamellar cristae and to the cytochrome P_{450} electron transfer chain associated with the steroid hydroxylating enzymes and located on the vesicular cristae (Allmann *et al.*, 1970; Wang *et al.*, 1974; Pfeiffer *et al.*, 1976). It is conceivable that the vesicular configuration of the mitochondrial cristae reflects not qualitative, but only quantitative, aspects of their enzymatic content. In fact, like the classic respiratory chain, the cytochrome P_{450} electron transfer chain requires an adequate steric arrangement for complete activity of the steroid hydroxylating enzymes, and the vesicular arrangement is the geometric configuration allowing the greatest concentration of cristal membranes in the unit mitochondrial volume (stereological demonstration of this was done by Nussdorfer *et al.*, 1977a). In support of this last view there are stereological and biochemical findings from Armato *et al.* (1978) showing that dedifferentiated human adrenocortical cells in primary tissue culture, containing mitochondria with lamelliform cristae, are able to secrete moderate amounts of cortisol in the growth medium and that the ACTH-induced reorganization of mitochondrial cristae into vesicles is coupled only with a significant increase in the hormone output. Similar results were reported in 19-day-old rat embryos by Manuelidis and Mulrow (1973).

In the mammalian zona fasciculata cells, Frühling and associates (1973) stereologically found an inverse correlation between the volume of the lipid compartment and the amount of SER tubules. The authors hypothesize that the SER-synthesized endogenous cholesterol is directly employed in the hormone synthesis, whereas exogenous cholesterol derived from the blood stream is esterified and stored in the lipid droplets (for a complete review of this point, see Nussdorfer *et al.*, 1978a). Stereological data from Frühling's group (Frühling *et al.*, 1973; Frühling, 1977) suggest that such an inverse correlation may subsist also in the zona glomerulosa, but further studies aimed at evaluating the differential capacity of the three adrenal zones for cholesterol endogenous synthesis are needed to settle this point.

Other subcellular features of the zona glomerulosa cells (the Golgi apparatus, lysosomes, microtubules and microfilaments, coated pits, and cell-to-cell attachments including septate-like zonulae adhaerentes) do not differ from those of the zona fasciculata cells (see Nussdorfer *et al.*, 1978a, for review). However, the following differences exist between these two cell types: zona glomerulosa cells of all the mammalian species examined, except the dog and the rabbit, contain RER profiles along with many free ribosomes, whereas true RER cisternae are reported only in the human zona fasciculata cells (Long and Jones, 1967b; Kawaoi, 1969); paracrystalline and lipid-like intramatrical inclusions, which are very abundant in the zona fasciculata cell mitochondria (see Nussdorfer *et al.*, 1978a), are only an exceptional finding in the zona glomerulosa cells; some zona glomerulosa mitochondria in the rat display a peculiar paracrystalline arrangement of otherwise normal tubular cristae. The significance of these differences between zona glomerulosa and zona fasciculata cells is completely unknown.

Another point is worth discussing: rat zona glomerulosa cells show structural differences according to their location in the zone, and quite analogous findings were reported in the rabbit. It appears that outer cells show a poorly differentiated appearance, while the inner ones display the typical features indicating their active engagement in steroid synthesis (i.e., mitochondria with tubular or vesicular cristae, well developed SER, and abundant lipid droplets). Whether or not these results can be extended to the zona glomerulosa of other mammalian species is still an open question. In any case, it is conceivable that most workers have observed and described only the outer zona glomerulosa cells since the gland connective capsule is the more easily recognizable landmark of this zone in the thin section for the electron microscope. This would explain why, in the zona glomerulosa cells of the various species, were found, almost exclusively, mitochondria with lamellar or plate-like cristae. It is to be emphasized that the settlement of this point will possibly contribute to the resolution of the problems concerning the significance of the zona glomerulosa in the adrenal cortex cytogenesis (see Section V).

The zona intermedia requires further accurate morphologic research: in fact, apart from the lack of lipid droplets, these cells seem to show noticeable pleomorphism among the species examined. The possible significance of the zona intermedia will be discussed in Section V.

The evidence reviewed in Section II,C excludes the possibility that a true histological zonation occurs in the interrenal gland of the lower vertebrates. However, the submammalian interrenal cells are able to respond both to the stimuli that are commonly thought to affect mammalian zona glomerulosa and those enhancing zona fasciculata activity (Section III,A). Whether the same interrenal cell is able to respond to both kinds of stimuli or whether there are two cell types, otherwise morphologically indistinguishable, which are each responsive to only one kind of stimulus, is a fascinating but still unresolved problem.

III. Fine Structure of the Hyperfunctioning and Hypofunctioning Zona Glomerulosa

In this section the ultrastructural changes induced in the zona glomerulosa by various experimental or pathological conditions will be reviewed. In addition, a brief account of the current concepts on the regulation of the growth and hormonal secretion of the zona glomerulosa will be provided, as a basis for an easier understanding of the experimental morphological data.

A. MULTIFACTORIAL REGULATION OF THE ZONA GLOMERULOSA FUNCTIONS

On this subject many excellent review articles have appeared, to which the readers can refer for comprehensive references and discussion (Gláz and Vecsei, 1971; Müller, 1971; Davis, 1975).

There are decisive proofs that the regulation of aldosterone secretion by the mammalian and sub-mammalian adrenal gland is chiefly mediated by the renin–angiotensin system, the plasma concentration of sodium and potassium, ACTH, and serotonin. Each of these adrenoglomerulotropic factors seems to promote aldosterone synthesis by a direct effect on the zona glomerulosa, as is well demonstrated by several *in vitro* experiments. According to Müller (1971) and Müller and Baumann (1974), angiotensin II, ACTH, and serotonin act only in the first steps of aldosterone synthesis (i.e., the conversion of cholesterol to pregnenolone), sodium deficiency acts only in the final steps (i.e., the conversion of DOC to aldosterone, via corticosterone), whereas potassium ions modulate both steps. This last finding was confirmed in bovine adrenal cell suspension by the use of various inhibitors of the steroid synthesis (McKenna *et al.,* 1978a). In contrast, Tan and Mulrow (1978) reported that the synthesis of 18OH-DOC is under the influence of ACTH but not of the renin–angiotensin system. Moore *et al.* (1978) demonstrated, however, that the synthesis of this steroid is regulated by ACTH, angiotensin II and K^+. McKenna and co-workers (1978b) showed that angiotensin II stimulates both pregnenolone formation and the conversion of DOC to aldosterone (for the aldosterone synthesic pathway and the subcellular localization of the enzymes involved, see Section IV,A).

Interrelationships among the various adrenoglomerulotropic factors *in vivo* cannot be disregarded since alterations of the Na/K balance may stimulate or inhibit the renin–angiotensin system (see Davis, 1975, for review), and ACTH was found to induce renin release by the kidney juxta-glomerular apparatus (Hauger–Klevene *et al.*, 1969, 1970; Winer *et al.*, 1971). On these grounds, the hypothesis of the predominance of the renin–angiotensin system in the physiological control of the aldosterone secretion has been advanced. However, on the basis of the evidence available in the human, dog, sheep and rat, Müller (1971) has seriously questioned this view. In addition, several investigators have proposed the unifying hypothesis that all the adrenoglomerulotropic factors act by increasing the intracellular concentration of potassium, which thus might be considered a "final common pathway" in the regulation of aldosterone production (Baumberg *et al.*, 1971; Davis, 1972; Boyd *et al.*, 1973). However, findings by Szalay *et al.* (1975) seem to contradict this concept: these authors, in fact, have determined, by electron probe X-ray microanalysis, the potassium contents in rat zona glomerulosa cells and have not shown appreciable changes in the ion concentration either in hyperaldosteronism induced by Na-deficiency, or in hypoaldosteronism due to Na-repletion.

Angiotensin II, Na-deficiency, K-repletion, and ACTH were found to maintain and stimulate zona glomerulosa growth (i.e., the volume of the zona glomerulosa and its cells) (see Sections III,B,C,D). Also, rat zona glomerulosa hyperplasia was found to be induced by chronic vasopressin administration (Isler, 1973; Payet and Isler, 1976). The mechanism of the mitogenic effect of the posterior pituitary hormone is not yet settled, although interrelationships between vasopressin and the renin–angiotensin system cannot be disregarded: by inducing changes in the body fluids, vasopressin may well elicit renin release, and vasopressin release was found to be induced by angiotensin II (Keil *et al.*, 1975). Payet and Isler (1976), however, have shown that renin, but not angiotensin II, significantly stimulates cell proliferation in the rat zona glomerulosa. In this connection, it must be recalled that chronic ACTH administration, along with the increase in the zona glomerulosa growth, markedly represses the conversion of [^{3}H]corticosterone to 18OH-DOC and aldosterone, thus transforming the secretion pattern of the zona glomerulosa into that typical of the zona fasciculata (Müller and Baumann, 1974).

B. Stimulation and Inhibition of the Renin–Angiotensin System

The ultrastructural investigations into the effects of chronic stimulation and inhibition of the renin–angiotensin system on the zona glomerulosa are very scarce. Moreover, morphological research on the acute effects of angiotensin II is completely lacking.

Fisher and Horvat (1971b) studied the effects of blood flow restriction to the left kidney in young female Wistar rats. Three weeks after the operation the adrenal weight was increased and the zona glomerulosa enlarged. Mitochondria were large and pale and occasionally contained elongated cristae that incompletely surrounded a homogeneous electron-dense material, SER was abundant, and the Golgi apparatus prominent. Hashida and Yunis (1972) examined the adrenal glands of two renovascular hypertensive children (4 and 11 years of age). They did not report obvious qualitative changes in mitochondria and lipid droplets; some mitochondria apparently contained an increased number of lamellar cristae, and sometimes the subcapsular zona glomerulosa cells displayed mitochondria with straight tubular cristae (500–800 Å in diameter) resembling those described in the rat by Giacomelli *et al.* (1965) (Section II, A,1). SER tubules were abundant and RER cisternae often aggregated in small stacks. Many membrane-bound electron-dense bodies (0.5–0.8 μm in diameter) tended to clump around lipid droplets and mitochondria.

Kasemsri and Nickerson (1976) made a stereological study of the adrenal glands of rats in which hypertension was induced, according to Deane and Masson (1951), by latex encapsulation of both kidneys for 4 and 6 weeks. The volume of the zona glomerulosa was increased only after 6 weeks. The volume of cells, nuclei, and lipid compartment was significantly enhanced, whereas the Golgi apparatus and the volume of the mitochondrial compartment did not vary. The SER surface area, but not that of the mitochondrial cristae, showed a marked increase. In Nussdorfer's laboratory (Rebuffat *et al.*, 1979) stereological studies were performed on the zona glomerulosa of male Wistar rats in which the renin–angiotensin system was stimulated by surgical stenosis of the left renal artery or inhibited by intraperitoneal administration of 5 mg/kg (twice a day) of timolol maleate (MK 950) for 6 consecutive days. For the effect of this last compound, see Graham *et al.* (1976).

Renovascular hypertension developed in about 10–15 days (160–170 mm Hg versus 100–105 mm Hg in the control sham-operated rats), and the animals were sacrificed 20 days after the operation. In comparison to the controls, plasma renin activity in the hypertensive animals was found to be significantly enhanced (30.4 ng/ml/hour versus 6.5 ng/ml/hour), and the aldosterone concentration significantly increased both in the peripheral plasma (360.5 pg/ml versus 240.2 pg/ml) and in the capsular adrenal homogenate (125.6 ng/100 mg versus 26.1 ng/100 mg). Timolol maleate-treated rats showed a significant decrease in the sistolic blood pressure (80–85 mm Hg) and in the plasma renin activity (2.7 ng/ml/hour). Plasma aldosterone concentration did not display evident changes as compared with the control value, whereas the intracellular concentration of the hormone was significantly decreased (12.4 ng/100 mg). Stereological findings in the zona glomerulosa of hypertensive rats closely agreed with those from Kasemsri and Nickerson (1976), except that the volume of the mitochondrial compartment and the surface area of the mitochondrial cristae were noticeably increased and the volume of the lipid compartment significantly decreased. Another striking ultrastructural finding was the presence of several clumps of electron-dense bodies resembling those described by Hashida and Yunis (1972) in the zona glomerulosa cells of renovascular hypertensive children. These granules did not show appreciable acid phosphatase activity, seemed to arise in the hypertrophic Golgi area, and were usually located at the juxta-sinusoidal poles of the cells (Fig. 10). Opposite stereological findings were observed in timolol maleate-treated hypotensive rats, with the exception that the volume of the lipid compartment did not display significant changes.

Noticeable disagreement subsists as to the effects of renal hypertension on the zona fasciculata cells. Fisher and Horvat (1971b) did not describe any significant variation, whereas Tsuchiyama *et al.* (1972) and Maruyama (1972) reported a conspicuous increase in SER and Golgi apparatus hypertrophy. Stereologically Kasemsri and Nickerson (1976) found quantitative structural changes analogous to those they described in zona glomerulosa cells, except that the SER surface area increase was not significant.

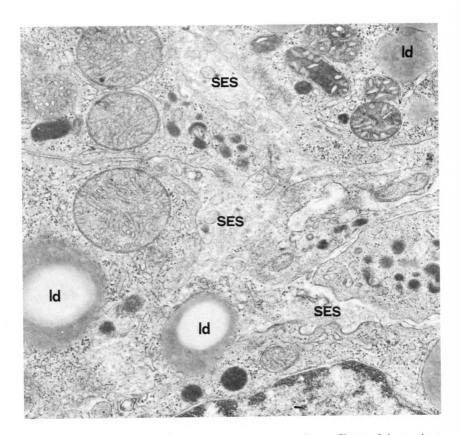

FIG. 10.   Zona glomerulosa cells from a renovascular hypertensive rat. Clumps of electron-dense granules are located at the cell periphery. ld, Lipid droplets; SES, subendothelial space. ×23,000.

C. Alteration of Na/K Balance

The morphological effects induced by chronic stimulation of the zona glomerulosa with a Na-deficient diet were described for the rat by Giacomelli *et al.* (1965), Fisher and Horvat (1971b), Smiciklas *et al.* (1971), Domoto *et al.* (1973), Palacios and Lafarga (1976), and Palacios *et al.* (1976); for the mouse by Shelton and Jones (1971); and for the opossum by Long and Jones (1970). The results were quite similar, though various periods of treatments were employed (from 2 days to 12 weeks).

Chronic sodium depletion provoked increase in width of the zona glomerulosa, which, according to Long and Jones (1970), Shelton and Jones (1971), and Smiciklas *et al.* (1971), was due to the increase both in the number and volume of cells. In this connection, Palacios *et al.* (1976) claimed that the onset of zona glomerulosa hypertrophy is due to the mitotic activation of its parenchymal cells. These authors also described binucleate cells in the subcapsular portion of the zone and suggested that amitosis could be a proliferative process that does not disturb the normal functioning of cellular synthetic pathways in states of secretory hyperactivity of the zona glomerulosa.

No evident mitochondrial changes were noted in the mouse by Shelton and Jones (1971); in the rat, however, after 2 and 4 days of sodium restriction, Domoto *et al.* (1973) reported that mitochondrial lamellar cristae are enlarged and transformed into tubulo-vesicular cristae. After 12 weeks of treatment Palacios *et al.* (1976) found an increase in the mitochondrial number. Giacomelli *et al.* (1965) described giant mitochondria (up to 4 μm in length) containing straight parallel cristal tubules and often showing discontinuities of their outer envelope through which the cristal tubules projected into the cytoplasm where they were closely apposed to the SER tubules. The increase in the number of mitochondria with straight tubular cristae was confirmed by Fisher and Horvat (1971b), but the ruptures of their limiting outer membrane seemed to be the effect of poor fixation. In the prolonged sodium-deprivation experiment, many mitochondria were found to contain lipid-like intramatrical inclusions, which were only rarely encountered in the normal animals (Section II,A) (Giacomelli *et al.*, 1965; Long and Jones, 1970; Domoto *et al.*, 1973).

The SER was noticeably increased (Giacomelli *et al.*, 1965; Fisher and Horvat, 1971b; Shelton and Jones, 1971; Smiciklas *et al.*, 1971; Domoto *et al.*, 1973) (Fig. 11), and sometimes, in the opossum, large collections of SER tubules were seen to enclose 1 or 2 mitochondria and a few ribosomes (Long and Jones, 1970). RER profiles and free ribosomes did not display evident variations, except in the mouse zona glomerulosa cells, in which Shelton and Jones (1971) described a transient RER increase: at the second day of treatment, many stacks of RER cisternae appeared in a juxta-nuclear location; after 7 days, RER profiles were decreased and sparse in the cytoplasm, and after 2–3 weeks, the number of

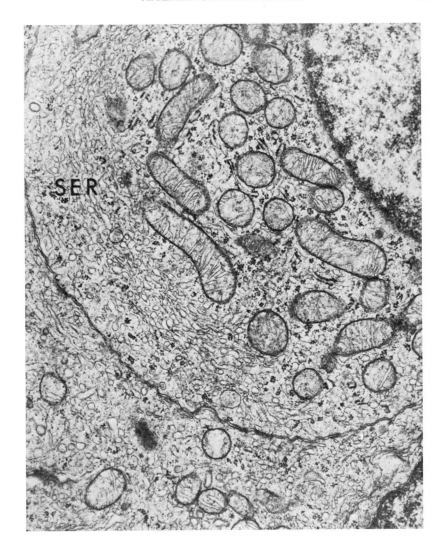

F_{IG}. 11. Zona glomerulosa cells from a sodium-depleted opossum. Elongated mitochondria contain lamellar cristae. SER characteristically shows a significant increase. ×17,500. (From Long and Jones, 1970.)

RER cisternae was restored. The authors stressed that RER hypertrophy precedes SER increase.

Lipid droplets were found to be decreased during the first phases of sodium deficiency (Giacomelli *et al.*, 1965; Shelton and Jones, 1971; Domoto *et al.*,

1973) but noticeably increased by prolonging the treatment (Shelton and Jones, 1971; Palacios et al., 1976). Long and Jones (1970), however, did not report evident lipid changes in the opossum. The Golgi apparatus was hypertrophied (Giacomelli et al., 1965; Long and Jones, 1970; Fisher and Horvat, 1971b; Shelton and Jones, 1971; Smiciklas et al., 1971; Domoto et al., 1973) and contained an increased number of coated vesicles (Palacios and Lafarga, 1976). Microvilli (Giacomelli et al., 1965; Shelton and Jones, 1971) and coated pits (Palacios and Lafarga, 1976) seemed to be increased in number. All these ultrastructural changes seem to be reversible, since the cells were found to recover completely in about 5 days after the suspension of the low-sodium diet (Palacios, 1978).

Smiciklas et al. (1971) also studied the effect of sodium restriction on pregnant rats in which sodium retention is attained through aldosterone-increased secretion by the zona glomerulosa. The ultrastructural changes reported above were intensified in these animals; in addition, several mitochondria containing lipid-like intramatrical deposits, many lysosomes, and some large autophagic vacuoles were observed. The authors suggested that these last ultrastructural features are signs of cell exhaustion.

The stimulating effects of chronic potassium loading on the zona glomerulosa morphology are quite analogous to those evoked by sodium restriction (Domoto et al., 1973; Lustyik et al., 1977). Mitochondria were increased in number and contained tubular cristae instead of lamellar ones, SER and free ribosomes were plentiful, the prominent Golgi apparatus contained several dense bodies, and the lipid compartment did not display evident changes. Kawai et al. (1978) described similar findings in the zona glomerulosa cells of rats chronically subjected to potassium loading and simultaneous sodium restriction.

The investigations on the morphology of the zona glomerulosa cells chronically inhibited by sodium repletion are rather scarce (Shelton and Jones, 1971; Hirano, 1976). In the rats a sodium-rich diet maintained for up to 4 weeks induced a rapid decrease in the zona glomerulosa width and in the number of mitochondrial cristae, atrophy of the Golgi apparatus, an increase in the number of lysosome-like dense bodies, and noticeable lipid depletion (Hirano, 1976). An analogous treatment did not induce striking changes in the mouse zona glomerulosa (Shelton and Jones, 1971): the changes included increase in lipid droplets, atrophy of the Golgi apparatus, decrease in the number of microvilli, and the appearance of β-glycogen particles and several small electron-dense granules, differing from typical lysosomes, at the cell periphery.

Different findings were reported in sodium-loaded rats, according to the route of salt administration (Nickerson and Molteni, 1972). After 16 weeks, rats on a high sodium diet did not display significant changes in the zona glomerulosa morphology, though plasma renin activity was decreased; as early as 7 days, high-sodium-drinking rats showed severe decrease in the zona glomerulosa width, along with Golgi hypertrophy, and increase in the number of caveolae and β-glycogen particles. Nickerson (1977) reconsidered this problem in a stereological study of the effects of high sodium intake (8% NaCl in the food for 11 weeks) on two strains of Sprague–Dawley rats, resistant or sensitive to the hypertensive effect of sodium loading; this study was made since adrenal gland seems to be involved in the pathogenesis of the hypertension (Iwai et al., 1969). Nickerson observed that the resistant strain did not show any change in the zona glomerulosa, whereas the sensitive rats displayed striking changes including increase in the cell volume and in the SER surface area. The treatment did not provoke significant variations either in mitochondria or in lipid droplets. Ultrastructural signs of hyperactivity (increase in the SER and mitochondrial cristae surface area, hypertrophy of the Golgi apparatus, and decrease in the volume of the lipid compartment) were present in the zona fasciculata cells. Nickerson claims that it is conceivable that the stress induced by the anomalous diet provokes increase in the ACTH release, which in turn stimulates both zona glomerulosa and zona fasciculata (Section III,D).

By selective inbreeding of Wistar rats, Okamoto and Aoki (1963) obtained a spontaneously hypertensive strain; Aoki (1963) demonstrated that adrenal cortex is essential for the development of the hypertension, which becomes evident at the twenty-first week of age. The fine morphology of the zona glomerulosa cells in this strain was stereologically described by Nickerson (1976). The volume of the zona glomerulosa is greater in the hypertensive than in the normotensive animals; the volume of cells and nuclei is less, while SER volume and surface are greater. No significant differences were observed in the volume of the mitochondrial compartment and in the surface area of mitochondrial cristae. The volume of the lipid compartment is significantly reduced and the Golgi apparatus is hypertrophic. Nickerson (1976) pointed out that it is unlikely that the renin–angiotensin system causes these changes in the hypertensive strain, inasmuch as Sokabe (1966) and Koletsky et al. (1970) showed significant decrease in the plasma renin activity, and Freeman et al. (1975) did not detect changes in the aldosterone secretion pattern.

Hirano (1976) described the effects of high sodium intake on the zona glomerulosa of a spontaneously hypertensive rat strain, but he used animals of 4 weeks of age in which the hypertension was not yet developed. Sodium repletion for 3 weeks provoked decrease in the volume of cells and lipid droplets as well as atrophy of the Golgi apparatus.

It should be stressed that a potassium-deficient diet for 10 weeks did not induce evident ultrastructural changes in the rat zona glomerulosa (Fisher and Horvat, 1971b); it is reasonable to suppose that such a procedure is not able to cause significant change in the zona glomerulosa intracellular potassium concentration, which by electron-probe microanalysis was found to be much higher than that in the zona fasciculata cells (Bacsy et al., 1973; Szalay et al., 1975).

Before concluding this subsection, it seems worthwhile to mention a few works dealing with the effects of the alteration of the electrolytic balance on the ultrastructure of the interrenal cells of lower vertebrates. Ogawa (1967) studied the interrenal cells of the *Carassius auratus* transferred in one-third sea water, and Berchtold (1973) those of the *Triturus cristatus* placed in 0.5% KCl solution; Bhattacharyya *et al.* (1975a,b) reported the effects of potassium loading in three avian species (*Anas platyrhynchos, Coturnix c. japonica,* and *Columba livia*). Berchtold and Bhattacharyya and associates reported very similar changes and agreed that there were no regional differences in the cell response to the stimulating procedure. The volumes of nuclei and nucleoli were increased; mitochondria showed an increase in number in the newt and in the Peking duck some giant organelles containing cristae arranged into a huge fenestrated plate were observed. SER was noticeably increased as were free ribosomes and polysomes; in the three species of birds several RER profiles were noted. Golgi apparatus was hypertrophic and contained many coated vesicles. In the Golgi area Ogawa (1967) described many electron-dense bodies.

D. ALTERATION OF THE HYPOPHYSEAL-ADRENAL AXIS

Since the first studies of Deane and Greep (1946), the "dogma" of the absolute independence of the zona glomerulosa morphology from the hypothalamo-hypophyseal axis has been established (see Deane, 1962; Idelman, 1970; and Long, 1975, for review). This contention was mainly based upon the observation that the ACTH- or hypophysectomy-induced increase or decrease, respectively, in the size of the adrenal gland, is not coupled with analogous variations in the zona glomerulosa width. According to the aforementioned dogma, many ultrastructural investigations did not report cytological changes in the zona glomerulosa following stimulation or inhibition of the hypophyseal-adrenal axis (Sabatini *et al.*, 1962; Nishikawa *et al.*, 1963; Idelman, 1970; Rhodin, 1971; Fujita, 1972; Buuck *et al.*, 1976).

However, this view has been questioned by several authors (Selye and Stone, 1950; Feldman, 1951; Lever, 1955; Bahn *et al.*, 1960), who maintained that hypophysectomy induces shrinkage of the zona glomerulosa, though not so marked as that occurring in the inner adrenal layers. On the assumption that these conflicting results may be due to the fact that a linear parameter (the thickness) is not adequate for assessing the zona glomerulosa volume, Nussdorfer and associates have studied the effect of hypophysectomy on the volume of the zona glomerulosa as evaluated by a morphometric technique (Nussdorfer *et al.*, 1973, 1974b). They found that 11 days after hypophysectomy a significant decrease in the volume of the rat zona glomerulosa occurred, in spite of its increase in width, and that this hypophysectomy-induced effect was reversed by a 5-day treatment with ACTH, cAMP, and cGMP. Nickerson and Brownie (1975) confirmed the atrophy of the rat zona glomerulosa 7 days after hypophysectomy but demonstrated that the volume of this zone was completely restored at the thirtieth postoperative day. The factor(s) involved in this phenomenon are still unknown since it does not seem that hypophysectomy exerts a stimulating effect on the renin–angiotensin system (Palkovits *et al.*, 1970; Rojo-Ortega *et al.*, 1972), nor does it affect the electrolytic balance (Knobil and Greep, 1958). The interrelationships between hypophysectomy and vasopressin release must be investigated, since vasopressin was found to elicit zona glomerulosa hyperplasia (Section III,A). Decrease and increase in the volume of the zona glomerulosa was also described by Nussdorfer *et al.* (1977d) after chronic administration of dexamethasone and ACTH to intact adult male rats.

The ultrastructural qualitative changes in the zona glomerulosa cells following activation or inhibition of the hypophyseal-adrenal axis are very scarce; nonetheless, hypertrophy of the Golgi apparatus in the zona glomerulosa cells of rats with transplanted ACTH-secreting tumor (Nickerson *et al.*, 1970) and of stressed monkeys (Penney and Brown, 1971) (Fig. 12), as well as an increased number of caveolae and microvilli in rats treated with ACTH and cyclic nucleotides (Nussdorfer *et al.*, 1973, 1974b), were reported.

Much more conspicuous are the ultrastructural quantitative changes appreciable only by stereological techniques (Nussdorfer *et al.*, 1973, 1974b, 1977d; Nickerson, 1975; Mazzocchi *et. al.*, 1977). In the mouse bearing a transplantable ACTH-secreting tumor, a significant increase was found in the volume of the zona glomerulosa cells, of the mitochondrial and lipid compartments, as well as in the surface area of SER and mitochondrial cristae (Nickerson, 1975). Nussdorfer *et al.* (1977d) confirmed Nickerson's findings, by treating intact rats with high doses of ACTH or dexamethasone up to 15 consecutive days. They showed that the volume of cells, of nuclei, of the mitochondrial compartment, and the surface area of SER and mitochondrial cristae increased or decreased linearly with the duration of ACTH or dexamethasone treatment, respectively. The volume of the lipid compartment displayed a transient decrease at the third day of ACTH treatment, and thereafter it increased linearly; in the dexamethasone-treated rats, this parameter had significantly increased at the third day of treatment, but from the third to the fifteenth day it remained in plateau.

Mazzocchi *et al.* (1977) made a stereological study of the effects of chronic treatment with ACTH and dexamethasone on the size and number of rat zona glomerulosa mitochondria. In the ACTH-treated rats, the average volume of individual mitochondria decreased significantly up to the sixth day of treatment and then increased from the sixth to the fifteenth day, whereas in dexamethasone-treated animals this parameter, after a small increase during the first 6 days of treatment, showed a significant decrease. The number of mitochondria per cell in ACTH-treated rats dramatically increased up to the sixth day of treatment, and then from the sixth to the fifteenth day it continued to

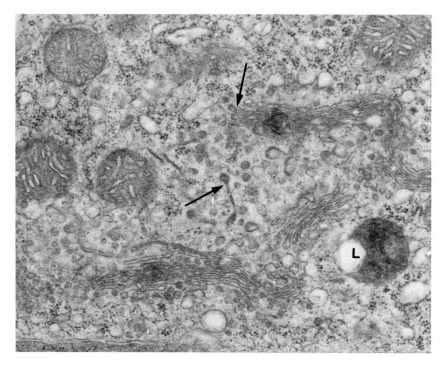

FIG. 12.   Cell from the zona glomerulosa of a hypophysectomized ACTH-treated rat. The Golgi apparatus shows a hypertrophic appearance. The arrows indicate some coated vesicles that seem to arise from the dilated endings of the cisternae. L, Lysosome. ×21,000. (From Nussdorfer *et al.*, 1973.)

increase, but only slightly. In contrast, the number of mitochondria per cell decreased as a function of the duration of dexamethasone treatment. The authors claimed that ACTH is involved in the maintenance and stimulation of the growth and proliferative activity of zona glomerulosa mitochondria.

Interesting findings were reported by Nickerson (1972) on the effects of hypophysectomy on the Mongolian gerbil zona glomerulosa. Two to four weeks after the operation a conspicuous increase in the width of the zona glomerulosa occurred, along with ultrastructural signs of cellular hyperactivity: increase in SER tubules, hypertrophy of the Golgi apparatus, and increase in the number of caveolae and coated vesicles. Unfortunately, Nickerson's observations were not supported by stereological evaluations. In any case, it must be recalled that differences in the regulation of the zona glomerulosa activity and growth among the various species have been reported (see Davis, 1967, 1975; and Müller, 1971, for review), and that the identification of the system(s) involved in the stimulation of the Mongolian gerbil zona glomerulosa is still an open problem (Nickerson, 1972).

E. Other Experimental Procedures

Some sporadic data concerning the effects of various experimental procedures will be reviewed here. Zona glomerulosa hyperactivity was obtained by spironolactone (aldactone) treatment; zona glomerulosa inhibition was achieved by treatment with mineralocorticoid hormones, carbenexolon, and heparin.

Aspects of zona glomerulosa hyperfunction were observed in both human patients bearing primary or secondary hyperaldosteronismus (Jenis and Hertzog, 1969; Symington, 1969; Davis and Medline, 1970; Kovacs et al., 1973) and experimental animals (Fisher and Horvat, 1971a; Rohrschneider et al., 1973) after spironolactone chronic administration. This drug, in fact, acts at the renal tubular level by competitively inhibiting aldosterone action and by eliciting an increased secretory response of the zona glomerulosa (see Glàz and Vecsei, 1971, for review).

The width of the zona glomerulosa was noticeably increased (Kovacs et al., 1973). The SER and RER were highly developed (Jenis and Hertzog, 1969; Davis and Medline, 1970; Kovacs et al., 1973). Typically the cells, especially those lying under the capsule (Davis and Medline, 1970; Kovacs et al., 1973) contained spherical laminated inclusions that were called spironolactone bodies (SB) (Fig. 13). They were sharply defined structures ranging from 2 to 20 μm in diameter and consisting of 2 to 20 concentric, smooth-surfaced, double-layered membranes arranged around a lipid-like electron-dense amorphous core. Sometimes the outermost membrane of the whorl retained ribosomes and was in continuity with SER tubules (Symington, 1969; Kovacs et al., 1973). According to Rohrschneider et al. (1973), SB are not specifically induced by spironolactone treatment since they were also found in the control dog zona glomerulosa cells, but this view was recently denied by Conn and Hinerman (1977), who found SB bodies only in cells actively producing aldosterone (i.e., in the cells of a Conn adenoma and not in the inactive zona glomerulosa cells proper). Although Fisher and Horvat (1971a) have suggested that SB originate in the rat from mitochondria, the bulk of the evidence indicates that they derive from SER (Symington, 1969; Davis and Medline, 1970; Kovacs et al., 1973; Conn and Hinerman, 1977). As to the functional significance of SB, Davis and Medline (1970) suggested that they are the expression of the spatial rearrangement of the newly formed spironolactone-induced SER membranes. Kovacs et al. (1973) and Rohrschneider et al. (1973), on the contrary, supposed that SB are the expression of a decreased SER catabolism. According to Hruban et al. (1972), the drug may be incorporated into newly formed SER membranes, thus interfering with

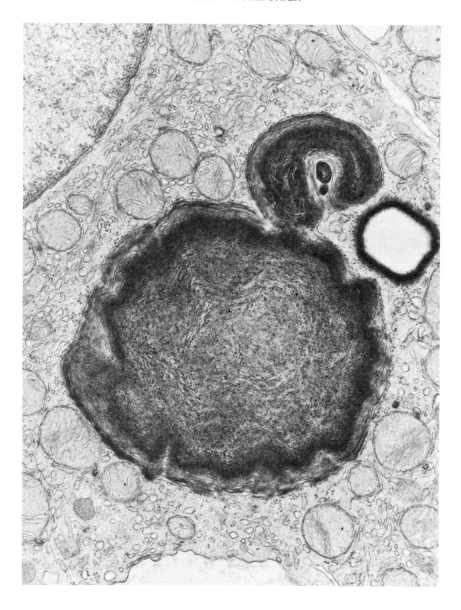

FIG. 13. Electron micrograph of a zona glomerulosa-like cell of an aldosterone-producing adenoma after spironolactone treatment of the patient. The cell contains a spironolactone body, along with abundant SER and round mitochondria with tubular and laminar cristae. ×26,000. (From Conn and Hinerman, 1977.)

their degradation by lysosomes. Conn and Hinerman (1977) claim that spironolactone, in addition to its peripheral effect, exerts a direct effect upon the zona glomerulosa cells by inhibiting the conversion of corticosterone to aldosterone. This contention would explain the specificity of SB in zona glomerulosa cells and permit one to hypothesize that these bodies are the morphologic reflection of a block in aldosterone biosynthesis.

The ultrastructural features of zona glomerulosa cells of rats chronically treated with *mineralocorticoid hormones* (aldosterone and DOC acetate) were described by Fisher and Horvat (1971b). The width of the zona glomerulosa was noticeably reduced; mitochondria were enlarged, more spherical in shape, and contained tubular and plate-like cristae extending across the major diameter of the organelle in parallel arrays (sometimes cristae appeared as coiled loops); the SER was decreased in amount; and the Golgi apparatus was small and fragmented.

Similar findings were obtained after chronic treatment of rats with *sodium carbenexolon* (Saeger and Mitschke, 1973), a drug that represses zona glomerulosa by its mineralocorticoid activity (Werning *et al.*, 1971). These authors also reported a decrease in number and volume of mitochondria. After withdrawal of the drug there was a complete restoration of the normal morphology. The specificity of the effects of mineralocorticoid hormones on the zona glomerulosa morphology is demonstrated by the fact that they did not induce changes in the zona fasciculata cells (Fisher and Horvat, 1971b); conversely, neither corticosterone in the rat (Nickerson, 1973) nor cortisol in the human (Mitschke and Saeger, 1973) were found to affect zona glomerulosa morphology.

Heparin and *heparinoids* were reported to decrease aldosterone synthesis in the rat, by partially blocking 18-hydroxylase activity (Sharma *et al.*, 1967; Glàz and Vecsei, 1971). Chronic heparin administration decreased the width of the zona glomerulosa (Glàz *et al.*, 1969; Levine *et al.*, 1972). The heparin-provoked fine structural changes were described by Lustyik *et al.* (1977). In accordance with the fact that heparin acts selectively on the mitochondrial 18-hydroxylase, the changes were mainly evidenced by mitochondria, which showed a decrease in number and contained only short plate-like cristae. A moderate increase in electron density of the lipid droplets was also noted.

F. Aldosterone-Secreting Tumors

Although primary hyperaldosteronismus (Conn's syndrome) is a well defined clinical entity, its histopathologic counterpart is not yet completely settled (see Symington, 1969; Neville and Mackay, 1972, for review). This syndrome, in fact, can be due to zona glomerulosa simple or nodular hyperplasia and to adrenocortical adenomas. Mackay (in Symington, 1969) showed that hyperplastic zona glomerulosa does not display evident structural differences from the normal one, except for a smaller average cell volume that is conceivably due to cell proliferation.

The ultrastructural investigations on the aldosterone-secreting neoplasms are very few, and these studies revealed that adrenocortical adenomas associated with increased levels of aldosterone do not possess a uniform ultrastructure, their cells possibly being similar to those of the zona glomerulosa or of the zona fasciculata; sometimes cells showing intermediate features between zona glomerulosa and zona fasciculata elements are also found (hybrid cells).

Cervòs-Navarro *et al.* (1965), Luse (1967), Reidbord and Fisher (1969), and Hashida and Yunis (1972) described adenomas composed of zona glomerulosa-like cells. Mitochondria were elongated, oval, or round, and contained lamellar cristae often grouped in the center of the organelle. Mitochondria never contained intramatrical lipid-like bodies. According to Reidbord and Fisher (1969), occasional mitochondria with vesicular cristae were also observed. SER was very abundant and lipid droplets numerous. Reidbord and Fisher (1969) claimed that an inverse correlation exists between the number of lipid droplets and the quantity of SER tubules. RER was present in small stacks (Reidbord and Fisher, 1969) or in a striking amount (Hashida and Yunis, 1972). Scanty glycogen rosettes and several dense bodies were seen; microvilli were very numerous.

Propst (1965) and Sommers and Terzakis (1970) described aldosterone-secreting tumors composed of zona fasciculata-like cells, which contained large nuclei with evident nucleoli, mitochondria with tubulo-vesicular cristae, and a very well developed SER. Kovacs *et al.* (1974) described a Conn's syndrome associated with an adrenal adenoma consisting of a mixture of cells resembling those of both zona fasciculata and zona reticularis. Mitochondria were pleomorphic and contained vesicular or tubular cristae; sometimes mitochondria with parallel tubular cristae running straight along the inner membrane, and giant organelles with tubular or vesicular cristae, as well as mitochondria with cristal loss, were observed; SER was hypertrophic; RER stacks and free ribosomes were plentiful; lipid droplets were very scarce; and the Golgi apparatus showed a focal increase. A large number of dense bodies were present, and microvilli projected into the SES.

IV. Morphological–Functional Correlations in Zona Glomerulosa Cells

In this section the morphologic features of the zona glomerulosa cells from normal and experimentally treated animals will be correlated with their functional activity on the grounds of our current knowledge of the biochemical pathways leading to aldosterone synthesis and of the possible mechanism(s) of action of adrenoglomerulotropic factors (see Section III,A).

In the following subsections, the subcellular topology of the enzymes involved in aldosterone synthesis, the presumed mechanism(s) of action of adrenoglomerulotropic factors, the possible involvement of the various subcellular organelles in the hormone synthesis, and the hypotheses concerning the mechanism of aldosterone release by zona glomerulosa cells will be discussed.

A. Subcellular Localization of the Enzymes of Aldosterone Synthesis

Several excellent reviews are available on this subject (Dorfman and Ungar, 1965; Glàz and Vecsei, 1971; Müller, 1971; Tamaoki, 1973; Samuels and Nelson, 1975; Sandor et al., 1976). The bulk of evidence clearly indicates that the enzymes of aldosterone synthesis are located in the SER and mitochondria. The various steps in the biosynthesis of aldosterone, as well as the most widely accepted subcellular topology of the enzymes involved, are depicted in Fig. 14.

Free cholesterol molecules enter the mitochondria where side chain-splitting, hydroxylating enzymes transform them into pregnenolone. Pregnenolone is then

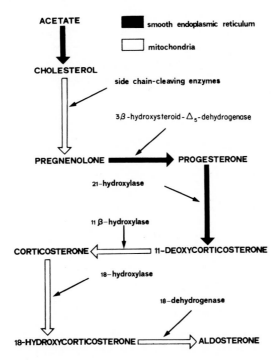

Fig. 14. Simplified scheme illustrating the pathways of synthesis of aldosterone and the subcellular topology of the enzymes involved.

transformed into progesterone by 3β-hydroxysteroid-$\Delta 5$-dehydrogenase located in the SER, where 21-hydroxylase also converts progesterone into 11-deoxycorticosterone. This intermediate product reenters the mitochondria to be transformed by 11β-hydroxylase into corticosterone, which in turn is converted to aldosterone by 18-hydroxylase and 18-hydroxy dehydrogenase. In contrast with the zona fasciculata, there are neither biochemical nor cytochemical investigations to suggest that, in the zona glomerulosa cells, 3β-hydroxysteroid-$\Delta 5$-dehydrogenase may be located not only in the SER, but also on the mitochondrial cristae (see Nussdorfer *et al.*, 1978a, for references).

From this brief account, it is clear that zona glomerulosa cells possess a more complex enzymatic steroidogenic pathway than that of the zona fasciculata cells, since corticosterone, which is the end product in the zona fasciculata cells in many species, is only an intermediate product in the zona glomerulosa. Some authors claim that, in the zona glomerulosa cells, two mitochondrial populations are present that contain the hydroxylating enzymatic systems leading to corticosterone production and the enzymes transforming corticosterone to aldosterone, respectively (Wassermann and Wassermann, 1974). This view, however, requires additional morphological and biochemical proofs before being accepted. Moreover, the subcellular topology of the enzymes involved in steroid synthesis indicates that the intermediate product molecules must switch between the SER and the mitochondria during their transformation into aldosterone: the intimate interrelationships existing among lipid droplets, mitochondria, and SER tubules (see Section II,A) may be the morphological counterpart of this process.

B. Mechanisms of Action of the Adrenoglomerulotropic Factors

There are very few reports dealing with the mechanism of action of adrenoglomerulotropic factors. Nevertheless, some evidence is available that ACTH (Shima *et al.*, 1971), potassium ions, and serotonin (Albano *et al.*, 1973) increase cAMP concentration in zona glomerulosa cells. However, according to Shima *et al.* (1978), angiotensin II does not seem to increase the cAMP level in the rat capsular adrenals; the mechanism underlying the effect of this last adrenoglomerulotropic factor appears to involve an increase in the intracellular calcium concentration and the activation of phosphodiesterase. It is well known that calcium ions regulate adrenal steroidogenesis (see Haksar and Péron, 1973, and Halkerston, 1975, for review).

It was also suggested that cAMP might function as an intracellular mediator of the glomerulotrophic action of ACTH (Nussdorfer *et al.*, 1973). Moreover, cGMP was also shown to stimulate zona glomerulosa growth (Nussdorfer *et al.*, 1974b). Therefore, the hypothesis that the various adrenoglomerulotropic factors act through activation of adenyl or guanyl cyclase cannot be excluded. Investigations into the effects of angiotensin II on the intracellular levels of cGMP in the capsular adrenals are needed.

If this were the case, the ultimate link coupling stimulus with response in zona glomerulosa cells consists in the activation of a protein kinase, which, in zona glomerulosa cells, would elicit the following effects (see Halkerston, 1975, for review):

1. Activation of glycogen phosphorylase, which, by promoting glycogenolysis, activates the pentose shunt, thus increasing the availability of NADPH, a rate-limiting coenzyme in steroidogenesis (review in Haynes, 1968, 1975);

2. Activation of cholesterol esterase, which hydrolyzes cholesterol esters stored in the lipid droplets (Frühling *et al.*, 1969; Moses *et al.*, 1969; Sand *et al.*, 1972), thus increasing the intracellular concentration of free cholesterol available for steroid synthesis (see Sandor *et al.*, 1976; and Nussdorfer *et al.*, 1978a, for comprehensive references);

3. Activation of cytoribosomal translation of "rapid turnover" proteins, which possibly affect steroidogenesis, mainly at the mitochondrial level (see Garren *et al.*, 1971; Haynes, 1975; and Sandor *et al.*, 1976, for review), or act as "carriers" in the transport of cholesterol (Kan *et al.*, 1972; Brownie and Paul, 1974; Lefèvre and Saez, 1977) and pregnenolone (Kream and Sauer, 1977; Strott, 1977) across the mitochondrial membranes;

4. Increase in the uptake of plasma cholesterol into adrenocortical cells (Dexter *et al.*, 1967a,b, 1970; Armato and Nussdorfer, 1972).

The four effects of activated protein kinase are conceivably involved in the rapid secretory effects of adrenoglomerulotropic factors. However, all these factors are able to exert trophic effects on the zona glomerulosa, and these effects also seem to be mediated by cyclic nucleotides. It was suggested that the mechanism underlying this action involves stimulation of the synthesis of both structural and enzymatic proteins at the transcriptional and translational levels (Mazzocchi and Nussdorfer, 1974; Nussdorfer and Mazzocchi, 1975).

C. The Functional Significance of the Ultrastructural Changes in Hyperfunctioning and Hypofunctioning Zona Glomerulosa Cells

From Section III, it is clear that contrasting ultrastructural changes are associated with the hyperfunction and hypofunction of zona glomerulosa cells. To summarize, long-term stimulating procedures induce an increase in the volume of cells, nuclei, and the mitochondrial compartment, as well as in the surface area of SER and mitochondrial cristae, and, at least in the more prolonged treatments, in the volume of the lipid compartment; the Golgi apparatus is hypertrophied. In addition, the hyperactive cells of aldosterone-secreting tumors show a noticeable hypertrophy of the SER membranes. Long-term suppression of the zona glomerulosa provokes decrease in the volume of cells, nuclei, and mitochondrial compartment, and in the surface area of SER and mitochondrial cristae; lipid droplets, after a transient increase, remain in plateau; the Golgi apparatus is atrophic and β-glycogen particles accumulate in the cytoplasm.

1. Endoplasmic Reticulum and Mitochondria

Stereological evaluations have shown that the most conspicuous subcellular changes, provoking cell hypertrophy and atrophy, concern SER and mitochondria, and this is easily understandable by considering that the enzymes of aldosterone synthesis are contained in these two organelles (Section IV,A).

Many biochemical and cytochemical investigations in vivo and in vitro showed that the activity of several of these enzymes is increased or decreased in stimulated or suppressed adrenocortical cells, respectively: side chain-cleaving enzymes (Kimura, 1969); 3β-hydroxysteroid-$\Delta 5$-dehydrogenase (Loveridge and Robertson, 1978); 21-hydroxylase (Milner and Villee, 1970; DeNicola, 1975); 11β-hydroxylase (Griffiths and Glick, 1966; Kowal, 1967, 1969; Kahri et al., 1970, 1976a; Kowal et al., 1970; Laury and McCarthy, 1970; Milner and Villee, 1970; DeNicola, 1973; DeNicola and Freire, 1973; Salmenperä and Kahri, 1976; Salmenperä et al., 1976; Vardolov and Weiss, 1978); and 18-hydroxylase (Kahri et al., 1970, 1976a; Salmenperä, 1976; Salmenperä and Kahri, 1976; Salmenperä et al., 1976). Furthermore, the quantity of cytochrome P_{450} in rat adrenocortical cells was found to be decreased after hypophysectomy and restored by ACTH 4 months after the operation (Pfeiffer et al., 1972).

According to Armato et al. (1974), it is reasonable to assume that, like the classic respiratory chain, the cytochrome P_{450} electron transfer chain requires an adequate steric arrangement for complete activity of the hydroxylating enzymes. Therefore, the increase in the surface area of SER membranes and of mitochondrial cristae can be interpreted as providing an increased framework of basic membrane into which newly synthesized enzymes of steroid synthesis can be inserted.

The increase in the surface area of the mitochondrial cristae, however, does not seem to involve the transformation of tubular cristae into vesicular ones in in vivo experiments. This transformation, occurring in in vitro cultured adrenocortical cells (see Kahri, 1966), allows a much more conspicuous increase in the surface area of the inner mitochondrial membranes, which, perhaps, permits the accommodation of more enzymatic moieties; stereological data showed that this transformation causes a 2- to 3-fold increase in the cristal surface area per unit mitochondrial volume (Nussdorfer et al., 1977a), which was coupled with only quantitative (but not qualitative) changes in the steroid secretion pattern (Armato et al., 1978). In in vivo experiments the increase in the mitochondrial cristal surface per cell in the zona glomerulosa seems to be obtained only by the increase in the average mitochondrial volume and number (Mazzocchi *et al.*, 1977).

As to the mechanism underlying the proliferation of SER and mitochondrial cristae, many lines of evidence are available that suggest that some of the adrenoglomerulotropic factors stimulate protein synthesis. The ACTH-elicited *in vitro* increase in the SER membranes and mitochondrial cristal surface areas was found to be inhibited by treating the cultures with specific inhibitors of nuclear and mitochondrial DNA-dependent protein synthesis: actinomycin and puromycin (Kahri, 1968a); chloramphenicol (Kahri, 1970; Milner, 1971); cycloheximide (Kahri, 1971); 5-bromodeoxyuridine (Kahri *et al.*, 1976b); and ethidium bromide (Milner, 1972; Salmenperä and Kahri, 1977). Mazzocchi *et al.* (1978b) showed that chloramphenicol *in vivo* blocks the ACTH-induced mitochondrial cristal surface area maintenance; high resolution autoradiographic studies demonstrated that ACTH and purine cyclic nucleotides enhance nuclear and mitochondrial RNA synthesis in zona glomerulosa cells of hypophysectomized rats (Mazzocchi and Nussdorfer, 1974; Nussdorfer and Mazzocchi, 1975).

On these grounds, the nuclear hypertrophy and atrophy in hyperfunctioning and hypofunctioning zona glomerulosa cells, respectively, can easily be explained, since the increase in nuclear size was coupled with an enhanced transcriptional process (Stark *et al.*, 1965; Palkovitz and Fischer, 1968). This is confirmed by the fact that actinomycin D, an antibiotic that blocks nuclear DNA-dependent RNA polymerase, inhibits the ACTH-elicited hypertrophy of the nucleus in rat adrenocortical cells cultured *in vitro* (Kahri, 1968a), a finding that was not observed after treatment with puromycin, cycloheximide, or chloramphenicol (Kahri, 1968a, 1970, 1971).

The activation of cytoplasmic translational processes by adrenoglomerulotropic factors can also be inferred from the data of Shelton and Jones (1971) (see Section III,C), who found that sodium deprivation induces a transient increase in RER before SER proliferation. In fact, many data indicate that SER membranes are built up in the RER and then transferred to (or transformed into) the SER in rapidly growing adrenocortical cells (McNutt and Jones, 1970; Black, 1972; Fujita and Ihara, 1973; Armato *et al.*, 1978; see also Nussdorfer *et al.*, 1978a, for review).

In conclusion, it is conceivable that adrenoglomerulotropic factors are involved in the maintenance of the integrity of the SER and mitochondrial compartment by balancing the rates of synthesis and degradation of SER and cristal membranes. For the zona glomerulosa cells there are no specific indications that, as in the zona fasciculata (see Nussdorfer *et al.*, 1978a, for review), ACTH or other adrenoglomerulotrophic factors increase the half-lives of cytoplasmic and mitochondrial proteins; this point requires further investigations. According to Mazzocchi *et al.* (1977), the decrease in the SER and the cristal surface area in the absence of the adrenoglomerulotropic factors can, thus, be the expression of the degradation processes in the presence of a block of the synthesis of new intracellular membranes.

Some research has shown that the mitochondrial number increases and decreases in the hyperfunctioning and hypofunctioning zona glomerulosa cells, respectively (see Section III,D). A stereological study of this subject was performed by Mazzocchi et al. (1977). It was found that ACTH controls the growth of the rat zona glomerulosa mitochondria by stimulating mitochondrial DNA duplication. The dexamethasone-provoked decrease in the number of mitochondria was interpreted as due to the fact that the degradation rate of mitochondria is not in balance with the proliferation rate of new organelles. Preliminary results from Nussdorfer's group suggested that, as in the zona fasciculata (Nussdorfer et al., 1977c), so in the zona glomerulosa, ACTH is able to prolong the half-life of mitochondria and dexamethasone reverses this effect (Nussdorfer and Mazzocchi, unpublished results). During the first days of zona glomerulosa suppression, mitochondrial gigantism was noted (Section III); Mazzocchi and co-workers (1977) proposed that this was caused by the fusion of preexisting organelles, a mechanism that is thought to be at play for repairing the deficiency in the synthesis of new mitochondrial membranes (Tandler et al., 1968; Canick and Purvis, 1972).

2. Lipid Droplets

Stereological data concerning the volume of the lipid compartment in chronically stimulated and suppressed zona glomerulosa cells do not seem to be in complete agreement. For instance, in the zona glomerulosa of hypertensive rats Kasemsri and Nickerson (1976) reported increase in the lipid droplets, whereas Nussdorfer and associates (see Section III,B) described lipid depletion. However, the bulk of evidence suggests that during the first phases of zona glomerulosa stimulation lipid depletion occurs, which is then followed by lipid increase if the treatment is prolonged.

It is well established that lipid droplets contain esterified cholesterol (Frühling et al., 1969; Moses et al., 1969; Sand et al., 1972), and it was suggested that only exogenous cholesterol taken up from the bloodstream is stored in the lipid compartment (Frühling et al., 1973). The rate of endogenous synthesis of cholesterol in adrenocortical cells seems to vary according to the species, at least as far as the whole adrenal gland is concerned (Werbin and Chaikoff, 1961; Goodman, 1965; Ichii et al., 1967; Chevalier et al., 1968), and it was proposed that the rate of this process is directly related to the amount of SER (Frühling et al., 1973), since cholesterol synthesis from acetate and glucose is known to occur in the SER (Christensen, 1965; Dorfman and Ungar, 1965; Olson, 1965; Chesterton, 1968).

Armato et al. (1974) pointed out that the volume of lipid compartment in adrenocortical cells is the expression of the balance of the rates of formation and utilization of lipid droplets, and, therefore, is the result of the following processes: (1) endogenous synthesis of cholesterol; (2) uptake of exogenous cholesterol and its esterification and storage in the lipid droplets; (3) transformation of esterified cholesterol stored in the lipid droplets into free cholesterol by cholesterol esterase; (4) utilization of exogenous and endogenous free cholesterol in steroid synthesis.

On these grounds, it is obvious that the findings concerning the behavior of the volume of the lipid compartment in stimulated or suppressed zona glomerulosa cells would vary according to the species and the experimental procedure employed. In fact, it seems quite well established that ACTH stimulates Processes 1 (Sharma *et al.*, 1972), 2, 3, and 4 (see Section IV,B), but nothing is known about the effects of other adrenoglomerulotropic factors. On the basis of these considerations, the decrease in the volume of lipid droplets at the onset of stimulating treatments can be interpreted as the morphological counterpart of the increased requirement of free cholesterol for utilization in the accelerated aldosterone synthesis (Processes 3 an 4); the cellular demand of free cholesterol would not be satisfied by Process 1, since by judging from the SER amount in the zona glomerulosa with respect to the zona fasciculata cells (Section II,D), zona glomerulosa cells do not possess a great capacity for endogenous cholesterol synthesis. In the prolonged stimulating treatments the volume of lipid droplets increases, and this may be correlated with the SER increase and the consequent presumable enhancement of Process 1, coupled with an increased uptake of exogenous cholesterol from the blood stream. Therefore, the discrepancies existing between stereological data from Nickerson's and Nussdorfer's groups (Section III,B) can be explained: Nickerson and co-workers observed the effects of the hypertensive treatment after 4–6 weeks, and Nussdorfer and associates, after 1–2 weeks from the operation (long- and short-term effects, respectively). The activation of the hypothalamo-hypophyseal axis seems to be much more rapid than the activation of the renin–angiotensin system in eliciting the aforementioned short- and long-term effects: in fact, the volume of the lipid compartment increases as early as 3 days after initiation of chronic ACTH administration (Nussdorfer *et al.*, 1977d). At present it is impossible to explain this fact; however, it can tentatively be suggested that angiotensin would stimulate process 2 to a lesser extent than ACTH.

A complete agreement subsists as to the effects of chronic suppression of zona glomerulosa cells on their lipid compartment. In this case, the volume of the lipid compartment shows a slight initial increase and then remains in plateau; this is easily explained if we consider that, in these experimental conditions, not only the processes leading to the utilization of lipid droplets, but also those provoking their accumulation, are blocked. In this connection, I wish to mention some preliminary findings concerning the effect of chloramphenicol on rat zona glomerulosa cells (Nussdorfer, Mazzocchi and Rebuffat, unpublished results).

As in the zona fasciculata cells (Mazzocchi *et al.*, 1978b), the chronic treatment with this antibiotic causes noticeable increase in the volume of the lipid compartment, which is easily explained by recognizing that the ACTH-induced maintenance of Process 2 at an adequate rate does not require protein synthesis (Dexter *et al.*, 1967a,b, 1970).

3. *Other Organelles and Cell Inclusions*

The Golgi apparatus hypertrophies in hyperfunctioning zona glomerulosa cells and shows atrophy in the suppressed cells (Section III). Specific papers on the role of the Golgi apparatus in the zona glomerulosa cells are not available at present, and therefore it may only be recalled here that some authors claim that this organelle does intervene, although in an unknown fashion, in steroid synthesis and secretion (Christensen and Gillim, 1969; Fawcett *et al.*, 1969). Whether or not the Golgi apparatus is involved in steroid sulfation–desulfation processes or in the binding of steroid intermediates to their carrier proteins is not settled at present (see Nussdorfer *et al.*, 1978a, for review). The possible role of the Golgi apparatus in the segregation of steroid hormones in secretory granules will be discussed in the next subsection.

Investigations into the changes in the lysosomes (and microbodies) in the hyperfunctioning or hypofunctioning zona glomerulosa cells were not performed; for a discussion of the possible involvement of these organelles in adrenocortical cell physiology, the reader is invited to consult two review articles (Nussdorfer *et al.*, 1978a; Idelman, 1978).

In the suppressed zona glomerulosa cells, a noticeable increase in the β-glycogen particles was observed. This change is in keeping with one of the proposed mechanisms of action of the adrenoglomerulotropic factors; namely, the activation of glycogen phosphorylase (Section IV,B). The blockade of glycogenolysis in the inhibited zona glomerulosa cells would obviously result in the intracytoplasmic clumping of β-glycogen particles.

D. The Mechanism of the Hormone Release by Zona Glomerulosa Cells

There are extremely few investigations into the mechanism of aldosterone release by the zona glomerulosa cells. The theories of holocrine, apocrine, and endoplasmocrine secretion of Rhodin (1971) do not seem to apply to this zone, though Brenner (1966) and Penney and Brown (1971) described parenchymal cell fragments in the sinusoid lumina of the monkey zona glomerulosa.

Perhaps the more widely accepted theory is that of "simple diffusion," which holds that steroid hormones (and aldosterone) diffuse freely throughout the aqueous phase of the cytoplasm and the lipid phase of the cellular membranes (Lever, 1955; Porter and Bonneville, 1967; Christensen and Gillim, 1969). Though aldosterone does not show high polarity, it is conceivable that this molecule must be modified prior to its release, and in this connection the possible involvement of the Golgi apparatus in the sulfation of steroid hormones (Fawcett *et al.*, 1969) and/or in their binding to a carrier protein should be recalled.

The view that an exocytotic mechanism underlies the hormone release by steroid-producing cells is now gaining more and more credit (Laychock and Rubin, 1974; Rubin *et al.*, 1974; Gemmell *et al.*, 1974, 1977a,b; Gemmell and Stacy, 1977; Pearce *et al.*, 1977; Nussdorfer *et al.*, 1978b), although demonstration of secretory granules has been very elusive as far as adrenocortical cells are concerned. Specific investigations on the zona glomerulosa are not available. However, electron-dense granules, not displaying conspicuous acid phosphatase activity, were found in the rabbit (Section II,A,6), in hypertensive children, and in rat zona glomerulosa cells (Section III,B).

As reviewed in Section II,A, rat zona glomerulosa cells possess all the organelles known to be involved in the segregation of secretory granules and in their exocytotic release: Golgi apparatus, microtubules, and microfilaments (see Winkler and Smith, 1975; and Normann, 1976, for review). Exocytosis is frequently coupled with pinocytosis in order to avoid the increase in the cell surface caused by the integration into the cell plasma membrane of the membrane of the granules during their extrusion (see Werb and Dingle, 1976, for review); some evidence indicates that, in the hyperfunctioning zona glomerulosa cells, an increased number of coated pits occurs at the plasma membrane (Nussdorfer *et al.*, 1973, 1974b; Palacios and Lafarga, 1976).

It is quite reasonable to assume that the difficulties encountered in visualizing secretory granules in adrenocortical cells can be ascribed to the lack of appreciable intracellular storage of secretion products (Holzbauer and Newport, 1969; Jaanus *et al.*, 1970), so that secretory granules are released as soon as they are formed (Nussdorfer *et al.*, 1978b). Since microtubules are thought to be involved in the exocytosis (see Normann, 1976, for review), Nussdorfer and his associates devised a method of blocking the release of aldosterone by rat zona glomerulosa cells, without impairing its synthesis, by treating the animals with vinblastine and colchicine, two antimicrotubular agents (Malawista *et al.*, 1968; Gemmell and Stacy, 1977).

This experimental procedure induces the appearance of several clumps of electron-dense granules in both the Golgi area and at the juxta-sinusoidal poles of zona glomerulosa cells (Fig. 15), as well as a 4- to 5-fold rise in the intracellular concentration of aldosterone in the rat capsular adrenals. The authors think that the electron opacity of these granules does not indicate their lysosomal nature, since they showed (1) scarce or absent acid phosphatase activity, and (2) positivity to hyperosmication. According to Friend and Brassil (1970), hyperosmication selectively fixes and stains steroid hormones. It is conceivable that in these granules aldosterone may be bound to a carrier protein in a "storage secretion

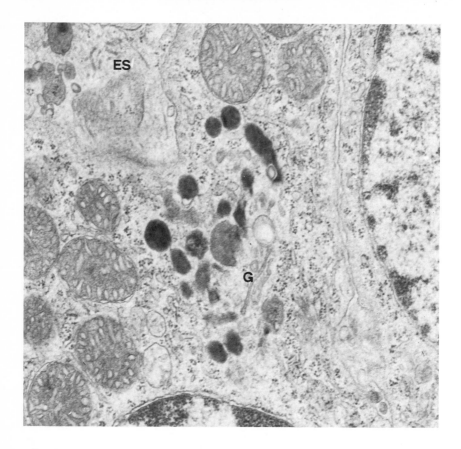

FIG. 15. Cells from the middle portion of the zona glomerulosa of a vinblastine-treated rat containing a clump of electron-dense granules in the Golgi area (G). ES, Extracellular space. ×28,000.

product," which prevents aldosterone diffusion throughout the granule membrane (see Rubin, 1974; Winkler and Smith, 1975, for review).

In this connection, it is to be recalled that some *in vitro* studies suggest that antimicrotubular agents enhance steroid synthesis and release by isolated normal adrenocortical cells (Ray and Strott, 1978) and Y-1 adrenal tumor cell line (Temple and Wolff, 1973). I think that these discrepancies with the *in vivo* results (Gemmell and Stacy, 1977; Nussdorfer *et al.*, 1978b) could be explained by recognizing that isolation procedures may have induced alterations in the cell plasma membrane, which can produce an aspecific release (diffusion) of the intracytoplasm-stored steroid hormones. This contention is also supported by the findings of O'Hare (1976), who did not show any change in steroid concentration in the growth medium of rat adrenocortical cells in primary monolayered culture after exposure to antimicrotubular agents; tissue culture, in fact, provides more physiological maintenance conditions for cells than isolation and suspension techniques, which presumably permit only a short-term survival. Morphological monitoring of dispersed adrenocortical cells would provide further insight into this problem.

V. Zona Glomerulosa and the Maintenance of the Adrenal Cortex

The great body of data on this subject was previously reviewed (Long, 1975; Idelman, 1978); therefore, in this section only a brief account will be provided, with emphasis on some new personal findings.

A. THEORIES ON THE CYTOGENESIS IN THE ADRENAL CORTEX

Two main theories on the cytogenesis in the mammalian adrenal cortex and on the functional significance of its three concentric zones are available: (1) the cell migration or escalatory theory, and (2) the zonal theory. A third one, the transformation field hypothesis advanced by Tonutti (1951) and Chester-Jones (1957) (see Idelman, 1978, for review), now has little credit and will be not discussed here. Furthermore, this last theory does not consider the problem of the parenchymal cell renewal.

The *cell migration theory,* first proposed by Gottschau (1883) and revived by Celestino da Costa (1951), postulates that new adrenocortical cells arise in the zona glomerulosa, migrate centripetally into the zona fasciculata and then degenerate in the zona reticularis. The fact that mitoses or ''S'' phase cells are almost exclusively present in the zona glomerulosa and in the outer portion of the zona fasciculata (Bachmann, 1954; Hunt and Hunt, 1964; Reiter and Pizzarello, 1966; Stöcker *et al.,* 1966; Reiter and Hoffman, 1967; Kwarecki, 1969), seems to favor this view. However, the migration of newly formed parenchymal cells labeled with [³H]thymidine or vital dyes from the outer to the inner layers of the normal adult adrenal cortex has so far eluded decisive demonstration (Diderholm and Hellman, 1960a,b; Walker and Rennels, 1961; Brenner, 1963; Ford and Young, 1963; Hunt and Hunt, 1964, 1966; Reiter and Pizzarello, 1966).

The finding of the independence of the zona glomerulosa from ACTH control (Deane and Greep, 1946; Greep and Deane, 1949a) and the evidence for each adrenal zone having a quite distinct steroid hormone pattern (''functional zonation'') (Stachenko and Giroud, 1959a,b, 1964; Sheppard *et al.,* 1963; and Vinson and Kenyon, 1978, for review) have resulted in serious questioning of the cell migration theory and provided the basis for the *zonal theory.* The latter theory proposes that each adrenal zone has a slow proliferative rate that is adequate to support its independent maintenance (Reiter and Pizzarello, 1966). The finding that one hour after intraperitoneal injection of [^{3}H]thymidine, zona reticularis displays a very low number of "S" phase cells in both the hamster (Reiter and Hoffman, 1967) and the adult rat (Pappritz et al., 1972) allowed some researchers to assume that, at least in these species, a local renewal occurs in the zona reticularis (Hunt and Hunt, 1966; Bertholet and Idelman, 1972). In any case, the number of mitoses decreases from the zona glomerulosa to the zona reticularis (see Idelman, 1978, for review).

Nevertheless, the following considerations cast some doubt on the absolute validity of the zonal theory: (1) The lack of a clear morphological zonation in the lower vertebrates (Section II,C) does not prevent interrenal cells from secreting mineralo- and glucocorticoid hormones (see Sandor et al., 1976, for review). (2) A great mass of evidence has challenged the view of the absolute independence of the zona glomerulosa from the hypothalamo-hypophyseal axis (Section III,A,D). (3) Adrenocortical cells, cultured in vitro in the absence of ACTH display the ultrastructural characteristics of the zona glomerulosa elements (e.g., mitochondria with lamellar or tubular cristae), whereas, if ACTH is added to the culture medium, they differentiate into zona fasciculata-like elements (e.g., mitochondria with vesicular cristae) (Kahri, 1966; Armato and Nussdorfer, 1972; Armato et al., 1974). (4) Müller and Baumann (1974) reported that chronic ACTH treatment shifts the secretion pattern of zona glomerulosa (capsular adrenal) into that of the zona fasciculata. (5) In experiments with enucleation of the adrenal gland or with the transplantation of the gland capsule, the residual zona glomerulosa was found to regenerate a histologically and functionally normal adrenal cortex (Greep and Deane, 1949b; Penney et al., 1963; Nickerson et al., 1969; Yago et al., 1972); furthermore, mitotic activation of the rat zona glomerulosa resulted from the chronic treatment with dimethylbenz(a)anthracene (Wheatley, 1967; Danz et al., 1973; Belloni et al., 1978b), which exerts a specific lytic effect on the inner adrenocortical zones (Huggins and Morii, 1961; Morii and Huggins, 1962; Horvath et al., 1969; Murad et al., 1973; Belloni et al., 1978b; Mazzocchi et al., 1978a).

The recognized finding that the greatest parenchymal mitotic activity occurs in the cells located between the zona glomerulosa and zona fasciculata, i.e., in the so-called zona intermedia (see Section II,B), allowed some investigators to hypothesize that new parenchymal cells arise in this "cambium" layer and then migrate centrifugally into the zona glomerulosa and centripetally into the zona fasciculata (see Idelman, 1978). For the reader's convenience this modification of the classic cell migration theory will be named here "proliferative intermediate zone hypothesis." The demonstration that the higher binding of ^{131}I-labeled ACTH (Golder and Boyns, 1971, 1972) and the pick of the ACTH-induced increase in the adenyl cyclase activity (Orenberg and Glick, 1972; Golder and Boyns, 1973) occurs in the zona intermedia, lends support to this hypothesis. It must be noted, however, that the presence of an evident zona intermedia is only an exceptional finding in the mammalian adrenal cortex (Section II,B).

B. The Normal Maintenance of the Adrenal Cortex

On the basis of the cytogenetic theories herein reviewed, it is possible to forecast the site of the pool of proliferating adrenocortical cells and of the cell deletion mechanisms. Furthermore, it must be stressed that in the steady state the maintenance of the adrenal cortex is due to the exact balance of the rates of cell proliferation and cell deletion.

According to the cell migration theory, cell proliferation occurs only in the zona glomerulosa and cell deletion exclusively in the zona reticularis (or also in the inner zona fasciculata). According to the "intermediate proliferative zone hypothesis," the growth of the gland is maintained by the double balance between the rate of cell proliferation in the zona intermedia and the rates of cell deletion in both the zona glomerulosa and the zona reticularis (or also in the inner zona fasciculata). According to the zonal theory, in the steady state, adrenocortical growth is maintained by the balance of the rate of cell proliferation and cell deletion occurring independently in each adrenal zone. Differences in the parenchymal cell turnover of the three adrenal zones would explain the reported zonal differences in the number of "S" phase cells. Since this parameter seems to decrease from the zona glomerulosa to the zona reticularis, it might be expected that in the steady state an analogous behavior would also be shown by the number of degenerate cells since cell removal would be maximum in the zona glomerulosa and minimum in the zona reticularis.

As far as I am aware, degenerate cells have never been found in the zona glomerulosa and in the outer half of the zona fasciculata, whereas signs indicating cell death occur more frequently in the zona reticularis (see Deane, 1962; and Idelman, 1978). Wyllie et al. (1973a,b) described a cell deletion mechanism, which they named *apoptosis,* in the zona reticularis of both neonatal and adult rat.

C. Supporting Evidence for the Cell Migration Theory
(Belloni et al., 1978a)

In the writer's opinion, the available data on the cytogenesis of the adrenal gland fit better with the cell migration theory, although a clear demonstration of centripetal cell migration from the zona glomerulosa is lacking. It is conceivable, however, that this was not achieved by previous workers because they employed inadequate experimental models: in fact, it seems well demonstrated that cell renewal in the adrenal cortex of normal animals is very slow, since parenchymal cells are in a very prolonged G_1 period (Ford and Young, 1963). This contention is supported by the observation that the unique convincing proofs for the existence of a centripetal cell migration from the zona glomerulosa were obtained in the rapidly growing glands of the prepubertal rats (Wright, 1971; Wright et al., 1973; Wright and Voncina, 1977) or of unilaterally adrenalectomized adult animals (Jones, 1967). Conversely, no data supporting centripetal migration were obtained in stressed or acutely ACTH-administered rats (Brenner, 1963; Machemer and Oehlert, 1964).

Numerous investigations have indicated that chronic ACTH treatment enhances DNA synthesis in adrenocortical cells both *in vivo* (Farese, 1968; Masui and Garren, 1970; Nussdorfer et al., 1974a; Saez et al., 1977) and *in vitro* (Armato et al., 1975, 1977; Salmenperä and Kahri, 1977), as well as the number of zona fasciculata cells (Nussdorfer et al., 1977b, 1978a, for review). To localize the pool of parenchymal cells whose mitotic activity is triggered by ACTH, Nussdorfer and co-workers have studied the effect of the chronic administration of ACTH (10 IU/kg) for 8 consecutive days on the number of labeled cells in the adrenal cortex of the [³H]thymidine-injected adult male rats (2.5 μCi/gm).

Quantitative autoradiographic data showed that chronic ACTH administration increases the number of "S" phase cells in the inner half of the zona glomerulosa and in the outer third of the zona fasciculata but does not induce any significant increase in the label uptake in the inner adrenal layers (Fig. 16). These data, coupled with the previously mentioned morphometric ones (Nussdorfer et al., 1977b), suggest that the ACTH-induced zona fasciculata hyperplasia involves the displacement of newly formed elements from the zona glomerulosa to the zona fasciculata.

To demonstrate this fact directly, we have studied the fate of ³H-labeled cells in rats under continuous ACTH administration. The following experimental schedule was employed: 42 adult male rats were treated with daily injections of 10 IU/kg of ACTH during the entire experimental period; at the eighth day the animals received i.p. injection of 2.5 μCi/gm of [³H]thymidine and were divided into seven groups, which were sacrificed after 2 hours (0 time), 3,6,12, 18, 24, and 30 days after the tracer injection. Quantitative autoradiographic data are depicted in Fig. 17. The number of "S" phase cells were observed to (1) in the zona glomerulosa, decrease parabolically in relation to the number of days elapsed from the tracer injection; (2) in the outer third of the zona fasciculata, increase up to the sixth day and then decrease parabolically; (3) in the middle third of the zona fasciculata, increase from the sixth to the twenty-fourth day and then significantly decrease; (4) in the inner third of the zona fasciculata, remain in plateau up to the twenty-fourth day and then show a significant increase; and finally (5) in the zona reticularis, display no significant change.

These findings clearly suggest that zona glomerulosa functions as a reservoir

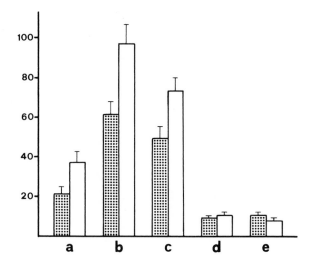

FIG. 16. Histogram demonstrating the effects of ACTH on the number of "S" phase paren-chymal cells per mm² of tissue area in the adult rat adrenal cortex. ▨, Control rats; □, ACTH-treated rats. (a) Outer half of the zona glomerulosa; (b) inner half of the zona glomerulosa; (c) outer third of the zona fasciculata; (d) middle and inner thirds of the zona fasciculata; (e) zona reticularis. Standard errors are indicated.

of newly formed cells, which, under appropriate stimulus, can be incorporated into the zona fasciculata. However, our experimental model failed to show migration of the labeled elements into the zona reticularis: whether this fact is due to insufficient experimental time or to an ACTH-induced elongation of the adrenocortical cell half-life, which blocks the passing of the labeled cells from the fascicular to the reticular compartment, requires further investigations. In this connection, it must be recalled that the question of whether migration is actively (Kahri, 1968b) or only passively due to the growth pressure in the zona glomerulosa, is still open.

Although the migration of newly formed cells from the zona glomerulosa to the zona fasciculata in chronically ACTH-treated rats seems quite well demon-strated, the localization of the pool of mitotic cells requires additional specula-tions. We suggest the following layout of the adrenocortical tissue, based on the classic assumption that parenchymal cells are arranged in cordonal units that form loops of various lengths in the zona glomerulosa (see Chester-Jones, 1976, for review). We propose that, in the rat adrenal, the long loops are the more numerous and their tips are in contact or enter into the zona fasciculata; the tips of the short loops are located in the middle portion of the zona glomerulosa, and only a few cords run straight without looping, their tips touching the connective capsule of the gland. We hypothesize that the "stem cells" are in the cordonal

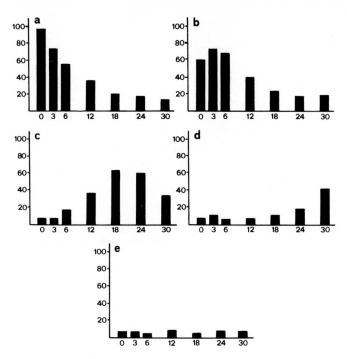

Fig. 17.   *Ordinate:* "S" phase parenchymal cells per mm²; Abscissa: number of days. Histograms illustrating the changes in the number of [³H]thymidine labeled cells in the various zones of the adrenal cortex of ACTH-treated rats as a function of the number of days elapsed from the tracer injection. (a) Zona glomerulosa; (b) outer third of the zona fasciculata; (c) middle third of the zona fasciculata; (d) inner third of the zona fasciculata; (e) zona reticularis. Standard errors are not indicated.

tips; therefore, the pool of proliferating cells in the rat adrenals would be located between the zona glomerulosa and zona fasciculata, i.e., in the zona intermedia. According to our model, the zona intermedia would consist of stem cells intermingled with transitional elements between the zona glomerulosa and zona fasciculata. The ultrastructural features of the zona intermedia cells (e.g., the absence of lipid droplets) may support this contention (Section II,B). Moreover, an evident zona intermedia cannot be found in that species in which the short loops are more numerous, and this consideration might explain the only sporadic appearance of this layer.

Investigations into the localization of the mitotic cell pool in the various mammalian species are needed, as well as the study of the ³H-labeled cell migration during the first days of ACTH administration in the rat adrenals: in fact, according to our hypothesis, the first part of the cell migration would happen inside the zona glomerulosa.

VI. Concluding Remarks

The preceding sections of this chapter indicate that in the last few years a conspicuous mass of research has considerably increased our knowledge of the cytophysiology of the adrenal zona glomerulosa. However, if the present article is compared with a previous one dealing with zona fasciculata (Nussdorfer *et al.*, 1978a), it appears that much work must still be done before our insight into zona glomerulosa cytophysiology can be comparable to that into zona fasciculata cytophysiology. The points needing further investigations were stressed in the five sections of this review.

I shall now take the opportunity to make some speculations on the zona glomerulosa cytophysiology and to emphasize some topics whose elucidation should be the task of future investigations.

The mechanism of action of the various factors modulating zona glomerulosa activity and growth is not yet clarified (Section IV,B). At variance with other adrenoglomerulotropic factors, angiotensin II does not activate adenyl cyclase. However, the possibility cannot be excluded that it acts on guanyl cyclase; previous papers, in fact, have indicated that both cAMP and cGMP elicit the growth of rat zona glomerulosa, but that only the contemporary administration of both cyclic nucleotides completely reverses the hypophysectomy-induced zona glomerulosa atrophy (Nussdorfer *et al.*, 1974b; Mazzocchi *et al.*, 1974). Only the effects of ACTH on the protein, RNA, and DNA synthesis by zona glomerulosa cells were investigated (Section IV,B,C): It is to be hoped that these studies will be extended to the other adrenoglomerulotropic factors. It must be recalled, however, that the investigations into zona glomerulosa cytophysiology present a basic difficulty owing to the possible interrelationships among the various adrenoglomerulotropic factors (Section III,A): the use of hypophysectomized animals treated with maintenance doses of ACTH and with timolol maleate; of timolol maleate-treated, sodium-repleted animals; and of sodium-repleted, hypophysectomized animals administered with maintenance doses of ACTH, would adequately permit, for instance, the study of the specific effects of Na/K balance, ACTH, or angiotensin II, respectively. An alternate approach to overcome this difficulty might involve *in vitro* experiments: however, I want to stress that only the use of tissue culture of capsular adrenals, according to the technique of Hornsby *et al.* (1974), will be suitable, since the isolated zona glomerulosa cells suspensions (Swallow and Sayers, 1969) do not assure adequate conditions of survival of the parenchymal elements and obviously do not allow the carrying out of long-term experiments.

It seems quite well demonstrated that, at least in some experimental conditions, newly formed zona glomerulosa cells migrate centripetally and incorporate into the zona fasciculata (Section V). Nussdorfer's group found that this process is accelerated by ACTH: it will be worthwhile to investigate whether the other adrenoglomerulotropic factors have the same effect.

The ACTH-induced centripetal migration of zona glomerulosa cells in the rat adrenals requires further speculation. The aldosterone synthesis is rather more complex than that of corticosterone: in fact, corticosterone, which is the end product in the rat zona fasciculata cells, appears to be only an intermediate product in the zona glomerulosa elements (Section IV,A). Therefore, the transformation of the steroid secretion pattern of the zona glomerulosa into that typical of the zona fasciculata seems to involve the loss of two mitochondrial enzymes (18-hydroxylase and 18OH-dehydrogenase). Müller and Baumann (1974) showed that this occurs in the rat capsular adrenals after chronic ACTH treatment; however, acute ACTH exposure is able to enhance the activity of 18-hydroxylase in adrenocortical cells cultured *in vitro* (Section IV,C). On these grounds the question arises: Do zona glomerulosa cells differentiate or dedifferentiate under ACTH chronic treatment? It is not yet possible to answer this question satisfactorily. However, I wish to stress that the biochemical investigations so far available did not discriminate clearly between the increase in the enzymatic activity that results from the activation of preexisting enzymes or that is due to *de novo* synthesis of enzymatic moieties. Therefore, it is conceivable that ACTH stimulates the *de novo* synthesis of all the enzymes participating in corticosterone synthesis, but not that of 18-hydroxylase (and 18OH-dehydrogenase), although it is able to activate all the enzymes of aldosterone synthesis: this would result in an increased output of corticosterone and in only a moderate amount of aldosterone release. This hypothesis implies that parenchymal cells in both the zona fasciculata and zona reticularis are able to secrete aldosterone, though in a moderate amount.

I think that efforts to express the enzymatic activity per single cell by using coupled biochemical and morphometric techniques (Neri *et al.*, 1978) will help us to gain insight into this problem.

REFERENCES

Albano, J. D. M., Brown, B. L., Ekins, R. P., Price, I., Tait, S. A. S., and Tait, J. F. (1973). *J. Endocrinol.* **58**, XI.
Allmann, D. W., Wakabayashi, T., Kornman, E. F., and Green, D. E. (1970). *J. Bioenerg.* **1**, 73.
Aoki, K. (1963). *Jpn Heart J.* **4**, 443.
Armato, U., and Nussdorfer, G. G. (1972). *Z. Zellforsch. Mikrosk. Anat.* **135**, 245.
Armato, U., Nussdorfer, G. G., Andreis, P. G., Mazzocchi, G., and Draghi, E. (1974). *Cell Tissue Res.* **155**. 155
Armato, U., Andreis, P. G., Draghi, E., and Meneghelli, V. (1975). *Horm. Res.* **6**, 105.
Armato, U., Andreis, P. G., and Draghi, E. (1977). *J. Endocrinol.* **72**, 97.
Armato, U., Nussdorfer, G. G., Neri, G., Draghi, E., Andreis, P. G., Mazzocchi, G., and Mantero, F. (1978). *Cell Tissue Res.* **190**, 187.
Bachmann, R. (1953). *Handb. Mikrosk. Anat. Mensch.* **6** (Pt. 5). "Die Nebenniere" (R. Bachmann, E. Scharrer and B. Scharrer, eds.), pp. 1–952. Springer-Verlag, Berlin and New York.
Bacsy, E., Szalay, K. Sz., Pantò, G., and Nagy, G. (1973). *Experientia* **29**, 485.

Bahn, R. C., Storino, H. E., and Smith, R. W. (1960). *Endocrinology* **66**, 403.

Baumberg, J. S., Davis, J. O., Johnson, J. A., and Wetty, R. T. (1971). *Am. J. Physiol.* **220**, 1094.

Belloni, A. S., Mazzocchi, G., Meneghelli, V., and Nussdorfer, G. G. (1978a). *Arch. Anat. Hist. Embr. Norm. Exp.* **61**, 195.

Belloni, A. S., Mazzocchi, G., Robba, C., Gambino, A. M., and Nussdorfer, G. G. (1978b). *Virchows Arch. B* **26**, 195.

Belt, W. D. (1960). *Anat. Rec.* **136**, 169.

Berchtold, J. P. (1973). *Z. Zellforsch. Mikrosk. Anat.* **139**, 47.

Berchtold, J. P. (1975). *C. R. Hebd. Seances Acad. Sci. (Paris)* **281**, 543.

Bertholet, J. Y., and Idelman, S. (1972). *J. Physiol. (Paris)* **64**, 101.

Bhattacharyya, T. K. (1975a). *Ann. Endocrinol. (Paris)* **36**, 21.

Bhattacharyya, T. K. (1975b). *Anat. Embryol.* **146**, 301.

Bhattacharyya, T. K., Calas, A., and Assenmacher, I. (1975a). *Gen. Comp. Endocrinol.* **26**, 115.

Bhattacharyya, T. K., Calas, A., and Assenmacher, I. (1975b). *Cell Tissue Res.* **160**, 219.

Black, V. (1972). *Am. J. Anat.* **135**, 381.

Bloodworth, J. M. B., and Powers, K. L. (1968). *J. Anat.* **102**, 457.

Boissin, J. (1967). *J. Physiol. (Paris)* **59**, 432.

Boyd, J., Mulrow, P. J., Palmore, W. P., and Silvo, P. (1973). *Circul. Res. Suppl.* **I** (to volumes 32–33), 39.

Brenner, R. M. (1963). *Am. J. Anat.* **112**, 81.

Brenner, R. M. (1966). *Am. J. Anat.* **119**, 429.

Brownie, A. C., and Paul, D. P. (1974). *Endocr. Res. Commun.* **1**, 321.

Buuck, R. J., Tharp, G. D., and Brumbaugh, J. A. (1976). *Cell Tissue Res.* **168**, 261.

Canick, J. A., and Purvis, J. L. (1972). *Exp. Mol. Pathol.* **16**, 79.

Celestino da Costa, A. (1951). *Ann. Endocrinol. (Paris)* **12**, 361.

Cervòs-Navarro, J., Tonutti, E., Garcia-Alvarez, F., Bayer, J. M., and Fritz, K. M. (1965). *Endokrinologie* **49**, 35.

Chester-Jones, I. (1957). "The Adrenal Cortex." Cambridge Univ. Press, London.

Chester-Jones, I. (1976). *J. Endocrinol.* **71**, 3P.

Chesterton, C. J. (1968). *J. Biol. Chem.* **243**, 1147.

Chevalier, F., D'Hollander, F., and Simmonet, F. (1968). *Biochim. Biophys. Acta* **164**, 339.

Christensen, A. K. (1965). *J. Cell Biol.* **26**, 911.

Christensen, A. K., and Gillim, S. W. (1969). *In* "The Gonads" (K. W. McKerns, ed.), pp. 415–488. North-Holland Publ., Amsterdam.

Conn, J. W., and Hinerman, D. L. (1977). *Metabolism* **26**, 1293.

Cotte, G., Michel-Bechet, M., Picard, D., and Haon, A. M. (1963). *Ann. Endocrinol. (Paris)* **24**, 1040.

Cronshaw, J., Holmes, W. N., and Loeb, S. L. (1974). *Anat. Rec.* **180**, 385.

Danz, M., Amlacher, E., Urban, H., and Bolck, F. (1973). *Exp. Pathol.* **8**, 122.

Davis, D. A., and Medline, N. H. (1970). *Am. J. Clin. Pathol.* **54**, 22.

Davis, J. O. (1967). *In* "The Adrenal Cortex" (A. B. Eisenstein, ed.), pp. 203–247. Churchill, London.

Davis, J. O. (1972). *New Engl. J. Med.* **286**, 100.

Davis, J. O. (1975). *Handb. Physiol., Sect. 7: Endocrinol.* **6**, 77–106.

Deane, H. W. (1962). *In* "Handbuch der Experimentellen Pharmakologie" (O. Eichler and A. Farah, eds.), Vol. 14, Pt. 1, pp. 1–185. Springer-Verlag, Berlin and New York.

Deane, H. W., and Greep, R. O. (1946). *Am. J. Anat.* **79**, 117.

Deane, H. W., and Masson, G. M. C. (1951). *J. Clin. Endocrinol.* **11**, 193.

Del Conte, E. (1972a). *Z. Zellforsch. Mikrosk. Anat.* **135**, 27.

Del Conte, E. (1972b). *Experientia* **28**, 451.

GASTONE G. NUSSDORFER

Del Conte, E. (1975). *Cell Tissue Res.* **157**, 493.

De Nicola, A. F. (1973). *Acta Physiol. Lat. Am.* **23**, 178.

De Nicola, A. E. (1975). *J. Steroid Biochem.* **6**, 1219.

De Nicola, A. F., and Freire, F. (1973). *J. Steroid Biochem.* **4**, 407.

De Robertis, E., and Sabatini, D. (1958). *J. Biophys. Biochem. Cytol.* **4**, 667.

Dexter, R. N., Fishman, L. N., Ney, R. L., and Liddle, G. W. (1967a). *Endocrinology* **81**, 1185.

Dexter, R. N., Fishman, L. N., Ney, R. L., and Liddle, G. W. (1967b). *J. Clin. Endocrinol. Metab.* **27**, 473.

Dexter, R. N., Fishman, L. H., and Ney, R. L. (1970). *Endocrinology* **87**, 836.

Diderholm, H., and Hellman, B. (1960a). *Acta Pathol. Microbiol. Scand.* **49**, 82.

Diderholm, H., and Hellman, B. (1960b). *Acta Physiol. Scand.* **50**, 197.

Domoto, D. T., Boyd, J. E., Mulrow, P. J., and Koshgarian, M. (1973). *Am. J. Pathol.* **72**, 433.

Dorfman, R. I., and Ungar, F. (1965). "Metabolism of Steroid Hormones." Academic Press, New York.

Farese, R. V. (1968). *In* "Function of the Adrenal Cortex" (K. W. McKerns, ed.), pp. 539–581. North-Holland Publ., Amsterdam.

Fawcett, D. W., Long, J., and Jones, A. L. (1969). *Recent Prog. Horm. Res.* **25**, 315.

Feldman, J. D. (1951). *Anat. Rec.* **109**, 41.

Fisher, E. R., and Horvat, B. (1971a). *Arch. Pathol.* **91**, 471.

Fisher, E. R., and Horvat, B. (1971b). *Arch. Pathol.* **92**, 172.

Ford, J. K., and Young, R. W. (1963). *Anat. Rec.* **146**, 125.

Freeman, R. H., Davis, J. O., Varsano-Aharon, N., Ulick, S., and Weinberger, M. H. (1975). *Circul. Res.* **37**, 66.

Friend, D. S., and Brassil, G. E. (1970). *J. Cell Biol.* **46**, 252.

Friend, D. S., and Gilula, N. B. (1972). *J. Cell Biol.* **53**, 148.

Frühling, J. (1977). Dissertation, Université Libre de Bruxelles.

Frühling, J., and Claude, A. (1968). *Eur. Reg. Conf. Electron Microsc. 4th, Rome. Pre-Congr. Abst.* **2**, 17.

Frühling, J., Meneghelli, V., and Claude, A. (1968). *J. Microsc. (Paris)* **7**, 705.

Frühling, J., Penasse, W., Sand, G., and Claude, A. (1969). *J. Microsc. (Paris)* **8**, 957.

Frühling, J., Sand, G., and Claude, A. (1971). *J. Microsc. (Paris)* **11**, 59.

Frühling, J., Sand, G., Penasse, W., Pecheux, F., and Claude, A. (1973). *J. Ultrastruct. Res.* **44**, 113.

Fujita, H. (1972). *Z. Zellforsch. Mikrosk. Anat.* **125**, 480.

Fujita, H., and Ihara, T. (1973). *Z. Anat. Entwicklunggesch.* **142**, 267.

Gabbiani, G., Chaponnier, C., and Lüscher, E. F. (1975). *Proc. Soc. Exp. Biol. Med.* **149**, 618.

Garren, L. D., Gill, G. N., Masui, H., and Walton, G. M. (1971). *Recent Prog. Horm. Res.* **27**, 433.

Gemmell, R. T., and Stacy, B. D. (1977). *J. Reprod. Fertil.* **49**, 115.

Gemmell, R. T., Stacy, B. D., and Thornburn, G. D. (1974). *Biol. Reprod.* **11**, 447.

Gemmell, R. T., Laychock, S. G., and Rubin, R. P. (1977a). *J. Cell Biol.* **72**, 209.

Gemmell, R. T., Stacy, B. D., and Nancarrow, C. D. (1977b). *Anat. Rec.* **189**, 161.

Giacomelli, F., Wiener, J., and Spiro, D. (1965). *J. Cell Biol.* **26**, 499.

Glàz, E., and Vecsei, P. (1971). "Aldosterone." Pergamon, Oxford.

Glàz, E., Kiss, R., Morvai, V., Péteri, M., Surag, K., Hajos, F., and Dauda, Gy. (1969). *Acta Endocrinol. (Copenhagen) Suppl.* **138**, 127.

Golder, M. P., and Boyns, A. R. (1971). *J. Endocrinol.* **49**, 649.

Golder, M. P., and Boyns, A. R. (1972). *J. Endocrinol.* **53**, 277.

Golder, M. P., and Boyns, A. R. (1973). *J. Endocrinol.* **56**, 471.

Goodman, D. S. (1965). *Physiol. Rev.* **45**, 747.

Gottschau, M. (1883). *Arch. Anat. Physiol. (Leipzig)* **83**, 412.

Graham, R. M., Oates, H. F., Weber, M. A., and Stokes, G. S. (1976). *Arch. Int. Pharmacodyn. Ther.* **219**, 205.

Greep, R. O., and Deane, H. W. (1949a). *Endocrinology* **45**, 42.

Greep, R. O., and Deane, H. W. (194b). *Ann. N.Y. Acad. Sci.* **50**, 596.

Griffiths, K., and Glick, D. (1966). *J. Endocrinol.* **35**, 1.

Haack, D. W., Abel, J. H., Jr., and Rhees, R. W. (1972). *Cytobiologie* **5**, 247.

Haksar, A., and Péron, F. G. (1973). *Biochim. Biophys. Acta* **313**, 363.

Halkerston, I. D. K. (1975). *Adv. Cyclic Nucleotide Res.* **6**, 99–136.

Hanke, W. (1978). *Gen. Comp. Clin. Endocrinol. Adrenal Cortex* **2**, 419–495.

Hardisty, M. W. (1972a). *In* "The Biology of Lamprey" (H. W. Hardisty and I. C. Potter, eds.), Vol. 2, pp. 171–192. Academic Press, New York.

Hardisty, M. W. (1972b). *Gen. Comp. Endocrinol.* **18**, 501.

Hardisty, M. W., and Baines, M. E. (1971). *Experientia* **27**, 1072.

Hashida, Y., and Yunis, E. J. (1972). *Human Pathol.* **3**, 301.

Hauger-Klevene, J. H., Brown, H., and Fleischer, N. (1969). *Proc. Soc. Exp. Biol. Med.* **131**, 539.

Hauger-Klevene, J. H., Brown, H., and Fleischer, N. (1970). *Acta Physiol. Lat. Am.* **20**, 373.

Haynes, R. C., Jr. (1968). *In* "Functions of the Adrenal Cortex" (K. W. McMerns, ed.), Vol. 1, pp. 583–600. North-Holland Publ., Amsterdam.

Haynes, R. C., Jr. (1975). *Handb. Physiol. Sec. 7: Endocrinol.* **6**, 69–76.

Hirano, T. (1976). *Acta Pathol. Jpn.* **26**, 589.

Holmes, W. N., and Phillips, J. G. (1976). *Gen. Comp. Clin. Endocrinol. Adrenal Cortex* **1**, 293–420.

Holzbauer, M., and Newport, H. (1969). *J. Physiol. (London)* **200**, 821.

Hornsby, P. J., O'Hare, M. J., and Neville, A. M. (1974). *Endocrinology* **95**, 1240.

Horvath, E., Kovacs, K., and Szabò, D. (1969). *J. Pathol.* **97**, 277.

Hruban, Z., Slesers, A., and Hopkins, E. (1972). *Lab. Invest.* **27**, 62.

Huggins, C., and Morii, S. (1961). *J. Exp. Med.* **114**, 741.

Hunt, T. E., and Hunt, E. A. (1964). *Anat. Rec.* **149**, 387.

Hunt, T. E., and Hunt, E. A. (1966). *Anat. Rec.* **156**, 361.

Ichii, S., Kobayashi, S., Yago, N., and Omata, S. (1967). *Endocrinol. Jpn.* **14**, 138.

Idelman, S. (1970). *Int. Rev. Cytol.* **27**, 181.

Idelman, S. (1978). *Gen. Comp. Clin. Endocrinol. Adrenal Cortex* **2**, 1–199.

Isler, H. (1973). *Anat. Rec.* **177**, 321.

Ito, T. (1959). *Arch. Histol. Jpn.* **18**, 179.

Iwai, J., Knudsen, K. D. Dahl, L. K., and Tassinari, L. (1969). *J. Exp. Med.* **129**, 663.

Jaanus, S., Rosenstein, M. J., and Rubin, R. P. (1970). *J. Physiol. (London)* **209**, 539.

Jenis, E. M., and Hertzog, R. W. (1969). *Arch. Pathol.* **88**, 530.

Jones, C. L. (1967). *Diss. Abstr.* **27**, 3373B.

Kadioglu, D., and Harrison, R. G. (1975). *J. Anat.* **120**, 179.

Kahri, A. I. (1966). *Acta Endocrinol. (Copenhagen) Suppl.* **108**, 1.

Kahri, A. I. (1968a). *J. Cell Biol.* **36**, 181.

Kahri, A. I., (1968b). *Acta Anat.* **71**, 67.

Kahri, A. I., (1970). *Am. J. Anat.* **127**, 130.

Kahri, A. I. (1971). *Anat. Rec.* **171**, 53.

Kahri, A. I., Pesonen, S., and Saure, A. (1970). Steroidologia **1**, 25.

Kahri, A. I., Huhtaniemi, I., and Salmenperä, M. (1976a). *Endocrinology* **98**, 33.

Kahri, A. I., Salmenperä, M., and Saure, A. (1976b). *J. Cell Biol.* **71**, 951.

Kan, K. W., Ritter, M. C., Ungar, F., and Dempsey, M. E. (1972). *Biochem. Biophys. Res. Commun.* **48**, 423.

Kasemsri, S., and Nickerson, P. A. (1976). *Am. J. Pathol.* **82**, 143.

Kawai, K., Sugihara, H., and Tsuchiyama, H. (1978). *Acta Pathol. Jpn.* **28**, 265.

Kawaoi, A. (1969). *Acta Pathol. Jpn.* **19,** 115.

Keil, L. C., Summy-Long, J., and Severs, W. B. (1975). *Endocrinology* **96,** 1063.

Kimura, T. (1969). *Endocrinology* **85,** 492.

Kjaerheim, Å. (1968a). *Z. Zellforsch. Mikrosk. Anat.* **91,** 429.

Kjaerheim, Å. (1968b). *Z. Zellforsch. Mikrosk. Anat.* **91,** 456.

Kjaerheim, Å. (1968c). *J. Microsc. (Paris)* **7,** 715.

Kjaerheim, Å. (1969). *Acta Anat.* **74,** 424.

Knobil, E., and Greep, R. O. (1958). *Endocrinology* **62,** 61.

Koletsky, S., Shook, P., and Rivera-Velez, J. (1970). *Proc. Soc. Exp. Biol. Med.* **134,** 1187.

Kondics, L. (1965). *Acta Morphol. Acad. Sci. Hung.* **13,** 233.

Kondics, L., and Kjaerheim, Å. (1966). *Z. Zellforsch. Mikrosk. Anat.* **70,** 81.

Kovacs, K., Horvath, E., and Singer, W. (1973). *J. Clin. Pathol.* **26,** 949.

Kovacs, K., Horvath, E., Delarue, N. C., and Laidlaw, J. C. (1974). *Horm. Res.* **5,** 47.

Kowal, J. (1967). *Clin. Res.* **15,** 262.

Kowal, J. (1969). *Endocrinology* **85,** 270.

Kowal, J., Simpson, E. R., and Estabrook, R. W. (1970). *J. Biol. Chem.* **245,** 2438.

Kream, B. E., and Sauer, L. A. (1977). *Endocrinology* **101,** 1318.

Kwarecki, K. (1969). *Folia Histochem. Cytochem.* **7,** 105.

Laury, L. W., and McCarthy, J. L. (1970). *Endocrinology* **87,** 1380.

Laychock, S. G., and Rubin, R. P. (1974). *Steroids* **24,** 177.

Lefèvre, A., and Saez, J. M. (1977). *C. R. Hebd. Seances Acad. Sci. (Paris)* **284,** 561.

Lever, J. D. (1955). *Am. J. Anat.* **97,** 409.

Levine, J. H., Laidlaw, J. C., and Ruse, J. L. (1972). *Can. J. Physiol. Pharmacol.* **50,** 270.

Lofts, B. (1978). *Gen. Comp. Clin. Endocrinol. Adrenal Cortex* **2,** 291-369.

Lofts, B., and Phillips, J. G. (1965). *J. Endocrinol.* **33,** 327.

Long, J. A. (1975). *Handb. Physiol., Sect. 7: Endocrinol.* **6,** 13-24.

Long, J. A., and Jones, A. L. (1967a). *Am. J. Anat.* **120,** 463.

Long, J. A., and Jones, A. L. (1967b). *Lab. Invest.* **17,** 355.

Long, J. A., and Jones, A. L. (1970). *Anat. Rec.* **166,** 1.

Loveridge, N., and Robertson, W. R. (1978). *J. Endocrinol.* **78,** 457.

Luse, S. A. (1967). *In* "The Adrenal Cortex" (A. B. Eisenstein, ed.), pp. 1–59. Churchill, London.

Lustyik, Gy., and Szabò, J. (1975). *Acta Morphol. Acad. Sci. Hung.* **23,** 1.

Lustyik, Gy., Szabò, J., Glàz, E., and Kiss, R. (1977). *Exp. Pathol.* **13,** 189.

Luthman, M. (1971). *Z. Zellforsch. Mikrosk. Anat.* **121,** 244.

Machemer, R., and Oehlert, W. (1964). *Endokrinologie* **46,** 77.

McKenna, T. J., Island, D. P., Nicholson, W. E., and Liddle, G. W. (1978a). *Endocrinology* **103,** 1411.

McKenna, T. J., Island, D. P., Nicholson, W. E., and Liddle, G. W. (1978b). *J. Steroid Biochem.* **9,** 967.

McNutt, N. S., and Jones, A. L. (1970). *Lab. Invest.* **22,** 513.

Magalhães, M. M. (1972). *J. Cell Biol.* **55,** 126.

Malamed, S. (1975). *Handb. Physiol. Sect. 7: Endocrinol.* **6,** 25-39.

Malawista, S. E., Sato, H., and Bensh, K. G. (1968). *Science* **160,** 770.

Manuelidis, L., and Mulrow, P. (1973). *Endocrinology* **93,** 1104.

Maruyama, T. (1972). *Jpn. Circul. J.* **36,** 363.

Masui, H., and Garren, L. D. (1970). *J. Biol. Chem.* **245,** 2627.

Mazzocchi, G., and Nussdorfer, G. G. (1974). *J. Endocrinol.* **63,** 581.

Mazzocchi, G., Nussdorfer, G. G., Rebuffat, P., Belloni, A. S., and Meneghelli, V. (1974). *Experientia* **30,** 87.

Mazzocchi, G., Robba, C., Rebuffat, P., Belloni, A. S., Gambino, A. M., and Nussdorfer, G. G. (1977). *Cell Tissue Res.* **181,** 287.

Mazzocchi, G., Belloni, A. S., Robba, C., and Nussdorfer, G. G. (1978a). *Virchows Arch. B* **28**, 97.

Mazzocchi, G., Neri, G., Robba, C., Belloni, A. S., and Nussdorfer, G. G. (1978b). *Am. J. Anat.* **153**, 34.

Milner, A. J. (1971). *Endocrinology* **88**, 66.

Milner, A. J. (1972). *J. Endocrinol.* **52**, 541.

Milner, A. J., and Villee, D. B. (1970). *Endocrinology* **87**, 596.

Mitschke, H., and Saeger, W. (1973). *Virchows Arch. A* **361**, 217.

Moore, T. J., Braley, L. M., and Williams, G. H. (1978). *Endocrinology* **103**, 152.

Morii, S., and Huggins, C. (1962). *Endocrinology* **71**, 972.

Moses, H. L., Davis, W. W., Rosenthal, A. S., and Garren, L. D. (1969). *Science* **163**, 1203.

Müller, J. (1971). "Regulation of Aldosterone Synthesis." Springer-Verlag, Berlin and New York.

Müller, J., and Baumann, K. (1974). *J. Steroid Biochem.* **5**, 795.

Murad, T. M., Leibach, J., and von Haam, E. (1973). *Exp. Mol. Pathol.* **18**, 305.

Neri, G., Gambino, A. M., Mazzocchi, G., and Nussdorfer, G. G. (1978). *Experientia* **34**, 815.

Neville, A. M., and MacKay, A. M. (1972). *Clin. Endocrinol. Metab.* **1**, 361–395.

Nickerson, P. A. (1971). *Anat. Rec.* **171**, 443.

Nickerson, P. A. (1972). *Tissue Cell* **4**, 677.

Nickerson, P. A. (1973). *Virchows Arch. B* **13**, 297.

Nickerson, P. A. (1975). *Am. J. Pathol.* **80**, 295.

Nickerson, P. A. (1976). *Am. J. Pathol.* **84**, 545.

Nickerson, P. A. (1977). *Lab. Invest.* **37**, 120.

Nickerson, P. A., and Brownie, A. C. (1975). *Endocrinol. Exp.* **9**, 187.

Nickerson, P. A., and Molteni, A. (1972). *Cytobiologie* **5**, 125.

Nickerson, P. A., Brownie, A. C., and Skelton, F. R. (1969). *Am. J. Pathol.* **57**, 335.

Nickerson, P. A., Brownie, A. C., and Molteni, A. (1970). *Lab. Invest.* **23**, 368.

Nishikawa, M., Murone, I., and Sato, T. (1963). *Endocrinology* **72**, 197.

Normann, T. C. (1976). *Int. Rev. Cytol.* **46**, 1.

Nussdorfer, G. G. (1970). *Z. Zellforsch. Mikrosk. Anat.* **103**, 382.

Nussdorfer, G. G., and Mazzocchi, G. (1975). *Horm, Res.* **6**, 20.

Nussdorfer, G. G., Mazzocchi, G., and Rebuffat, P. (1973). *Endocrinology* **92**, 141.

Nussdorfer, G. G., Mazzocchi, G., and Gottardo, G. (1974a). *Cell Tissue Res.* **151**, 281.

Nussdorfer, G. G., Mazzocchi, G., and Rebuffat, P. (1974b). *Endocrinology* **94**, 602.

Nussdorfer, G. G., Armato, U., Mazzocchi, G., Andreis, P. G., and Draghi, E. (1977a). *Cell Tissue Res.* **182**, 145.

Nussdorfer, G. G., Mazzocchi, G., Belloni, A. S., and Robba, C. (1977b). *Acta Anat.* **99**, 359.

Nussdorfer, G. G., Mazzocchi, G., Robba, C., Belloni, A. S., and Gambino, A. M. (1977c). *Anat. Rec.* **188**, 125.

Nussdorfer, G. G., Mazzocchi, G., Robba, C., Belloni, A. S., and Rebuffat, P. (1977d). *Acta Endocrinol. (Copenhagen)* **85**, 608.

Nussdorfer, G. G., Mazzocchi, G., and Meneghelli, V. (1978a). *Int. Rev. Cytol.* **55**, 291.

Nussdorfer, G. G., Mazzocchi, G., Neri, G., and Robba, C. (1978b). *Cell Tissue Res.* **189**, 403.

Ogawa, M. (1967). *Z. Zellforsch. Mikrosk. Anat.* **81**, 174.

O'Hare, M. J. (1976). *Experientia* **32**, 251.

Okamoto, K., and Aoki, K. (1963). *Jpn. Circul. J.* **27**, 282.

Olson, J. A. (1965). *Ergeb. Physiol. Chem. Exp. Pharmacol.* **56**, 173.

Orenberg, E. K., and Glick, D. (1972). *J. Histochem. Cytochem.* **20**, 923.

Palacios, G. (1978). *Experientia* **34**, 653.

Palacios, G., and Lafarga, M. (1976). *Experientia* **32**, 381.

Palacios, G., Lafarga, M., and Perez, R. (1976). *Experientia* **32**, 909.

Palkovitz, M., and Fischer, J. (1968). "Karyometric Investigations." Akadémiai Kiadò, Budapest.

Palkovitz, M., de Jong, W., van der Wal, B., and de Wied, D. (1970). *J. Endocrinol.* **47**, 243.

Pappritz, G., Trieb, G., and Dhom, G. (1972). *Z. Zellforsch. Mikrosk. Anat.* **126**, 421.

Payet, N., and Isler, H. (1976). *Cell Tissue Res.* **172**, 93.

Pearce, R. B., Cronshaw, J., and Holmes, W. N. (1977). *Cell Tissue Res.* **183**, 203.

Pearce, R. B., Cronshaw, J., and Holmes, W. N. (1978). *Cell Tissue Res.* **192**, 363.

Péczely, P. (1972). *Acta Biol. Acad. Sci. Hung.* **23**, 11.

Penney, D. P., and Brown, G. M. (1971). *J. Morphol.* **134**, 447.

Penney, D. P., Patt, D. I., and Dixon, W. C. (1963). *Anat. Rec.* **146**, 319.

Pfeiffer, D. R., Chu, J. W., Kuo, T. H., Chan, S. W., Kimura, T., and Tchen, T. T. (1972). *Biochem. Biophys. Res. Commun.* **48**, 486.

Pfeiffer, D. R., Kuo, T. H., and Tchen, T. T. (1976). *Arch. Biochem. Biophys.* **176**, 556.

Porter, K. R., and Bonneville, M. A. (1967). "Fine Structure of Cells and Tissues," pp. 75–76. Lea & Febiger, Philadelphia, Pennsylvania.

Propst, A. (1965). *Beitr. Pathol.* **131**, 1.

Propst, A., and Müller, O. (1966). *Z. Zellforsch. Mikrosk. Anat.* **75**, 404.

Ray, P., and Strott, C. A. (1978). *Endocrinology* **103**, 1281.

Rebuffat, P., Belloni, A. S., Mazzocchi, G., Vassanelli, P., and Nussdorfer, G. G. (1979). *J. Anat.* **129**, 561.

Reidbord, H., and Fisher, E. R. (1969). *Arch. Pathol.* **88**, 155.

Reiter, R. J., and Hoffman, R. A. (1967). *J. Anat.* **101**, 723.

Reiter, R. J., and Pizzarello, D. J. (1966). *Tex. Rep. Biol. Med.* **24**, 189.

Rhodin, J. A. G. (1971). *J. Ultrastruct. Res.* **34**, 23.

Rohr, H. P., Oberholzer, M., Bartsch, G., and Keller, M. (1975). *Int. Rev. Exp. Pathol.* **15**, 233.

Rohr, H. P., Bartsch, G., Eichenberger, P., Rasser, Y., Kaiser, C., and Keller, M. (1976). *J. Ultrastruct. Res.* **54**, 11.

Rohrschneider, I., Schinko, I., and Schmiedek, P. (1973). *Res. Exp. Med.* **159**, 321.

Rojo-Ortega, J. M., Chrétien, M., and Genest, J. (1972). *Horm. Metab. Res.* **4**, 500.

Romagnoli, P. (1972). *Arch. Ital. Anat. Embriol.* **77**, 87.

Rubin, R. P. (1974). "Calcium and the Secretory Process." Plenum, New York.

Rubin, R. P., Sheid, B., McCauley, R., and Laychock, S. G. (1974). *Endocrinology* **96**, 370.

Sabatini, D. D., De Robertis, E. D. P., and Bleichmar, H. B. (1962). *Endocrinology* **70**, 390.

Saeger, W., and Mitschke, H. (1973). *Virchows Arch. A* **358**, 45.

Saez, J. M., Morera, A. M., and Gallet, D. (1977). *Endocrinology* **100**, 1268.

Salmenperä, M. (1976). *J. Ultrastruct. Res.* **56**, 642.

Salmenperä, M., and Kahri, A. I. (1976). *Acta Endocrinol. (Copenhagen)* **83**, 781.

Salmenperä, M., and Kahri, A. I. (1977). *Exp. Cell Res.* **104**, 223.

Salmenperä, M., Kahri, A. I., and Saure, A. (1976). *J. Endocrinol.* **70**, 215.

Samuels, L. T., and Nelson, D. H. (1975). *Handb. Physiol. Sect. 7: Endocrinol.* **6**, 55–68.

Sand, G., Frühling, J., Penasse, W., and Claude, A. (1972). *J. Microsc. (Paris)* **15**, 41.

Sand, G., Frühling, J., and Platten-Godfroid, C. (1973). *J. Microsc. (Paris)* **17**, 283.

Sandor, T., Fazekas, A. G., and Robinson, B. H. (1976). *Gen. Comp. Clin. Endocrinol. Adrenal Cortex* **1**, 25–142.

Selye, M., and Stone, H. (1950). *In* "On the Experimental Morphology of the Adrenal Cortex." Thomas, Springfield, Illinois.

Sharma, D. C., Nerenberg, C. A., and Dorfman, R. I. (1967). *Biochemistry* **6**, 3472.

Sharma, R. K., Hashimoto, K., and Kitabchi, A. E. (1972). *Endocrinology* **91**, 994.

Shelton, J. M., and Jones, A. L. (1971). *Anat. Rec.* **170**, 147.

Sheppard, H., Swenson, R., and Mowles, T. F. (1963). *Endocrinology* **73**, 819.

Sheridan, M. N., and Belt, W. D. (1964). *Anat. Rec.* **149**, 73.

Sheridan, M. N., Belt, W. D., and Hartman, F. A. (1963). *Acta Anat.* **53**, 55.

Shima, S., Mitzunaga, M., and Nakao, T. (1971). *Endocrinology* **88**, 465.

Shima, S., Kawashima, Y., and Hirai, M. (1978). *Endocrinology* **103**, 1361.

Smiciklas, H. A., Pike, R. L., and Schraer, H. (1971). *J. Nutrition* **101**, 1045.

Sokabe, H. (1966). *Jpn. J. Physiol.* **16**, 380.

Sommers, S. C., and Terzakis, J. A. (1970). *Am. J. Clin. Pathol.* **54**, 303.

Stachenko, J., and Giroud, C. J. P. (1959a). *Endocrinology* **64**, 730.

Stachenko, J., and Giroud, C. J. P. (1959b). *Endocrinology* **64**, 743.

Stachenko, J., and Giroud, C. J. P. (1964). *Can. J. Biochem.* **42**, 1777.

Stark, E., Palkovitz, M., Fachet, J., and Hajtman, B. (1965). *Acta Med. Acad. Sci. Hung.* **21**, 263.

Stöcker, E., Kabus, K., and Dhom, G. (1966). *Z. Zellforsch. Mikrosk. Anat.* **72**, 1.

Strott, C. A. (1977). *J. Biol. Chem.* **252**, 464.

Swallow, R. L., and Sayers, G. (1969). *Proc. Soc. Exp. Biol. Med.* **131**, 1.

Symington, T. (1969). "Functional Pathology of the Human Adrenal Gland." Livingstone, Edinburgh.

Szalay, K. S., Bàcsy, E., and Stark, E. (1975). *Acta Endocrinol. (Copenhagen)* **80**, 114.

Tamaoki, B. I. (1973). *J. Steroid Biochem.* **4**, 89.

Tan, S. Y., and Mulrow, P. J. (1978). *Endocrinology* **102**, 1113.

Tandler, B., Erlandson, R. A., and Wynder, E. L. (1968). *Am. J. Pathol.* **52**, 69.

Taylor, A. A., Davis, J. O., Breitenbach, R. P. and Hartroft, P. M. (1970). *Gen. Comp. Endocrinol.* **14**, 321.

Taylor, J. D., Honn, K. V., and Chavin, W. (1975). *Gen. Comp. Endocrinol.* **27**, 358.

Temple, R., and Wolff, J. (1973). *J. Biol. Chem.* **248**, 2691.

Tonutti, E. (1951). *Endokrinologie* **28**, 1.

Tsuchiyama, H., Sughihara, H., and Kawai, K. (1972). In "Spontaneous Hypertension" (K. Okamoto, ed.), pp. 177-184. Igaku Shoin, Tokyo.

Unsicker, K. (1971). *Z. Zellforsch. Mikrosk. Anat.* **116**, 151.

Unsicker, K. (1973). *Z. Zellforsch. Mikrosk. Anat.* **146**, 385.

Vardolov, L., and Weiss, M. (1978). *J. Steroid Biochem.* **9**, 47.

Varna, M. M. (1977). *Gen. Comp. Endocrinol.* **33**, 61.

Vinson, G. P., and Kenyon, C. J. (1978). *Gen. Comp. Clin. Endocrinol. Adrenal Cortex* **2**, 201-264.

Walker, B. E., and Rennels, E. G. (1961). *Endocrinology* **68**, 365.

Wang, H. P., Pfeiffer, D. K., Kimura, T., and Tchen, T. T. (1974). *Biochem. Biophys. Res. Commun.* **57**, 93.

Wassermann, D., and Wassermann, M. (1974). *Cell Tissue Res.* **149**, 235.

Werb, Z., and Dingle, J. T. (1976). *Lisosomes Biol. Pathol.* **5**, 127.

Werbin, S. C., and Chaikoff, I. L. (1961). *Arch. Biochem. Biophys.* **93**, 476.

Werning, C., Stadelmann, O., Miederer, S. E., Vetter, W., Schweikert, H. U., Stiel, D., and Siegenthaler, W. (1971). *Klin. Wschr.* **49**, 1285.

Wheatley, D. N. (1967). *J. Anat.* **101**, 223.

Wheatley, D. N. (1968). *J. Anat.* **103**, 151.

Winer, N., Chokshi, D. S., and Walkenhorst, W. G. (1971). *Circul. Res.* **29**, 239.

Winkler, H., and Smith, A. D. (1975). *Handb. Physiol. Sect. 7: Endocrinol.* **6**, 321-339.

Wright, N. A. (1971). *J. Endocrinol.* **49**, 599.

Wright, N. A., and Voncina, D. (1977). *J. Anat.* **123**, 147.

Wright, N. A., Voncina, D., and Morley, A. R. (1973). *J. Endocrinol.* **59**, 451.

Wyllie, A. H., Kerr, J. F. R., and Currie, A. R. (1973a). *J. Pathol.* **111**, 255.

Wyllie, A. H., Kerr, J. F. R., Macaskill, I. A. M., and Currie, A. R. (1973b). *J. Pathol.* **111**, 85.

Yago, N., Seki, M., Sekiyama, S., Kobayashi, S., Kurokawa, H., Iwai, Y., Sato, F., and Shiragai, A. (1972). *J. Cell Biol.* **52**, 503.

Yonetsu, T. (1966). *Endocrinol. Jpn.* **13**, 269.
Yoshimura, F., Harumiya, K., Suzuki, N., and Totsuka, S. (1968). *Endocrinol. Jpn.* **15**, 20.
Youson, J. H. (1972). *Gen. Comp. Endocrinol.* **19**, 56.
Youson, J. H. (1973). *Can. J. Zool.* **51**, 796.

NOTE ADDED IN PROOF. A number of relevant papers confirming or extending statements made in the text have recently appeared.

Fine structure of normally functioning zona glomerulosa. The ultrastructure of the human [Zwierzina, W. D. (1979). *Acta Anat.* **103**, 409] and rabbit zona glomerulosa [Mazzocchi, G., Belloni, A. S., Rebuffat, P., Robba, C., Neri, G., and Nussdorfer, G. G. (1979). *Cell Tissue Res.* **201**, 165] were carefully described. The three-dimensional organization of the adrenal cortex of rat, pig, and cat was studied by scanning electron microscopy [Motta, P., Muto, M., and Fujita, T. (1979). *Cell Tissue Res.* **196**, 23]. The zonation in the duck interrenals was morphometrically confirmed [Pearce, R. B., Cronshaw, J., and Holmes, W. N. (1979). *Cell Tissue Res.* **196**, 429], and its functional significance discussed [Klingbeil, C. K., Holmes, W. N., Pearce, R. B., and Cronshaw, J. (1979). *Cell Tissue Res.* **201**, 23].

Fine structure of hyperfunctioning and hypofunctioning zona glomerulosa. The factors regulating zona glomerulosa growth and secretion were further investigated *in vivo* [Müller, J. (1978). *Endocrinology* **103**, 2061; Komor, J., and Müller, J. (1979). *Acta Endocrinol. (Copenhagen)* **90**, 680] and *in vitro* [Aguillera, G., and Catt, K. J. (1979). *Endocrinology* **104**, 1046]. The interrelationships occurring between the renin–angiotensin system and ACTH were studied [Kralem, Z., Rosenthal, T., Rotzak, R., and Lunenfeld, B. (1979). *Acta Endocrinol. (Copenhagen)* **91**, 657; Lefebvre, J., Dewailly, D., Racadot, A., Fossati, P., and Linquette, M. (1979). *Ann. Endocrinol. (Paris)* **40**, 433], and it was demonstrated that in man the presence of ACTH is necessary for a full corticosteroid response to angiotensin II and that infused angiotensin II inhibits ACTH secretion [Mason, P. A., Fraser, R., Semple, P. F., and Morton, J. J. (1979). *J. Steroid Biochem.* **10**, 235]. The effects of low potassium diet on the rat zona glomerulosa cells [Kawai, K., Sugihara, H., and Tsuchiyama, M. (1979). *Acta Pathol. Jpn.* **29**, 351] and the hyperactivity of interrenal cells during long-term adaptation to seawater of the *Anguilla rostrata* [Bhattacharyya, T. K., and Butler, D. G. (1979). *Anat. Rec.* **193**, 213] were investigated by the electron microscope. An accurate description of the ultrastructure of 25 adrenal adenomas causing primary aldosteronism was done [Kano, K., Sato, S., and Hama, H. (1979). *Virchows Arch. A* **384**, 93], and the structure of the pathological human zona glomerulosa was reviewed [Neville, A. M., and O'Hare, M. J. (1979). *In* "The Adrenal Gland" (V. H. T. James, ed.), pp. 1–65. Raven Press, New York].

Morphological–functional correlations in zona glomerulosa cells. The mechanisms of action of the adrenoglomerulotropic factors were investigated. The ACTH-induced increase in the cAMP production by zona glomerulosa cells [Shima, S., Kawashima, Y., and Hirai, M. (1979). *Endocrinol. Jpn.* **26**, 219], and the role of cytochrome P_{450} in the action of sodium depletion on aldosterone biosynthesis in rats [Kramer, R. E., Gallant, S., and Brownie, A. C. (1979). *J. Biol. Chem.* **254**, 3953] were described. The ACTH-elicited increase in the uptake of cholesterol by adrenocortical cells [Watanuki, M., and Hall, P. H. (1979). *FEBS Lett.* **101**, 239] and in the activity of sterol ester hydrolases [Trzeciak, W. H., Mason, J. I., and Boyd, G. S. (1979). *FEBS Lett.* **102**, 13] was confirmed.

Many papers furthered the hypothesis that an exocytotic mechanism is involved in the hormone release by steroid-producing cells [Gemmell, R. T., and Stacy, B. D. (1979). *J. Reprod. Fertil.* **57**, 87; Gemmell, R. T., and Stacy, B. D. (1979). *Cell Tissue Res.* **197**, 413; Gemmell, R. T., and Stacy, B. D. (1979). *Am. J. Anat.* **155**, 1; Nussdorfer, G. G., Neri, G., and Gambino, A. M. (1978). *J. Steroid Biochem.* **9**, 835; Nussdorfer, G. G., Mazzocchi, G., Robba, C., and Rebuffat, P. (1979). *Anat. Anz.* **145**, 319; Sawyer, H. R., Abel, J. H., Jr., McClellan, M. C., Schmitz, M., and Niswender, G. D. (1979). *Endocrinology* **104**, 476].

Subject Index

A

Actin-myosin regulation, interaction via Ca^{2+}
  activating effect of tropomyosin, 292–293
  kinase-phosphatase regulation, 298–299
  myosin dephosphorylation, 295–298
  myosin light chain kinase, 293–295
  myosin phosphorylation, 283–291
Actomyosin, of smooth muscle, 266–274
Adrenal zona glomerulosa
  fine structure in hyper- and hypofunction
    aldosterone-secreting tumors, 341–342
    alteration of Na/K balance, 332–336
    multifactorial regulation of growth and se-
      cretion, 328–329
    other experimental procedures, 339–341
    stimulation and inhibition of
      hypothalamo-hypophyseal-adrenal axis,
      336–339
    stimulation and inhibition of the renin-
      angiotensin system, 329–331
  fine structure in normal function, 308
    general remarks, 325–328
    mammalian zona glomerulosa, 309–321
    mammalian zona intermedia, 321–323
    zona glomerulosa in lower vertebrates,
      323–325
  maintenance of adrenal cortex and normal
    maintenance, 355
  supporting evidence for cell migration
    theory, 355–359
  theories on cytogenesis in adrenal cortex,
    353–355
  morphological-functional correlations, 342–
    343
    functional significance of ultrastructural
      changes, 345–350
    mechanism of action of adrenoglomerulo-
      tropic factors, 344–345
    mechanism of hormone release, 350–353
    subcellular localization of enzymes of al-
      dosterone synthesis, 343–344
Adrenoglomerulotropic factors, mechanism of
  action, 344–345
Aldosterone
  secretion by tumors, 341–342
  synthesis, localization of enzymes for,
    343–344

B

Bleaching, conformational changes during,
  150–152

C

Cells, multicentriolar, 100–102
Centriolar complex, mitotic, organization of,
  82–84
Centriole
  function, nucleic acids and, 98–100
  movements during mitosis, 84–85
  multiple, 100–102
  systems lacking, 85–90
  two functional states of, 90–91
Choline, uptake of, 230–232
Chromatin, mitosis and, 29–31
Contractile material, of smooth muscle, 263–
  266

D

Dense bodies, and attachment plaques, of
  smooth muscle, 262–263
Disc membranes
  chemical composition, 111–120
  dynamic aspects of
    conformational changes during bleaching,
      150–152
    lipid fluidity, 149–150
    renewal of membrane, 153–156
    rhodopsin mobility, 148–149
  isolation of, 109–111

H

Hexoses, uptake of, 213–214
  malignant transformation and, 227–230
  nutritional and hormonal regulation of, 223–
    227
  phosphorylation and, 217–219
  transport and, 214–217
  transport and phosphorylation in tandem,
    219–223
Hormone, adrenal zona glomerulosa,
  mechanism of release, 350–353
Hypothalamo-hypophyseal-adrenal axis, stimulation and inhibition of, 336–339

369

I

Immunofluorescence, indirect, use in mitotic investigations, 91–96

K

Kinase-phosphatase, regulation, 298–299

L

Lipid, disc membrane, fluidity of, 149–150
Lipid bilayer, organization of rhodopsin molecules into
  size and nature of intramembranous particles, 143–148
  tangential, 139–142
  transverse, 124–138
Lower vertebrates, adrenal zona glomerulosa in, 323–325

M

Mammals, adrenal zona glomerulosa, fine structure in normal function, 309–321
Mammals, adrenal zona intermedia, fine structure in normal function, 321–323
Matrix, mitosis and, 25–26
Membranes, mitosis and, 20–25
Microtubules
  cytoplasmic, mitosis and, 53–54
  nucleation, *in vitro* studies, 96–98
Mitosis
  centriolar complex organization, 82–84
  centriolar movements during, 84–85
  characteristics of
    chromatin, 29–31
    control mechanisms, 54–55
    general, 5–20
    involvement of cytoplasmic microtubules, 53–54
    matrix, 25–26
    membranes, 20–25
    nucleolus, 26–29
    polar structures, 31–37
    spindle, 37–53
    summary, 55
  evolution of, 55
    hypothetical early mitotic systems, 58–63

    phylogenetic position of extant mitoses, 63–65
    prokaryote ancestors, 56–58
  as indicator of phylogeny, 65–66
    advantages, 66
    disadvantages, 66–67
    tentative relationships, 67–69
  indirect immunofluorescence and, 91–96
Myosin
  dephosphorylation and relaxation, 295–298
  light chain kinase, 293–295
  phosphorylation
    Ca^{2+} regulation and, 283–291
    of smooth muscle, 277–280

N

Nucleobases, uptake of, 207–208
Nucleic acid(s), centriolar function and, 98–100
Nucleolus, mitosis and, 26–29
Nucleoside kinase, nucleoside uptake and, 195–200
Nucleosides, uptake of
  effect of transport inhibitors, 207
  nonmediated permeation and, 203–204
  nucleoside kinases, 195–200
  nucleoside transport and phosphorylation in tandem, 200–203
  nucleoside transport system, 192–195
  physiological regulation of, 204–206

O

Outer segments
  isolation of, 109–111
  structure, 109

P

Permeation, nonmediated, contribution to nucleoside uptake, 203–204
Phosphorylation
  dissociation from transport, 190–191
  model of, 176–177
  nucleoside transport and, 200–203
Photoreceptor membranes, aspects of organization
  chemical composition of disc membranes, 111–120

general morphology, 108–109
isolation of outer segments and disc mem-
    branes, 109–111
outer segment structure, 109
rhodopsin properties, 121–124
Phylogeny, mitosis as indicator of, 65–69
Polar structures, mitosis and, 31–37
Protein(s), in addition to actomyosin, of smooth
    muscle, 280–283
Purines
    phosphoribosylation of, 210–211
    transport and phosphoribosylation in tandem,
        211–212
    transport of, 208–210
    uptake, physiological regulation, 212–213

R

Renin-angiotensin system, stimulation and inhi-
    bition of, 329–331
Rhodopsin
    mobility of, 148–149
    properties of, 121–124
Rhodopsin molecules, organization into lipid
    bilayer
    size and nature of intramembranous particles,
        143–148
    tangential, 139–142
    transverse, 124–138

S

Skeletin filaments, of smooth muscle, 256–262
Smooth muscle
    architecture of contractile apparatus
        contractile apparatus, 263–266
        dense bodies and attachment plaques,
            262–263
        general development of ultrastructural
            studies, 245–249

general morphology, 242
    organization of contractile material, 242–
        245
    ten nm skeletin filaments, 256–262
    thick filament architecture and structural
        polarity, 249–256
contractile proteins
    actomyosin, 266–274
    additional proteins in actomyosin, 280–283
    myosin, 277–280
    thin filaments, 274–277
Sodium/potassium balance, alteration of, 332–
    336
Spindle, mitosis and, 37–53

T

Thick filaments, architecture and structural po-
    larity, 249–256
Thin filaments, of smooth muscle, 274–277
Transport
    inhibitors, effect on nucleoside uptake, 207
    model of, 174–176
Transport and phosphorylation, theoretical and
        methodological considerations
    dissociating transport from phosphorylation,
        190–191
    kinetic behavior of transport and phosphoryla-
        tion operating in tandem, 177–190
    nomenclature, 173
    phosphorylation models, 176–177
    transport models, 174–176
Tropomyosin, activating effect of, 292–293
Tumors, aldosterone-secreting, 341–342

V

Vitamins, uptake of, 230–232

Contents of Previous Volumes

Volume 1

Some Historical Features in Cell Biology—
ARTHUR HUGUES
Nuclear Reproduction—C. LEONARD HUSKINS
Enzymic Capacities and Their Relation to Cell
Nutrition in Animals—GEORGE W. KIDDER
The Application of Freezing and Drying Tech-
niques in Cytology—L. G. E. BELL
Enzymatic processes in Cell Membrane Pene-
tration—TH. ROSENBERG AND W. WILBRANDT
Bacterial Cytology—K. A. BISSET
Protoplast Surface Enzymes and Absorption of
Sugar—R. BROWN
Reproduction of Bacteriophage—A. D. HER-
SHEY
The Folding and Unfolding of Protein Molecules
as a Basis of Osmotic Work—R. J. GOLD-
ACRE
Nucleo-Cytoplasmic Relations in Amphibian
Development—G. FRANK-HAUSER
Structural Agents in Mitosis—M. M. SWANN
Factors Which Control the Staining of Tissue
Sections with Acid and Basic Dyes—MARCUS
SINGER
The Behavior of Spermatozoa in the Neighbor-
hood of Eggs—LORD ROTHSCHILD
The Cytology of Mammalian Epidermis and
Sebaceous Glands—WILLIAM MONTAGNA
The Electron-Microscopic Investigation of Tis-
sue Sections—L. H. BRETSCHNEIDER
The Histochemistry of Esterases—G. GOMORI
AUTHOR INDEX—SUBJECT INDEX

Volume 2

Quantitative Aspects of Nuclear Nucleopro-
teins—HEWSON SWIFT
Ascorbic Acid and Its Intracellular Localization,
with Special Reference to Plants—J. CHAYEN
Aspects of Bacteria as Cells and as Organisms—
STUART MUDD AND EDWARD D. DE LAMATER
Ion Secretion in Plants—J. F. SUTCLIFFE
Multienzyme Sequences in Soluble Extracts—
HENRY R. MAHLER
The Nature and Specificity of the Feulgen Nu-
clear Reaction—M. A. LESSLER

Quantitative Histochemistry of Phosphatases—
WILLIAM L. DOYLE
Alkaline Phosphatase of the Nucleus—M.
CHÈVREMONT AND H. FIRKET
Gustatory and Olfactory Epithelia—A. F.
BARADI AND G. H. BOURNE
Growth and Differentiation of Explanted
Tissues—P. J. GAILLARD
Electron Microscopy of Tissue Sections—A. J.
DALTON
A Redox Pump for the Biological Performance
of Osmotic Work, and Its Relation to the
Kinetics of Free Ion Diffusion across
Membranes—E. J. CONWAY
A Critical Survey of Current Approaches in
Quantitative Histo- and Cytochemistry—
DAVID GLICK
Nucleo-cytoplasmic Relationships in the De-
velopment of Acetabularia—J. HAMMERLING
Report of Conference of Tissue Culture Workers
Held at Cooperstown, New York—D. J.
HETHERINGTON
AUTHOR INDEX—SUBJECT INDEX

Volume 3

The Nutrition of Animal Cells—CHARITY WAY-
MOUTH
Caryometric Studies of Tissue Cultures—OTTO
BUCHER
The Properties of Urethan Considered in Rela-
tion to Its Action on Mitosis—IVOR CORNMAN
Composition and Structure of Giant Chromo-
somes—ALEXANDER L. DOUNCE
How Many Chromosomes in Mammalian Soma-
tic Cells?—R. A. BEATTY
The Significance of Enzyme Studies on Isolated
Cell Nuclei—ALEXANDER L. DOUNCE
The Use of Differential Centrifugation in the
Study of Tissue Enzymes—CHR. DE DUVE
AND J. BERTHET
Enzymatic Aspects of Embryonic Differentia-
tion—TRYGGVE GUSTAFSON
Azo Dye Methods in Enzyme Histochemistry—
A. G. EVERSON PEARSE
Microscopic Studies in Living Mammals with
Transparent Chamber Methods—ROY G.
WILLIAMS

The Mast Cell—G. ASBOE-HANSEN
Elastic Tissue—EDWARDS W. DEMPSEY AND ALBERT I. LANSING
The Composition of the Nerve Cell Studied with New Methods—SVEN-OLOE BRATTGÅRD AND HOLGER HYDEN
AUTHOR INDEX—SUBJECT INDEX

Volume 4

Cytochemical Micrurgy—M. J. KOPAC
Amoebocytes—L. E. WAGGE
Problems of Fixation in Cytology, Histology, and Histochemistry—M. WOLMAN
Bacterial Cytology—ALFRED MARSHAK
Histochemistry of Bacteria—R. VENDRELY
Recent Studies on Plant Mitochondria—DAVID P. HACKETT
The Structure of Chloroplasts—K. MÜHLE-THALER
Histochemistry of Nucleic Acids—N. B. KUR-NICK
Structure and Chemistry of Nucleoli—W. S. VINCENT
On Goblet Cells, Especially of the Intestine of Some Mammalian Species—HARALD MOE
Localization of Cholinesterases at Neuromuscular Junctions—R. COUTEAUX
Evidence for a Redox Pump in the Active Transport of Cations—E. J. CONWAY
AUTHOR INDEX—SUBJECT INDEX

Volume 5

Histochemistry with Labeled Antibody—ALBERT H. COONS
The Chemical Composition of the Bacterial Cell Wall—C. S. CUMMINS
Theories of Enzyme Adaptation in Microorganisms—J. MANDELSTAM
The Cytochondria of Cardiac and Skeletal Muscle—JOHN W. HARMON
The Mitochondria of the Neuron—WARREN ANDREW
The Results of Cytophotometry in the Study of the Deoxyribonucleic Acid (DNA) Content of the Nucleus—R. VENDRELY AND C. VENDRELY

Protoplasmic Contractility in Relation to Gel Structure: Temperature-Pressure Experiments on Cytokinesis and Amoeboid Movement—DOUGLAS MARSLAND
Intracellular pH—PETER C. CALDWELL
The Activity of Enzymes in Metabolism and Transport in the Red Cell—T. A. PRANKERD
Uptake and Transfer of Macromolecules by Cells with Special Reference to Growth and Development—A. M. SCHECHTMAN
Cell Secretion: A Study of Pancreas and Salivary Glands—L. C. J. JUNQUEIRA AND G. C. HIRSCH
The Acrosome Reaction—JEAN C. DAN
Cytology of Spermatogenesis—VISHWA NATH
The Ultrastructure of Cells, as Revealed by the Electron Microscope—FRITIOF S. SJÖSTRAND
AUTHOR INDEX—SUBJECT INDEX

Volume 6

The Antigen System of *Paramecium aurelia*—G. H. BEALE
The Chromosome Cytology of the Ascites Tumors of Rats, with Special Reference to the Concept of the Stemline Cell—SAJIRO MAKINO
The Structure of the Golgi Apparatus—ARTHUR W. POLLISTER AND PRISCHIA F. POLLISTER
An Analysis of the Process of Fertilization and Activation of the Egg—A. MONROY
The Role of the Electron Microscope in Virus Research—ROBLEY C. WILLIAMS
The Histochemistry of Polysaccharides—ARTHUR J. HALE
The Dynamic Cytology of the Thyroid Gland—J. GROSS
Recent Histochemical Results of Studies on Embryos of Some Birds and Mammals—ELIO BORGHESE
Carbohydrate Metabolism and Embryonic Determination—R. J. O'CONNOR
Enzymatic and Metabolic Studies on Isolated Nuclei—G. SIEBERT AND R. M. S. SMELLIE
Recent Approaches of the Cytochemical Study of Mammalian Tissues—GEORGE H. HOGEBOOM, EDWARD L. KUFF, AND WALTER C. SCHNEIDER
The Kinetics of the Penetration of Nonelectro-

lytes into the Mammalian Erythrocyte—
FREDA BOWYER
AUTHOR INDEX—SUBJECT INDEX
CUMULATIVE SUBJECT INDEX (VOLUMES 1-5)

Volume 7

Some Biological Aspects of Experimental
Radiology: A Historical Review—F. G.
SPEAR
The Effect of Carcinogens, Hormones, and
Vitamins on Organ Cultures—ILSE LASNITZKI
Recent Advances in the Study of the
Kinetochore—A. LIMA-DE-FARIA
Autoradiographic Studies with S³⁵-Sulfate—D.
D. DZIEWIATKOWSKI
The Structure of the Mammalian Sper-
matozoon—DON W. FAWCETT
The Lymphocyte—O. A. TROWELL
The Structure and Innervation of Lamellibranch
Muscle—J. BOWDEN
Hypothalamo-neurohypophysial   Neurosecre-
tion—J. C. SLOPER
Cell Contact—PAUL WEISS
The Ergastoplasm: Its History, Ultrastructure,
and Biochemistry—FRANÇOISE HAGUENAU
Anatomy of Kidney Tubules—JOHANNES RHO-
DIN
Structure and Innervation of the Inner Ear Sen-
sory Epithelia—HANS ENGSTRÖM AND JAN
WERSÄLL
The Isolation of Living Cells from Animal
Tissues—L. M. RINALDINI
AUTHOR INDEX—SUBJECT INDEX

Volume 8

The Structure of Cytoplasm—CHARLES OBER-
ING
Wall Organization in Plant Cells—R. D. PRES-
TON
Submicroscopic Morphology of the Synapse—
EDUARDO DE ROBERTIS
The Cell Surface of *Paramecium*—C. F. EHRET
AND E. L. POWERS
The Mammalian Reticulocyte—LEAH MIRIAM
LOWENSTEIN
The Physiology of Chromatophores—MILTON
FINGERMAN

The Fibrous Components of Connective Tissue.
with Special Reference to the Elastic
Fiber—DAVID A. HALL
Experimental Heterotopic Ossification—J. B.
BRIDGES
A Survey of Metabolic Studies on Isolated
Mammalian Nuclei—D. B. ROODYN
Trace Elements in Cellular Function—BERT L.
VALLEE AND FREDERIC L. HOCH
Osmotic Properties of Living Cells—D. A. T.
DICK
Sodium and Potassium Movements in Nerve,
Muscle, and Red Cells—I. M. GLYNN
Pinocytosis—H. HOLTER
AUTHOR INDEX—SUBJECT INDEX

Volume 9

The Influence of Cultural Conditions on Bacte-
rial Cytology—J. F. WILKINSON AND J. P.
DUGUID
Organizational Patterns within Chromosomes—
BERWIND P. KAUFMANN, HELEN GAY, AND
MARGARET R. MCDONALD
Enzymic Processes in Cells—JAY BOYD BEST
The Adhesion of Cells—LEONARD WEISS
Physiological and Pathological Changes in
Mitochondrial Morphology—CH. ROUILLER
The Study of Drug Effects at the Cytological
Level—G. B. WILSON
Histochemistry of Lipids in Oogenesis—VISHWA
NATH
Cyto-Embryology of Echinoderms and Am-
phibia—KUTSUMA DAN
The Cytochemistry of Nonenzyme Proteins—
RONALD R. COWDEN
AUTHOR INDEX—SUBJECT INDEX

Volume 10

The Chemistry of Shiff's Reagent—FREDERICK
H. KASTEN
Spontaneous and Chemically Induced Chromo-
some Breaks—ARUN KUMAR SHARMA AND
ARCHANA SHARMA
The Ultrastructure of the Nucleus and Nucleo-
cytoplasmic Relations—SAUL WISCHNITZWE
The Mechanics and Mechanism of Cleavage—
LEWIS WOLPERT

The Growth of the Liver with Special Reference to Mammals—F. DOLJANSKI

Cytology Studies on the Affinity of the Carcinogenic Azo Dyes for Cytoplasmic Components—YOSHIMA NAGATANI

Epidermal Cells in Culture—A. GEDEON MATOLTSY

AUTHOR INDEX—SUBJECT INDEX

CUMULATIVE SUBJECT INDEX (VOLUMES 1-9)

Volume 11

Electron Microscopic Analysis of the Secretion Mechanism—K. KUROSUMI

The Fine Structure of Insect Sense Organs—ELEANOR H. SLIFER

Cytology of the Developing Eye—ALFRED J. COULOMBRE

The Photoreceptor Structures—J. J. WOLKEN

Use of Inhibiting Agents in Studies on Fertilization Mechanisms—CHARLES B. METZ

The Growth-Duplication Cycle of the Cell—D. M. PRESCOTT

Histochemistry of Ossification—ROMULO L. CABRINI

Cinematography, Indispensable Tool for Cytology—C. M. POMERAT

AUTHOR INDEX—SUBJECT INDEX

Volume 12

Sex Chromatin and Human Chromosomes—JOHN L. HAMERTON

Chromosomal Evolution in Cell Populations—T. C. HSU

Chromosome Structure with Special Reference to the Role of Metal Ions—DALE M. SEFFENSEN

Electron Microscopy of Human White Blood Cells and Their Stem Cells—MARCEL BESSIS AND JEAN-PAUL THIERY

In Vivo Implantation as a Technique in Skeletal Biology—WILLIAM J. L. FELTS

The Nature and Stability of Nerve Myelin—J. B. FINEAN

Fertilization of Mammalian Eggs *in Vitro*—C. R. AUSTIN

Physiology of Fertilization in Fish Eggs—TOKI-O YAMAMOTO

AUTHOR INDEX—SUBJECT INDEX

Volume 13

The Coding Hypothesis—MARTYNAS YCAS

Chromosome Reproduction—J. HERBERT TAYLOR

Sequential Gene Action, Protein Synthesis, and Cellular Differentiation—REED A. FLICKINGER

The Composition of the Mitochondrial Membrane in Relation to Its Structure and Function—ERIC G. BALL AND CLIFFE D. JOEL

Pathways of Metabolism in Nucleate and Anucleate Erythrocytes—H. A. SCHWEIGER

Some Recent Developments in the Field of Alkali Cation Transport—W. WILBRANDT

Chromosome Aberrations Induced by Ionizing Radiations—H. J. EVANS

Cytochemistry of Protozoa, with Particular Reference to the Golgi Apparatus and the Mitochondria—VISHWA NATH AND G. P. DUTTA

Cell Renewal—FELIX BERTALANFFY AND CHOSEN LAU

AUTHOR INDEX—SUBJECT INDEX

Volume 14

Inhibition of Cell Division: A Critical and Experimental Analysis—SEYMOUR GELFANT

Electron Microscopy of Plant Protoplasm—R. BUVAT

Cytophysiology and Cytochemistry of the Organ of Corti: A Cytochemical Theory of Hearing—J. A. VINNIKOV AND L. K. TITOVA

Connective Tissue and Serum Proteins—R. E. MANCINI

The Biology and Chemistry of the Cell Walls of Higher Plants, Algae, and Fungi—D. H. NORTHCOTE

Development of Drug Resistance by Staphylococci *in Vitro* and *in Vivo*—MARY BARBER

Cytological and Cytochemical Effects of Agents

Implicated in Various Pathological Conditions: The Effect of Viruses and of Cigarette Smoke on the Cell and Its Nucleic Acid—CECILIE LEUCHTENBERGER AND RUDOLF LEUCHTENBERGER
The Tissue Mast Wall—DOUGLAS E. SMITH
AUTHOR INDEX—SUBJECT INDEX

Volume 15

The Nature of Lampbrush Chromosomes—H. G. CALLAN
The Intracellular Transfer of Genetic Information—J. L. SIRLIN
Mechanisms of Gametic Approach in Plants—LEONARD MACHLIS AND ERIKA RAWITSCHER-KUNKEL
The Cellular Basis of Morphogenesis and Sea Urchin Development—T. GUSTAFSON AND L. WOLPERT
Plant Tissue Culture in Relation to Development Cytology—CARL R. PARTANEN
Regeneration of Mammalian Liver—NANCY L. R. BUCHER
Collagen Formation and Fibrogenesis with Special Reference to the Role of Ascorbic Acid—BERNARD S. GOULD
The Behavior of Mast Cells in Anaphylaxis—IVAN MOTA
Lipid Absorption—ROBERT M. WOTTON
AUTHOR INDEX—SUBJECT INDEX

Volume 16

Ribosomal Functions Related to Protein Synthesis—TORE HULTIN
Physiology and Cytology of Chloroplast Formation and "Loss" in *Euglena*—M. GRENSON
Cell Structures and Their Significance for Ameboid Movement—K. E. WOHLFARTH-BOTTERMAN
Microbeam and Partial Cell Irradiation—C. L. SMITH
Nuclear-Cytoplasmic Interaction with Ionizing Radiation—M. A. LESSLER
In Vivo Studies of Myelinated Nerve Fibers—CARL CASKEY SPEIDEL

Respiratory Tissue: Structure, Histophysiology, Cytodynamics. Part 1: Review and Basic Cytomorphology—FELIX D. BERTALANFFY
AUTHOR INDEX—SUBJECT INDEX

Volume 17

The Growth of Plant Cell Walls—K. WILSON
Reproduction and Heredity in Trypanosomes: A Critical Review Dealing Mainly with the African Species in the Mammalian Host—P. J. WALKER
The Blood Platelet: Electron Microscopic Studies—J. F. DAVID-FERREIRA
The Histochemistry of Mucopolysaccharides—ROBERT C. CURRAN
Respiratory Tissue Structure, Histophysiology. Cytodynamics. Part II. New Approaches and Interpretations—FELIX D. BERTALANFFY
The Cells of the Adenohypophysis and Their Functional Significance—MARC HERLANT
AUTHOR INDEX—SUBJECT INDEX

Volume 18

The Cell of Langerhans—A. S. BREATHNACH
The Structure of the Mammalian Egg—ROBERT HADEK
Cytoplasmic Inclusions in Oogenesis—M. D. L. SRIVASTAVA
The Classification and Partial Tabulation of Enzyme Studies on Subcellular Fractions Isolated by Differential Centrifuging—D. B. ROODYN
Histochemical Localization of Enzyme Activities by Substrate Film Methods: Ribonucleases, Deoxyribonucleases, Proteases, Amylase, and Hyaluronidase—R. DAOUST
Cytoplasmic Deoxyribonucleic Acid—P. B. GAHAN AND J. CHAYEN
Malignant Transformation of Cells *in Vitro*—KATHERINE K. SANFORD
Deuterium Isotope Effects in Cytology—E. FLAUMENHAFT, S. BOSE, H. I. CRESPI, AND J. J. KATZ
The Use of Heavy Metal Salts as Electron

Stains—C. RICHARD ZOBEL AND MICHAEL
BEER
AUTHOR INDEX—SUBJECT INDEX

Volume 19

"Metabolic" DNA: A Cytochemical Study—H.
ROELS
The Significance of the Sex Chromatin—MUR-
RAY L. BARR
Some Functions of the Nucleus—J. M. MITCHI-
SON
Synaptic Morphology on the Normal and De-
generating Nervous System—E. G. GRAY
AND R. W. GUILLERY
Neurosecretion—W. BARGMANN
Some Aspects of Muscle Regeneration—E. H.
BETZ, H. FIRKET, AND M. REZNIK
The Gibberellins as Hormones—P. W. BRIAN
Phototaxis in Plants—WOLFGANG HAUPT
Phosphorus Metabolism in Plants—K. S.
ROWAN
AUTHOR INDEX—SUBJECT INDEX

Volume 20

The Chemical Organization of the Plasma Mem-
brane of Animal Cells—A. H. MADDY
Subunits of Chloroplast Structure and Quantum
Conversion in Photosynthesis—RODERIC B.
PARK
Control of Chloroplast Structure by Light—LES-
TER PACKER AND PAUL-ANDRÉ SIEGEN-
THALER
The Role of Potassium and Sodium Ions as Stud-
ied in Mammalian Brain—H. HILLMAN
Triggering of Ovulation by Coitus in the Rat—
CLAUDE ARON, GITTA ASCH, AND JAQUELINE
ROOS
Cytology and Cytophysiology of Non-
Melanophore Pigment Cells—JOSEPH T.
BAGNARA
The Fine Structure and Histochemistry of Prosta-
tic Glands in Relation to Sex Hormones
—DAVID BRANDES
Cerebellar Enzymology—LUCIE ARVY
AUTHOR INDEX—SUBJECT INDEX

Volume 21

Histochemistry of Lysosomes—P. B. GAHAN
Physiological Clocks—R. L. BRAHMACHARY
Ciliary Movement and Coordination in Ciliates—
BELA PARDUCA
Electromyography: Its Structural and Neural
Basis—JOHN V. BASMAJIAN
Cytochemical Studies with Acridine Orange and
the Influence of Dye Contaminants in the
Staining of Nucleic Acids—FREDERICK H.
KASTEN
Experimental Cytology of the Shoot Apical Cells
during Vegetative Growth and Flowering—A.
NOUGARÈDE
Nature and Origin of Perisynaptic Cells of the
Motor End Plate—T. R. SHANTHAVEERAPPA
AND G. H. BOURNE
AUTHOR INDEX—SUBJECT INDEX

Volume 22

Current Techniques in Biomedical Electron Mi-
croscopy—SAUL WISCHNITZER
The Cellular Morphology of Tissue Repair—R.
M. H. McMINN
Structural Organization and Embryonic Dif-
ferentiation—GAJANAN V. SHERBET AND M.
S. LAKSHMI
The Dynamism of Cell Division during Early
Cleavage Stages of the Egg—N. FAUTREZ-
FIRLEFYN AND J. FAUTREZ
Lymphopoiesis in the Thymus and Other Tis-
sues: Functional Implications—N. B. EVERETT
AND RUTH W. TYLER (CAFFREY)
Structure and Organization of the Myoneural
Junction—C. COËRS
The Ecdysial Glands of Arthropods—WILLIAM S.
HERMAN
Cytokinins in Plants—B. I. SAHAI SRIVASTAVA
AUTHOR INDEX—SUBJECT INDEX
CUMULATIVE SUBJECT INDEX (VOLUMES 1–21)

Volume 23

Transformationlike Phenomenon in Somatic Cells
—J. M. OLENOV

Recent Developments in the Theory of Control and regulation of Cellular Processes—ROBERT ROSEN

Contractile Properties of Protein Threads from Sea Urchin Eggs in Relation to Cell Division—HIKOICHI SAKAI

Electron Microscopic Morphology of Oogenesis —ARNE NORREVANG

Dynamic Aspects of Phospholipids during Protein Secretion—LOWELL E. HOKIN

The Golgi Apparatus: Structure and Function—H. W. BEAMS AND R. G. KESSEL

The Chromosomal Basis of Sex Determination—KENNETH R. LEWIS AND BERNARD JOHN

AUTHOR INDEX—SUBJECT INDEX

Volume 24

Synchronous Cell Differentiation—GEORGE M. PADILLA AND IVAN L. CAMERON

Mast Cells in the Nervous System—YNGVE OLSSON

Development Phases in Intermitosis and the Preparation for Mitosis of Mammalian Cells in Vitro—BLAGOJE A. NEŠKOVIĆ

Antimitotic Substances—GUY DEYSSON

The Form and Function of the Sieve Tube: A Problem in Reconciliation—P. E. WEATHERLEY AND R. P. C. JOHNSON

Analysis of Antibody Staining Patterns Obtained with Striated Myofibrils in Fluorescence Microscopy and Electron Microscopy—FRANK A. PEPE

Cytology of Intestinal Epithelial Cells—PETER G. TONER

Liquid Junction Potentials and Their Effects on Potential Measurements in Biology Systems—P. C. CALDWELL

AUTHOR INDEX—SUBJECT INDEX

Volume 25

Cytoplasmic Control over the Nuclear Events of Cell Reproduction—NOEL DE TERRA

Coordination of the Rhythm of Beat in Some Ciliary Systems—M. A. SLEIGH

The Significance of the Structural and

Functional Similarities of Bacteria and Mitochondria—SYLVAN NASS

The Effects of Steroid Hormones on Macrophage Activity—B. VERNON-ROBERTS

The Fine Structure of Malaria Parasites—MARIA A. RUDZINSKA

The Growth of Liver Parenchymal Nuclei and Its Endocrine Regulation—RITA CARRIERE

Strandedness of Chromosomes—SHELDON WOLFF

Isozymes: Classification, Frequency, and Significance—CHARLES R. SHAW

The Enzymes of the Embryonic Nephron—LUCIE ARVY

Protein Metabolism in Nerve Cells—B. DROZ

Freeze-Etching—HANS MOOR

AUTHOR INDEX—SUBJECT INDEX

Volume 26

A New Model for the Living Cell: A Summary of the Theory and Recent Experimental Evidence in Its Support—GILBERT N. LING

The Cell Periphery—LEONARD WEISS

Mitochondrial DNA: Physicochemical Properties, Replication, and Genetic Function—P. BORST AND A. M. KROON

Metabolism and Enucleated Cells—KONRAD KECK

Stereological Principles for Morphometry in Electron Microscopic Cytology—EWALD R. WEIBEL

Some Possible Roles for Isozymic Substitutions during Cold Hardening in Plants—D. W. A. ROBERTS

AUTHOR INDEX—SUBJECT INDEX

Volume 27

Wound-Healing in Higher Plants—JACQUES LIPETZ

Chloroplasts as Symbiotic Organelles—DENNIS L. TAYLOR

The Annulate Lamellae—SAUL WISCHNITZER

Gametogenesis and Egg Fertilization in Planarians—G. BENAZZI LENTATI

Ultrastructure of the Mammalian Adrenal Cortex—SIMON IDELMAN

The Fine Structure of the Mammalian Lymphoreticular System—IAN CARR

Immunoenzyme Technique: Enzymes as Markers for the Localization of Antigens and Antibodies—STRATIS AVRAMEAS

AUTHOR INDEX—SUBJECT INDEX

Volume 28

The Cortical and Subcortical Cytoplasm of *Lymnaea* Egg—CHRISTIAAN P. RAVEN

The Environment and Function of Invertebrate Nerve Cells—J. E. TREHERNE AND R. B. MORETON

Virus Uptake, Cell Wall Regeneration, and Virus Multiplication in Isolated Plant Protoplasts—E. C. COCKING

The Meiotic Behavior of the *Drosophila* Oocyte—ROBERT C. KING

The Nucleus: Action of Chemical and Physical Agents—RENÉ SIMARD

The Origin of Bone Cells—MAUREEN OWEN

Regeneration and Differentiation of Sieve Tube Elements—WILLIAM P. JACOBS

Cells, Solutes, and Growth: Salt Accumulation in Plants Reexamined—F. C. STEWARD AND R. L. MOTT

AUTHOR INDEX—SUBJECT INDEX

Volume 29

Gram Staining and Its Molecular Mechanism—B. B. BISWAS, P. S. BASU, AND M. K. PAL

The Surface Coats of Animal Cells—A. MARTÍNEZ-PALOMO

Carbohydrates in Cell Surfaces—RICHARD J. WINZLER

Differential Gene Activation in Isolated Chromosomes—MARKUS LEZZI

Intraribosomal Environment of the Nascent Peptide Chain—HIDEKO KAJI

Location and Measurement of Enzymes in Single Cells by Isotopic Methods Part I—E. A. BARNARD

Location and Measurement of Enzymes in Single Cells by Isotopic Methods Part II—G. C. BUDD

Neuronal and Glial Perikarya Preparations: An Appraisal of Present Methods—PATRICIA V. JOHNSTON AND BETTY I. ROOTS

Functional Electron Microscopy of the Hypothalamic Median Eminence—HIDESHI KOBAYASHI, TOKUZO MATSUI, AND SUSUMI ISHII

Early Development in Callus Cultures—MICHAEL M. YEOMAN

AUTHOR INDEX—SUBJECT INDEX

Volume 30

High-pressure Studies in Cell Biology—ARTHUR M. ZIMMERMAN

Micrurgical Studies with Large Free-Living Amebas—K. W. JEON AND J. F. DANIELLI

The Practice and Application of Electron Microscope Autoradiography—J. JACOB

Applications of Scanning Electron Microscopy in Biology—K. E. CARR

Acid Mucopolysaccharides in Calcified Tissues—SHINJIRO KOBAYASHI

AUTHOR INDEX—SUBJECT INDEX

CUMULATIVE SUBJECT INDEX (VOLUMES 1–29)

Volume 31

Studies on Freeze-Etching of Cell Membranes—KURT MÜHLETHALER

Recent Developments in Light and Electron Microscope Radioautography—G. C. BUDD

Morphological and Histochemical Aspects of Glycoproteins at the Surface of Animal Cells—A. RAMBOURG

DNA Biosynthesis—H. S. JANSZ, D. VAN DER MEI, AND G. M. ZANDVLIET

Cytokinesis in Animal Cells—R. RAPPAPORT

The Control of Cell Division in Ocular Lens—C. V. HARDING, J. R. REDDAN, N. J. UNAKAR, AND M. BAGCHI

The Cytokinins—HANS KENDE

Cytophysiology of the Teleost Pituitary—MARTIN SAGE AND HOWARD A. BERN

AUTHOR INDEX—SUBJECT INDEX

Volume 32

Highly Repetitive Sequences of DNA in Chromosomes—W. G. FLAMM
The Origin of the Wide Species Variation in Nuclear DNA Content—H. REES AND R. N. JONES
Polarized Intracellular Particle Transport: Saltatory Movements and Cytoplasmic Streaming —LIONEL I. REBHUN
The Kinetoplast of the Hemoflagellates—LARRY SIMPSON
Transport across the Intestinal Mucosal Cell: Hierarchies of Function—D. S. PARSONS AND C. A. R. BOYD
Wound Healing and Regeneration in the Crab *Paratelphusa hydrodromous*—RITA G. ADIYODI
The Use of Ferritin-Conjugated Antibodies in Electron Microscopy—COUNCILMAN MORGAN
Metabolic DNA in Ciliated Protozoa, Salivary Gland Chromosomes, and Mammalian Cells —S. R. PELC
AUTHOR INDEX—SUBJECT INDEX

Volume 33

Visualization of RNA Synthesis on Chromosomes—O. L. MILLER, JR. AND BARBARA A. HAMKALO
Cell Disjunction ("Mitosis") in Somatic Cell Reproduction—ELAINE G. DIACUMAKOS, SCOTT HOLLAND, AND PAULINE PECORA
Neuronal Microtubules, Neurofilaments, and Microfilaments—RAYMOND B. WUERKER AND JOEL B. KIRKPATRIC
Lymphocyte Interactions in Antibody Responses —J. F. A. P. MILLER
Laser Microbeams for Partial Cell Irradiation—MICHAEL W. BERNS AND CHRISTIAN SALET
Mechanisms of Virus-Induced Cell Fusion— GEORGE POSTE
Freeze-Etching of Bacteria—CHARLES C. REMSEN AND STANLEY W. WATSON
The Cytophysiology of Mammalian Adipose Cells—BERNARD G. SLAVIN
AUTHOR INDEX—SUBJECT INDEX

Volume 34

The Submicroscopic Morphology of the Interphase Nucleus—SAUL WISCHNITZER
The Energy State and Structure of Isolated Chloroplasts: The Oxidative Reactions Involving the Water-Splitting Step of Photosynthesis—ROBERT L. HEATH
Transport in *Neurospora*—GENE A. SCARBOROUGH
Mechanisms of Ion Transport through Plant Cell Membranes—EMANUEL ERSTEIN
Cell Motility: Mechanisms in Protoplasmic Streaming and Ameboid Movement—H. KOMNICK, W. STOCKEM, AND K. E. WOHLEFARTH-BOTTERMANN
The Gliointerstitial System of Molluscs— GHISLAIN NICAISE
Colchicine-Sensitive Microtubules—LYNN MARGULIS
AUTHOR INDEX—SUBJECT INDEX

Volume 35

The Structure of Mammalian Chromosomes— ELTON STUBBLEFIELD
Synthetic Activity of Polytene Chromosomes— HANS D. BERENDES
Mechanisms of Chromosome Synapsis at Meiotic Prophase—PETER B. MOENS
Structural Aspects of Ribosomes—N. NANNINGA
Comparative Ultrastructure of the Cerebrospinal Fluid-Contacting Neurons—B. VIGH AND I. VIGH-TEICHMANN
Maturation-Inducing Substances in Starfishes— HARUO KANATANI
The *Limonium* Salt Gland: A Biophysical and Structural Study—A. E. HILL AND B. S. HILL
Toxic Oxygen Effects—HAROLD M. SWARTZ
AUTHOR INDEX—SUBJECT INDEX

Volume 36

Molecular Hybridization of DNA and RNA *in Situ*—WOLFGANG HENNIG
The Relationship between the Plasmalemma and Plant Cell Wall—JEAN-CLAUDE ROLAND

Recent Advances in the Cytochemistry and Ultrastructure of Cytoplasmic Inclusions in Mastigophora and Opalinata (Protozoa)—G. P. DUTTA

Chloroplasts and Algae as Symbionts in Molluscs—LEONARD MUSCATINE AND RICHARD W. GREENE

The Macrophage—SAIMON GORDON AND ZANVIL A. COHN

Degeneration and Regeneration of Neurosecretory Systems—HORST-DIETER DELLMANN

AUTHOR INDEX—SUBJECT INDEX

Volume 37

Units of DNA Replication in Chromosomes of Eukaroytes—J. HERBERT TAYLOR

Viruses and Evolution—D. C. REANNEY

Electron Microscope Studies on Spermiogenesis in Various Animal Species—GONPACHIRO YASUZUMI

Morphology, Histochemistry, and Biochemistry of Human Oogenesis and Ovulation—SARDUL S. GURAYA

Functional Morphology of the Distal Lung—KAYE H. KILBURN

Comparative Studies of the Juxtaglomerular Apparatus—HIROFUMI SOKABE AND MIZUHO OGAWA

The Ultrastructure of the Local Cellular Reaction to Neoplasia—IAN CARR AND J. C. E. UNDERWOOD

Scanning Electron Microscopy in the Ultrastructural Analysis of the Mammalian Cerebral Ventricular System—D. E. SCOTT, G. P. KOZLOWSKI, AND M. N. SHERIDAN

AUTHOR INDEX—SUBJECT INDEX

Volume 38

Genetic Engineering and Life Synthesis: An Introduction to the Review by R. Widdus and C. Ault—JAMES F. DANIELLI

Progress in Research Related to Genetic Engineering and Life Synthesis—ROY WIDDUS AND CHARLES R. AULT

The Genetics of C-Type RNA Tumor Viruses—J. A. WYKE

Three-Dimensional Reconstruction from Projections: A Review of Algorithms—RICHARD GORDON AND GABOR T. HERMAN

The Cytophysiology of Thyroid Cells—VLADIMIR R. PANTIĆ

The Mechanisms of Neural Tube Formation—PERRY KARFUNKEL

The Behavior of the XY Pair in Mammals—ALBERTO J. SOLARI

Fine-Structural Aspects of Morphogenesis in *Acetabularia*—G. WERZ

Cell Separation by Gradient Centrifugation—R. HARWOOD

SUBJECT INDEX

Volume 39

Androgen Receptors in the Nonhistone Protein Fractions of Prostatic Chromatin—TUNG YUE WANG AND LEROY M. NYBERG

Nucleocytoplasmic Interactions in Development of Amphibian Hybrids—STEPHEN SUBTELNY

The Interactions of Lectins with Animal Cell Surfaces—GARTH L. NICOLSON

Structure and Functions of Intercellular Junctions—L. ANDREW STAEHELIN

Recent Advances in Cytochemistry and Ultrastructure of Cytoplasmic Inclusions in Ciliophora (Protozoa)—G. P. DUTTA

Structure and Development of the Renal Glomerulus as Revealed by Scanning Electron Microscopy—FRANCO SPINELLI

Recent Progress with Laser Microbeams—MICHAEL W. BERNS

The Problem of Germ Cell Determinants—H. W. BEAMS AND R. G. KESSEL

SUBJECT INDEX

Volume 40

B-Chromosome Systems in Flowering Plants and Animal Species—R. N. JONES

The Intracellular Neutral SH-Dependent Protease Associated with Inflammatory Reactions—HIDEO HAYASHI

The Specificity of Pituitary Cells and Regulation of Their Activities—VLADIMIR R. PANTIC

Fine Structure of the Thyroid Gland—HISAO FUJITA

Postnatal Gliogenesis in the Mammalian Brain —A. PRIVAT

Three-Dimensional Reconstruction from Serial Sections—RANDLE W. WARE AND VINCENT LoPRESTI

SUBJECT INDEX

Volume 41

The Attachment of the Bacterial Chromosome to the Cell Membrane—PAUL J. LEIBOWITZ AND MOSELIO SCHAECHTER

Regulation of the Lactose Operon in *Escherichia coli* by cAMP—G. CARPENTER AND B. H. SELLS

Regulation of Microtubules in *Tetrahymena*—NORMAN E. WILLIAMS

Cellular Receptors and Mechanisms of Action of Steroid Hormones—SHUTSUNG LIAO

A Cell Culture Approach to the Study of Anterior Pituitary Cells—A. TIXIER-VIDAL, D. GOURDJI, AND C. TOUGARD

Immunohistochemical Demonstration of Neurophysin in the Hypothalamoneurohypophysial System—W. B. WATKINS

The Visual System of the Horseshoe Crab *Limulus polyphemus*—WOLF H. FAHRENBACH

SUBJECT INDEX

Volume 42

Regulators of Cell Division: Endogenous Mitotic Inhibitors of Mammalian Cells—BISMARCK B. LOZZIO, CARMEN B. LOZZIO, ELENA G. BAMBERGER, AND STEPHEN V. LAIR

Ultrastructure of Mammalian Chromosome Aberrations—B. R. BRINKLEY AND WALTER N. HITTELMAN

Computer Processing of Electron Micrographs: A Nonmathematical Account—P. W. HAWKES

Cyclic Changes in the Fine Structure of the Epithelial Cells of Human Endometrium —MILDRED GORDON

The Ultrastructure of the Organ of Corti—ROBERT S. KIMURA

Endocrine Cells of the Gastric Mucosa—ENRICO SOLCIA, CARLO CAPELLA, GABRIELE VASSALLO, AND ROBERTO BUFFA

Membrane Transport of Purine and Pyrimidine Bases and Nucleosides in Animal Cells—RICHARD D. BERLIN AND JANET M. OLIVER

SUBJECT INDEX

Volume 43

The Evolutionary Origin of the Mitochondrion: A Nonsymbiotic Model—HENRY R. MAHLER AND RUDOLF A. RAFF

Biochemical Studies of Mitochondrial Transcription and Translation—C. SACCONE AND E. QUAGLIARIELLO

The Evolution of the Mitotic Spindle—DONNA F. KUBAI

Germ Plasma and the Differentiation of the Germ Cell Line—E. M. EDDY

Gene Expression in Cultured Mammalian Cells—RODY P. COX AND JAMES C. KING

Morphology and Cytology of the Accessory Sex Glands in Invertebrates—K. G. ADIYODI AND R. G. ADIYODI

SUBJECT INDEX

Volume 44

The Nucleolar Structure—SIBDAS GHOSH

The Function of the Nucleolus in the Expression of Genetic Information: Studies with Hybrid Animal Cells—E. SIDEBOTTOM AND I. I. DEÁK

Phylogenetic Diversity of the Proteins Regulating Muscular Contraction—WILLIAM LEHMAN

Cell Size and Nuclear DNA Content in Vertebrates—HENRYK SZARSKI

Ultrastructural Localization of DNA in Ultrathin Tissue Sections—ALAIN GAUTIER

Cytological Basis for Permanent Vaginal Changes in Mice Treated Neonatally with Steroid Hormones—NOBORU TAKASUGI

On the Morphogenesis of the Cell Wall of Staphylococci—PETER GIESBRECHT, JÖRG WECKE, AND BERNHARD REINICKE

Cyclic AMP and Cell Behavior in Cultured Cells—MARK C. WILLINGHAM

Recent Advances in the Morphology, Histochemistry, and Biochemistry of Steroid-Synthesizing Cellular Sites in the Nonmammalian Vertebrate Ovary—SARDUL S. GURAYA

SUBJECT INDEX

Volume 45

Approaches to the Analysis of Fidelity of DNA Repair in Mammalian Cells—MICHAEL W. LIEBERMAN

The Variable Condition of Euchromatin and Heterochromatin—FRIEDRICH BACK

Effects of 5-Bromodeoxyuridine on Tumorigenicity, Immunogenicity, Virus Production, Plasminogen Activator, and Melanogenesis of Mouse Melanoma Cells—SELMA SILAGI

Mitosis in Fungi—MELVIN S. FULLER

Small Lymphocyte and Transitional Cell Populations of the Bone Marrow; Their Role in the Mediation of Immune and Hemopoietic Progenitor Cell Functions—CORNELIUS ROSSE

The Structure and Properties of the Cell Surface Coat—J. H. LUFT

Uptake and Transport Activity of the Median Eminence of the Hypothalamus—K. M. KNIGGE, S. A. JOSEPH, J. R. SLADEK, M. F. NOTTER, M. MORRIS, D. K. SUNDBERG, M. A. HOLZWARTH, G. E. HOFFMAN, AND L. O'BRIEN

SUBJECT INDEX

Volume 46

Neurosecretion by Exocytosis—TOM CHRISTIAN NORMANN

Genetic and Morphogenetic Factors in Hemoglobin Synthesis during Higher Vertebrate Development: An Approach to Cell Differentiation Mechanisms—VICTOR NIGON AND JACQUELINE GODET

Cytophysiology of Corpuscles of Stannius—V. G. KRISHNAMURTHY

Ultrastructure of Human Bone Marrow Cell Maturation—J. BRETON-GORIUS AND F. REYES

Evolution and Function of Calcium-Binding Proteins—R. H. KRETSINGER

SUBJECT INDEX

Volume 47

Responses of Mammary Cells to Hormones—M. R. BANERJEE

Recent Advances in the Morphology, Histochemistry, and Biochemistry of Steroid-Synthesizing Cellular Sites in the Testes of Non-mammalian Vertebrates—SARDUL S. GURAYA

Epithelial-Stromal Interactions in Development of the Urogenital Tract—GERALD R. CUNHA

Chemical Nature and Systematization of Substances Regulating Animal Tissue Growth—VICTOR A. KONYSHEV

Structure and Function of the Choroid Plexus and Other Sites of Cerebrospinal Fluid Formation—THOMAS H. MILHORAT

The Control of Gene Expression in Somatic Cell Hybrids—H. P. BERNHARD

Precursor Cells of Mechanocytes—ALEXANDER J. FRIEDENSTEIN

SUBJECT INDEX

Volume 48

Mechanisms of Chromatin Activation and Repression—NORMAN MACLEAN AND VAUGHAN A. HILDER

Origin and Ultrastructure of Cells *in Vitro*—L. M. FRANKS AND PATRICIA D. WILSON

Electrophysiology of the Neurosecretory Cell—KINJI YAGI AND SHIZUKO IWASAKI

Reparative Processes in Mammalian Wound Healing: The Role of Contractile Phenomena—GIULIO GABBIANI AND DENYS MONTADON

Smooth Endoplasmic Reticulum in Rat Hepatocytes during Glycogen Deposition and Depletion—ROBERT R. CARDELL, JR.

Potential and Limitations of Enzyme Cytochemistry: Studies of the Intracellular Digestive Apparatus of Cells in Tissue Culture—M. HÜNDGEN

Uptake of Foreign Genetic Material by Plant Protoplasts—E. C. COCKING

The Bursa of Fabricius and Immunoglobulin Synthesis—BRUCE GLICK
SUBJECT INDEX

Volume 49

Cyclic Nucleotides, Calcium, and Cell Division—LIONEL I. REBHUN
Spontaneous and Induced Sister Chromatid Exchanges as Revealed by the BUdR-Labeling Method—HATAO KATO
Structural, Electrophysiological, Biochemical, and Pharmacological Properties of Neuroblastoma-Glioma Cell Hybrids in Cell Culture —B. HAMPRECHT
Cellular Dynamics in Invertebrate Neurosecretory Systems—ALLAN BERLIND
Cytophysiology of the Avian Adrenal Medulla—ASOK GHOSH
Chloride Cells and Chloride Epithelia of Aquatic Insects—H. KOMNICK
Cytosomes (Yellow Pigment Granules) of Molluscs as Cell Organelles of Anoxic Energy Production—IMRE ZS.-NAGY
SUBJECT INDEX

Volume 50

Cell Surface Enzymes: Effects on Mitotic Activity and Cell Adhesion—H. BRUCE BOSMANN
New Aspects of the Ultrastructure of Frog Rod Outer Segments—JÜRGEN ROSENKRANZ
Mechanisms of Morphogenesis in Cell Cultures —J. M. VASILIEV AND I. M. GELFAND
Cell Polyploidy: Its Relation to Tissue Growth and Functions—W. YA. BRODSKY AND I. V. URYVAEVA
Action of Testosterone on the Differentiation and Secretory Activity of a Target Organ: The Submaxillary Gland of the Mouse—MONIQUE CHRÉTIEN
SUBJECT INDEX

Volume 51

Circulating Nucleic Acids in Higher Organisms —MAURICE STROUN, PHILIPPE ANKER, PIERRE MAURICE, AND PTER B. GAHAN

Recent Advances in the Morphology, Histochemistry, and Biochemistry of the Developing Mammalian Ovary—SARDUL S. GURAYA
Morphological Modulations in Helical Muscles (Aschelminthes and Annelida)—GIULIO LANZAVECCHIA
Interrelations of the Proliferation and Differentiation Processes during Cardiac Myogenesis and Regeneration—PAVEL P. RUMYANTSEV
The Kurloff Cell—PETER A. REVELL
Circadian Rhythms in Unicellular Organisms: An Endeavor to Explain the Molecular Mechanism—HANS-GEORG SCHWEIGER AND MANFRED SCHWEIGER
SUBJECT INDEX

Volume 52

Cytophysiology of Thyroid Parafollicular Cells—ELADIO A. NUNEZ AND MICHAEL D. GERSHON
Cytophysiology of the Amphibian Thyroid Gland through Larval Development and Metamorphosis—ELIANE REGARD
The Macrophage as a Secretory Cell—ROY C. PAGE, PHILIP DAVIES, AND A. C. ALLISON
Biogenesis of the Photochemical Apparatus —TIMOTHY TREFFRY
Extrusive Organelles in Protists—KLAUS HAUSMANN
Lectins—JAY C. BROWN AND RICHARD C. HUNT
SUBJECT INDEX

Volume 53

Regular Arrays of Macromolecules on Bacterial Cell Walls: Structure, Chemistry, Assembly, and Function—UWE B. SLEYTR
Cellular Adhesiveness and Extracellular Substrata—FREDERICK GRINNELL
Chemosensory Responses of Swimming Algae and Protozoa—M. LEVANDOWSKY AND D. C. R. HAUSER
Morphology, Biochemistry, and Genetics of Plastid Development in *Euglena gracilis*—V. NIGON AND P. HEIZMANN

Plant Embryological Investigations and Fluorescence Microscopy: An Assessment of Integration—R. N. KAPIL AND S. C. TIWARI

The Cytochemical Approach to Hormone Assay —J. CHAYEN

SUBJECT INDEX

Volume 54

Microtubule Assembly and Nucleation—MARC W. KIRSCHNER

The Mammalian Sperm Surface: Studies with Specific Labeling Techniques—JAMES K. KOEHLER

The Glutathione Status of Cells—NECHAMA S. KOSOWER AND EDWARD M. KOSOWER

Cells and Senescence—ROBERT ROSEN

Immunocytology of Pituitary Cells from Teleost Fishes—E. FOLLÉNIUS, J. DOERR-SCHOTT, AND M. P. DUBOIS

Follicular Atresia in the Ovaries of Nonmammalian Vertebrates—SRINIVAS K. SAIDAPUR

Hypothalamic Neuroanatomy: Steroid Hormone Binding and Patterns of Axonal Projections—DONALD W. PFAFF AND LILY C. A. CONRAD

Ancient Locomotion: Prokaryotic Motility Systems—LELENG P. TO AND LYNN MARGULIS

An Enzyme Profile of the Nuclear Envelope—I. B. ZBARSKY

SUBJECT INDEX

Volume 55

Chromatin Structure and Gene Transcription: Nucleosomes Permit a New Synthesis— THORU PEDERSON

The Isolated Mitotic Apparatus and Chromosome Motion—H. SAKAI

Contact Inhibition of Locomotion: A Reappraisal—JOAN E. M. HEAYSMAN

Morphological Correlates of Electrical and Other Interactions through Low-Resistance Pathways between Neurons of the Vertebrate Central Nervous System—C. SOTELO AND H. KORN

Biological and Biochemical Effects of Phenylalanine Analogs—D. N. WHEATLEY

Recent Advances in the Morphology, Histochemistry, Biochemistry, and Physiology of Interstitial Gland Cells of Mammalian Ovary—SARDUL S. GURAYA

Correlation of Morphometry and Stereology with Biochemical Analysis of Cell Fractions —R. P. BOLENDER

Cytophysiology of the Adrenal Zona Fasciculata—GASTONE G. NUSSDORFER, GIUSEPPINA MAZZOCCHI, AND VIRGILIO MENEGHELLI

SUBJECT INDEX

Volume 56

Synapses of Cephalopods—COLLETTE DUCROS

Scanning Electron Microscope Studies on the Development of the Nervous System in Vivo and in Vitro—K. MELLER

Cytoplasmic Structure and Contractility in Amoeboid Cells—D. LANSING TAYLOR AND JOHN S. CONDEELIS

Methods of Measuring Intracellular Calcium— ANTHONY H. CASWELL

Electron Microscope Autoradiography of Calcified Tissues—ROBERT M. FRANK

Some Aspects of Double-Stranded Hairpin Structures in Heterogeneous Nuclear RNA— HIROTO NAORA

Microchemistry of Microdissected Hypothalamic Nuclear Areas—M. PALKOVITS

SUBJECT INDEX

Volume 57

The Corpora Allata of Insects—PIERRE CASSIER

Kinetic Analysis of Cellular Populations by Means of the Quantitative Radioautography— J.-C. BISCONTE

Cellular Mechanisms of Insect Photoreception— F. G. GRIBAKIN

Oocyte Maturation—YOSHIO MASUI AND HUGH J. CLARKE

The Chromaffin and Chromaffin-like Cells in the Autonomic Nervous System—JACQUES TAXI

The Synapses of the Nervous System—A. A. MANINA

SUBJECT INDEX

Volume 58

Functional Aspects of Satellite DNA and Heterochromatin—BERNARD JOHN AND GEORGE L. GABOR MIKLOS

Determination of Subcellular Elemental Concentration through Ultrahigh Resolution Electron Microprobe Analysis—THOMAS E. HUTCHINSON

The Chromaffin Granule and Possible Mechanisms of Exocytosis—HARVEY B. POLLARD, CHRISTOPHER J. PAZOLES, CARL E. CREUTZ, AND OREN ZINDER

The Golgi Apparatus, the Plasma Membrane, and Functional Integration—W. G. WHALEY AND MARIANNE DAUWALDER

Genetic Control of Meiosis—I. N. GOLUBOVSKAYA

Hypothalamic Neurons in Cell Culture—A. TIXIER-VIDAL AND F. DE VITRY

The Subfornical Organ—H. DIETER DELLMANN AND JOHN B. SIMPSON

SUBJECT INDEX

Volume 59

The Control of Microtubule Assembly in Vivo—ELIZABETH C. RAFF

Membrane-Coating Granules—A. F. HAYWARD

Innervation of the Gastrointestinal Tract—GIORGIO GABELLA

Effects of Irradiation on Germ Cells and Embryonic Development in Teleosts—NOBUO EGAMI AND KEN-ICHI IJIRI

Recent Advances in the Morphology, Cytochemistry, and Function of Balbiani's Vitelline Body in Animal Oocytes—SARDUL S. GURAYA

Cultivation of Isolated Protoplasts and Hybridization of Somatic Plant Cells—RAISA G. BUTENKO

SUBJECT INDEX

Volume 60

Transfer RNA-like Structure in Viral Genomes—TIMOTHY C. HALL

Cytoplasmic and Cell Surface Deoxyribonucleic Acids with Consideration of Their Origin—BEVAN L. REID AND ALEXANDER J. CHARLSON

Biochemistry of the Mitotic Spindle—CHRISTIAN PETZELT

Alternatives to Classical Mitosis in Hemopoietic Tissues of Vertebrates—VIBEKE E. ENGELEBERT

Fluidity of Cell Membranes—Current Concepts and Trends—M. SHINITZKY AND P. HENKART

Macrophage-Lymphocyte Interactions in Immune Induction—MARC FELDMANN, ALAN ROSENTHAL, AND PETER ERB

Immunohistochemistry of Luteinizing Hormone-Releasing Hormone-Producing Neurons of the Vertebrates—JULIEN BARRY

Cell Reparation of Non-DNA Injury—V. YA. ALEXANDROV

Ultrastructure of the Carotid Body in the Mammals—ALAIN VERNA

The Cytology and Cytochemistry of the Wool Follicle—DONALD F. G. ORWIN

SUBJECT INDEX

Volume 61

The Association of DNA and RNA with Membranes—MARY PAT MOYER

Electron Cytochemical Stains Based on Metal Chelation—DAVID E. ALLEN AND DOUGLAS D. PERRIN

Cell Electrophoresis—THOMAS G. PRETLOW, II AND THERESA P. PRETLOW

The Wall of the Growing Plant Cell: Its Three-Dimensional Organization—JEAN-CLAUDE ROLAND AND BRIGITTE VIAN

Biochemistry and Metabolism of Basement Membranes—NICHOLAS A. KEFALIDES, ROBERT ALPER, AND CHARLES C. CLARK

The Effects of Chemicals and Radiations within the Cell: An Ultrastructural and Micrurgical Study Using Amoeba proteus as a Single-Cell Model—M. J. ORD

Growth, Reproduction, and Differentiation in Acanthamoeba—THOMAS J. BYERS

SUBJECT INDEX

Volume 62

Calcification in Plants—ALLAN PENTECOST

Cellular Microinjection by Cell Fusion: Technique and Applications in Biology and Medicine—MITSURU FURUSAWA

Cytology, Physiology, and Biochemistry of Germination of Fern Spores—V. RAGHAVAN

Immunocytochemical Localization of the Vertebrate Cyclic Nonapeptide Neurohypophyseal Hormones and Neurophysins—K. DIERICKX

Recent Progress in the Morphology, Histochemistry, Biochemistry, and Physiology of Developing and Maturing Mammalian Testis—SARDUL S. GURAYA

Transitional Cells of Hemopoietic Tissues: Origin, Structure, and Development Potential—JOSEPH M. YOFFEY

Human Chromosomal Heteromorphisms: Nature and Clinical Significance—RAM S. VERMA AND HARVEY DOSIK

SUBJECT INDEX

Volume 63

Physarum polycephalum: A Review of a Model System Using a Structure–Function Approach—EUGENE M. GOODMAN

Microtubules in Cultured Cells: Indirect Immunofluorescent Staining with Tubulin Antibody—B. BRINKLEY, S. FISTEL, J. M. MARCUM, AND R. L. PARDUE

Septate and Scalariform Junctions in Arthropods—CÉCILE NOIROT-TIMOTHÉE AND CHARLES NOIROT

The Cytology of Salivary Glands—CARLIN A. PINKSTAFF

Development of the Vertebrate Cornea—ELIZABETH D. HAY

Scanning Electron Microscopy of the Primate Sperm—KENNETH G. GOULD

Cortical Granules of Mammalian Eggs—BELA J. GULYAS

SUBJECT INDEX